buas

Principles and Practices of Method Validation

Principles and Practices of Method Validation

Edited by

A. Fajgelj
International Atomic Energy Agency, Agency's Laboratories, Seibersdorf, Austria

Á. Ambrus
Food and Agricultural Organisation/International Atomic Energy Agency Agricultural and Biotechnology Laboratory, Seibersdorf, Austria

RS•C
ROYAL SOCIETY OF CHEMISTRY

The proceedings of the Joint AOAC/FAO/IAEA/IUPAC International Workshop on the Principles and Practices of Method Validation held in November 1999 in Budapest, Hungary.

Special Publication No. 256

ISBN 0-85404-783-2

A catalogue record for this book is available from the British Library

Published by The Royal Society of Chemistry,
Thomas Graham House, Science Park, Milton Road,
Cambridge CB4 0WF, UK

For further information see our web site at www.rsc.org

Printed by MPG Books Ltd, Bodmin, Cornwall, UK

Foreword

Co-operative efforts over the past twenty years by the AOAC International, ISO, IUPAC, and other international organisations, have resulted in a number of harmonised protocols related to various aspects of chemical analysis. These protocols have gained wide international acceptance and, thereby, have contributed to the reliability and comparability of analytical results.

The Food and Agriculture Organisation of the United Nations and the International Atomic Energy Agency, through the Joint FAO/IAEA Programme, have recently joined with the AOAC International and IUPAC in further developing international guidelines on harmonisation of quality assurance requirements for analytical laboratories and the validation of acceptable analytical methods. This co-operative activity is aimed at supporting the implementation of national legislation and international agreements in order to both enhance food safety and facilitate international trade. To further these aims, an AOAC/FAO/IAEA/IUPAC International Workshop on Principles and Practices of Method Validation was convened from 4 to 6 November 1999, in Budapest, Hungary.

The Workshop provided a forum for analytical chemists and representatives of government agencies, standards organisations and accreditation bodies involved in method validation to share their experience and to contribute to the development of guidelines for the validation of analytical methods in general, and specifically for the determination of pesticide and veterinary drug residues in food.

An AOAC/FAO/IAEA/IUPAC Expert Consultation, held after the Workshop, took into consideration the recommendations of the Workshop in drafting the *Guidelines for Single-Laboratory Validation of Analytical Methods for Trace-level Concentrations of Organic Chemicals.*

The minimum data requirements, as developed during the Workshop and specified in the *Guidelines,* contribute to making method validation practical, timely and cost effective. It is hoped that they will be adopted by relevant International and National Organisations, thereby contributing to improvements in food safety and the reduction in barriers to international food trade.

John W. Jost	E. James Bradford	James Dargie
Executive Director,	Executive Director	Director
IUPAC	AOAC International	Joint FAO/IAEA
		Division
		of Nuclear Techniques
		in Food and Agriculture

Introduction

This book contains lectures presented at the international workshop on 'Principles and Practices of Method Validation', held from 4 to 6 November 1999 in Budapest, Hungary. The International Association of Official Analytical Chemists (AOAC International), Food and Agriculture Organisation (FAO), International Atomic Energy Agency (IAEA), the International Union of Pure and Applied Chemistry (IUPAC) and the Plant Health and Soil Conservation Station of Budapest co-operated in the organisation of this event.

The organisation of this workshop has a certain historical background. Namely, in the 1980s a strong co-operation between various international organisations (AOAC, IDF, FAO, WHO, ISO, IUPAC *etc.*) dealing with method validation or using the results of analytical methods was established with the aim of harmonising a number of existing protocols on method performance studies and to define criteria for the assessment of method performance characteristics. As a result, in 1988 a 'Protocol for the Design, Conduct and Interpretation of Method Performance Studies' was published under the auspices of IUPAC. In its original or in the 1995 revised version this protocol was for many years a basis for acceptance of analytical methods applied for legislative, trading, health related and similar purposes. This is especially valid for the food analysis sector. According to this protocol, newly developed methods were tested in a number of laboratories and the performance characteristics established with the help of statistical tools. However, with the fast development of new analytical techniques it became practically impossible to organise interlaboratory collaborative studies that would cover all of the analyte/matrix/method combinations. The exclusive use of interlaboratory method performance studies for acceptance of analytical methods became questionable in the sense of its practicability and due to the high costs involved. In parallel, the quality control and quality assurance requirements for testing laboratories have been more strictly enforced over the last decade, and they are well described in standards or guides, *e.g.* ISO 17025 (previously ISO Guide 25), EN 45001, GLP guidelines, *etc.* One of the requirements is that analytical methods, even standardised methods, have to be tested in the laboratory prior to being used for routine purposes. Together with the requirement of ISO 17025 that the results need to be accompanied by the associated measurement uncertainty, this compels the laboratories to perform a fitness for purpose test. This actually means that a partial or full method validation is performed. Such a validation is known as 'in-house method validation'. More than 15 different guidance documents on 'in-house method validation' have been prepared in recent years by various organisations. However, these general guidelines for methods of chemical analysis do not address the specific problems of trace analysis, and they are impractical for the validation of methods used for the determination of trace level organic chemicals.

To assist laboratories performing the analysis of trace organic contaminants, the FAO/IAEA Secretariat, through the FAO/IAEA Training and Reference Centre for Food and Pesticide Control, initiated a programme for the elaboration of Guidelines for the validation of analytical methods in a single laboratory, and prepared a working document in co-operation with a number of experienced and practising analysts. The formal basis for organisation of this workshop was the recommendations of the FAO/IAEA Consultants' Meeting on Validation of Analytical Methods for Food Control, held in Vienna in 1997, and the IUPAC project on 'Preparation and Harmonisation of International Guidelines for In-house Method Validation'. Its aim was to review the current practices of method validation and to discuss possible future development in this field. The workshop recognised the concerns regarding validation of methods expressed by the Vienna consultation. It endorsed the need for guidance on validation of methods within a single laboratory, whether for use only within that laboratory or as a precursor to an inter-laboratory validation. Various points were also raised at the workshop in regard to the requirements for method validation in general, and specifically on two draft documents: 'Practical procedures to validate method performance and results of analysis of pesticide and veterinary drug residues, and trace organic contaminants in food', prepared by the FAO/IAEA Secretariat, and the IUPAC technical report on 'Harmonised guidelines for the in-house validation of methods of analysis'.

During the workshop a number of lectures and posters were presented, and their full text is published in this book. Most of the contributions in this book relate to the validation of new methods for pesticide residue analyses in various foodstuffs and water. Further, the effects of sample processing and storage on the stability of pesticide residues are discussed, as well as the measurement uncertainty resulting from these processes. Different chromatographic methods are discussed, including estimation of various effects, *e.g.* matrix-induced effects, the influence of the equipment set-up, *etc.* In addition to the methods used for routine purposes, validation of analytical data in research and development environment is also presented.

A part of the book is dedicated to the legislation covering the EU Guidance on residue analytical methods, and to the extensive review of the existing 'in-house method validation documentation'. Finally, 'Guidelines for single-laboratory validation of analytical methods for trace-level concentrations of organic chemicals' are included. These guidelines have been extensively discussed in the second part of the workshop and finalised by a FAO/IAEA/AOAC Expert Consultation held after the workshop.

As many times indicated during the workshop, the process required to elaborate all technical details and to change the 'philosophy' and consequently the international legislation related to acceptance of method validation and the acceptance of analytical methods for trading purposes might take some years. In this process the workshop and these proceedings are an important milestone. We hope that practical examples and discussions presented in this book will be a useful source of technical information for students, researchers, and scientists who are directly dealing with method validation. On the other hand, for all those who are dealing with method validation from the regulative and legislative point of view this book should serve as a source of up-dated information.

We are grateful to the Hungarian Organising Committee and especially to Mr. János Szabó, Dr. László Győrfi and Ms. Gabriella Sz. Kükedi for the excellent conditions provided for the workshop, and to Professor Ernő Pungor, Chairman of the Hungarian IUPAC National Committee for his support.

Aleš Fajgelj   Árpád Ambrus

Contents

The Potential Use of Quality Control Data to Validate Pesticide Residue Method Performance

William Horwitz

CENTER FOR FOOD SAFETY AND APPLIED NUTRITION HFS-500, US FOOD AND DRUG
ADMINISTRATION, WASHINGTON DC 20204, USA

Full scale interlaboratory (collaborative) studies are becoming too expensive and time-consuming to support their use as the only way to validate methods of analysis. Furthermore, reliable estimates of method performance parameters, such as accuracy, precision, and limits of applicability, cannot be achieved by individual collaborative studies at the concentration levels of 0.01-1 mg.kg^{-1}, the region of interest for residue analysis, when the expected random error among laboratories is of the order of 20-30% of the mean. Performance data from proficiency studies of tens of thousands of control determinations accumulated over the past decade from individual and multiple laboratories are being examined to determine their potential as a substitute for interlaboratory performance data. An analysis of variance indicates that as much as 80% of the total variability of pesticide residue analysis is "random error." If this is the case, proficiency data may be substituted for method-performance data, when recovery is acceptable, because the individual factors of analyte, method, matrix, laboratory, and time contribute little to overall variability.

1 INTRODUCTION

A full-scale method performance study, utilizing the IUPAC-AOAC harmonized protocol,[1] requires a minimum of 8 laboratories to analyze at least 5 materials related to the analyte-matrix-concentration combinations of interest by the proposed method. Although such a study is probably the most pertinent and reliable way of demonstrating the performance of a method with a specific test sample, several alternatives are also available. The Youden pair technique (split level design)[2] conducted at several relevant concentration levels is the model favored by the extensive water analysis program of the U.S. Environmental Protection Agency (EPA). In Europe, the "uncertainty" budget approach has been advocated, particularly in conjunction with laboratory accreditation, although this design has been modified during the past few years to become more similar to the method-performance model.[3]

All of these models are expensive and time-consuming. What few calculations that have been conducted independently suggest that they do not provide the same performance parameters with the same method from the same data and information. Proficiency testing, also advocated by the FAO/WHO Codex Alimentarius Program as a way of demonstrating equivalency in laboratory results,[4] has been suggested as still another potential substitute for the method-performance trials. The purpose of this paper is to explore this possibility.

2 AVAILABLE PROFICIENCY STUDIES

Data from several large scale proficiency programs have been made available to us in order to examine their potential use as a substitute for method performance in the determination of pesticide residues in food. As of late 1999, data from three programs have been examined. These include the following:

2.1 Total Diet Program (FDA-TD)

The Total Diet Program of the U.S. Food and Drug Administration (FDA), Kansas City Laboratory, has been in continuous operation for almost 40 years. The recoveries of approximately 7000 control determinations of analytes added to individual foods in the pesticide residue portion of the 4 annual market baskets analyzed during 1991-1997 were available as databases in EXCEL4. This is a within-laboratory program.

2.2 Food Analysis Performance Assessment Schemes (FAPAS)

This is an international fee-based program conducted by the Ministry of Agriculture, Fisheries, and Food of the United Kingdom.[5] Each test material is thoroughly homogenized to a smooth paste and aliquots of stock standard solutions of pesticides of known purity are added with additional mixing. The fortified test material is measured into screw top glass jars, tested for homogeneity, and stored at $-20°$ until shipped. More than 100 laboratories participated in many of the tests. Participants were required to submit their results within 8 weeks for their values to be included in a report that was distributed. The pesticide residue reports (Nos. 1901-1905), uncorrected for recovery, and percent recovery of about 1100 concurrent controls that were submitted, were the data used in the present examination.

2.3   State of California Quality Assurance Program (CA)

The State of California has been monitoring the quality of their Residue Enforcement Program since 1988, utilizing their headquarters and field laboratories and an occasional contract laboratory. The laboratories are supplied centrally with a control spiking solution containing 3-8 pesticides quarterly. The pesticides are selected from those encountered routinely in the state. The laboratories selected a test sample daily to be fortified with the control solution. Both test samples, spiked and unspiked, were conducted through the entire analytical procedure of extraction, isolation, and measurement. The commodity selected for fortification could be new, one that had shown problems, or was the subject of a special application investigation. The results are submitted to the Quality Assurance Unit of the State Laboratory for collation and about 15000 data points are reviewed here.

3 PROCEDURE

The data was usually available as percent recovery for each control sample (fortified analyte/food) from relatively long time periods as in the case of the FDA-TD and the CA programs and from many laboratories in the case of the FAPAS program. These records were calculated to an overall pooled recovery and relative standard deviation for each analyte/food combination and, when available, for substantially different characteristics (laboratory, method, type of food). To keep the examination within reasonable limits, only those combinations with at least 8 values were reviewed. The HORRAT values were also calculated for each group from the following formulae:

$$\text{Relative standard deviation} = \text{RSD} = s \times 100/\bar{x}, \tag{1}$$

where s is the standard deviation *within*-laboratory from FDA-TD and *among*-laboratories for the FAPAS and CA programs.

$$\text{Predicted RSD} = \text{PRSD} = 2\ C^{(-0.1505)}, \tag{2}$$

where C is the concentration of the pesticide added, as a decimal fraction (1 mg.kg^{-1} = 10^{-6}). Equation (2) is the so-called "Horwitz-curve,"[6] and

$$\text{HORRAT} = \text{RSD/PRSD}. \tag{3}$$

A HORRAT value of about 0.5-0.7 is expected from within-laboratory studies (FDA-TD) and about 1 from among-laboratories studies (FAPAS and CA),[7] although values in the interval from about 0.5 to 2 times the expected value may be acceptable.

3.1 Outlier Removal

A concurrent investigation suggests that all values outside of a recovery of 100±50% be removed as beyond acceptable limits for the concentration levels examined in these studies (about 0.01-1 $\times$ 10^{-6}). Such limits usually result in the removal of an acceptably small fraction (i.e., <5%) of results. These limits also happen to approximate the 3-sigma quality control limits that would be calculated from the Horwitz formula (1) for a concentration between 0.5 and 1.0×10^{-6}.

4 RESULTS

Considering all of the studies as a group, the overall recovery of pesticide residues in general is about 90%. The among-laboratory precision is about 15%. These values have remained approximately constant since the introduction of multiresidue methods.[8] The FDA-TD program utilizes a within-laboratory model whose variability is expected to be one half to two thirds that of the among-laboratories models used in the other two programs. Each of the programs exhibit some special features discussed below.

4.1 FDA-TD

To keep this database within reasonable bounds, Table 1 summarizes the salient information only from those analytes with approximately 100 or more records and related entries. For this database, outliers, defined as recoveries >150% or <50%, are removed prior to calculating the statistics. In most cases few values had to be removed. If a substantial number of values had to be removed, the method(s) is considered inappropriate for the analyte.

Table 1 gives a summary of the analytical characteristics of the major analytes included in the FDA-TD market baskets over a recent 5-year period where a substantial number of values were available for examination. A more detailed examination of all of the data will be presented separately. In general, for most of the pesticides, the differences in extraction and measurement methods that are made to accommodate high or low fat, moisture, and sugar contents of various foods and the use of different types of columns, conditions and detectors make little difference in the overall recoveries. The extensive details that would be required to list each different method or condition used would require too much space and are not pertinent to the present summary. Those pesticides that deviate substantially in recovery and precision (ETU, herbicides) are known to produce problems in routine analytical work or an inapplicable method may have been used (e.g., pentachlorophenol).

The acceptable HORRAT values in this data set average about 0.5, which is somewhat better than would be expected for multiple-analyst, within-laboratory data. The typical overall analytical variability is so large (ca 10%) at these low levels (20-500 $\mu g.kg^{-1}$) that statistical tests of significance are generally meaningless. In most cases there were insufficient numbers of values to provide reliable statistics for individual analyte/food combinations, aside from the general categories of high and low fat and moisture with some analytes.

4.2 FAPAS

This is a fee-based program operated to meet the ISO requirements for accreditation. Although a list of about 20 potential analytes is usually supplied to participants, no more than about 6 pesticides are present in any single test sample. About half of the participants voluntarily supplied control data in addition to the required program data. This resulted in the availability of analyte/commodity data from test material analyzed both as a known and as an unknown. This fortuity permitted applying a correlation coefficient (r) calculation to the two types of data with the unexpected result of an average r $\approx$ 0, i.e., the analysis of a test sample as a known does not correlate with its measurement as an unknown. Another interesting finding was that the analyte thiabendazole analyzed by HPLC by 15 laboratories showed definite evidence of censoring the data when analyzed as a known (matrix recovery) compared to the similar analyses analyzed as an unknown (spike recovery), as shown in Figure 1. The values exhibit a much tighter cluster when reported as a known, as compared to the wider cluster when reported as an unknown. Although rare, such reports occasionally do appear in the literature.[9] Until the questions raised by these findings are resolved, control values reported on known additions must be viewed with skepticism.

Table 1. *Performance parameters of major analytes in the FDA Total Diet Database*

Analyte[a]	No. of Values	Concn. $\times 10^9$	Ave. Recy %	SD	HORRAT
Benomyl (0)	167	610	80.6	6.7	0.48
Thiabendazole (0)	131	220	80.5	6.8	0.42
Carbaryl (0)	198	50	99.0	10.9	0.44
Carbofuran (0)	85	50	99.5	9.3	0.37
Isoprocarb (0)	82	50	95.2	10.1	0.42
Methomyl (0)	136	50	95.9	11.1	0.46
Ethylenethiourea (9)	258	50	88.6	16.2	0.73
Diazinon (0) FF	105	16	79.9	14.0	0.59
Diazinon (1) NFF	152	14	96.1	11.1	0.38
Dieldrin (0) FF	252	20	91.6	10.6	0.40
Dieldrin (1) NFF	152	20	91.5	11.8	0.45
Parathion (0) FF	251	22	84.1	10.3	0.43
Parathion (0) NFF	124	20	95.2	8.7	0.32
Parathion (0) NFF	157	20	99.7	10.3	0.36
Parathion (0) NFF	29	20	94.6	12.0	0.44
Parathion (0) NFF	21	20	95.6	15.0	0.54.
Parathion (0) NFF	21	20	96.5	12.9	0.46
Propargite (0) NFF	17	100	93.6	16.9	0.80
Propargite (0) NFF	29	400	93.3	18.0	1.05
Propargite (0) NFF	90	400	94.8	16.0	0.92
2,4,5-T (0)	89	68	65.7	41.0	2.60
2,4-D (0)	88	123	55.8	36.0	2.94
Pentachlorophenol (0)	91	32	62.8	30.1	1.78
Diuron (3)	155	50	92.6	15.1	0.65
Metoxuron (1)	92	50	97.4	19.4	0.79
Monuron (2)	64	50	87.2	14.8	0.68
Nebruron (23)	97	50	85.4	15.4	0.72

Values in parentheses are number of outliers <50% or >150%; FF = fatty foods; NFF = nonfatty foods.
The same analyte on different lines indicate different methods.

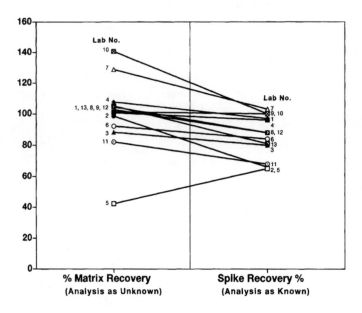

Figure 1. *Control data for thiabendazole examined as an unknown ("% Matrix Recovery") on left and as a known ("% Spike Recovery") on right. The lines connect the two values from the same laboratory, designated by an arbitrary number. Note the compression of values when analyzed as a known compared to the variability when analyzed as an unknown.*

4.3 CA Program

Almost 15000 values from about 65 analytes in numerous foods, analyzed by typically 5 laboratories, are available from this pioneer quality control program that has been operating for over a decade. A preliminary analysis of the available data is given in Table 1. The original data was first checked for the effect of outlier removal, using limits of $100\pm40\%$ (maximum removal), $100\pm50\%$, and $100\pm60\%$ (minimum removal). Although, on an overall basis, outlier removal has a negligible impact on the statistical parameters calculated from 15000 records, they may affect calculations of the analyte/food/laboratory combinations.

Some preliminary conclusions are: (1) Most of the outliers (values outside $100\pm50\%$) are from the smaller analyte/matrix combinations (<8 values). (2) Limits of $100\pm50\%$ strike a reasonable balance between excessive and restrained outlier removal. (3) A small percent of values are reported as "0." On review, these may be found to reflect clerical errors (failure to record a value, recording an incorrect value, or not adhering to a scheduled protocol) rather than an analytical failure. The implication of their presence is considerably greater if the data is used to support method performance rather than routine laboratory quality control.

Table 2 provides the initial analysis of variance of all the data from the CA program, with values outside of $100\pm40\%$, $100\pm50\%$, and $100\pm60\%$ recovery removed as "outliers." With no values removed, 82% of the variance is "random." Only 18% of the variance is attributable to specific factors, primarily analyte and method. Food and laboratory make a negligible contribution. If "outliers" are removed, about 75% of the variance is "random," but which outlier-removal procedure is used is immaterial. Only 25% of the variance is attributable to specific factors, with food and laboratory again making a negligible contribution.

Table 2   *Analysis of Variance by Major Variables of the 15000 Records From the CA Program by % of Total Variance*

Parameter	All Data	$100\pm40\%$	$100\pm50\%$	$100\pm60\%$
Analyte	9.4	12.6	12.8	13.0
Food	0.5	0.7	0.7	0.9
Method	7.8	10.3	10.4	10.3
Lab.	0.2	1.4	1.3	0.6
Error	82.1	75.0	74.9	75.2

5   DISCUSSION

The analysis of variance shows that the major factor in the variability of pesticide analysis is "random error" and therefore is irreducible under the conditions of these studies. This conclusion is reinforced by the constancy of the precision and recovery of pesticide residue analysis over the last quarter century. The improvements that have been made in the direction of better columns and instrumentation require relaxing control of operating conditions to permit optimizing resolution, sensitivity, and peak sharpness and shape, and minimizing baseline interference. Such general directions provided under the title "system suitability tests" at best can only maintain current among-laboratory precision as characterized by equation (2). System suitability requirements have attained a high degree of refinement in the pharmaceutical analysis and are recognized in the official compendia.[10]

The necessity for the use of broad control limits (i.e, $\pm50\%$) for individual analyses or for numerous replicates to reduce variability, decreases the value of quality control specifications to monitor analyst or method performance. The tendency of analysts to censor their values, as illustrated in Figure 1 by the tighter clusters of results when controls are analyzed as knowns (right side) as compared with results from similar analyses when conducted as unknowns (left side), must be overcome. The use of automation, at least in the chromatographic, measurement, and calculation steps, may surmount the natural tendency of analysts to provide expected rather than actual values.

Imposition of general specifications is likely to be self-defeating because they would have to be sufficiently broad to overcome the local systematic errors of individual laboratories. But, necessarily, imposition of local specifications would not apply to other laboratories.

6   CONCLUSION

Most collaborative studies emphasize estimation of precision.   As long as bias (trueness, accuracy) does not make a substantial contribution to overall differences from "true values," it can be tolerated, and no adjustment is made. Under such circumstances, where analyte, matrix, method, laboratory, and time contribute little to the overall effect, proficiency studies can provide a reasonable estimate of performance parameters, if the natural tendency of analysts to censor their results can be avoided.

References

1. W. Horwitz, "Protocol for the Design, Conduct and Interpretation of Method-Performance Studies; Revised 9 May 1994" Pure & Appl. Chem. 67, 331-343(1995); J. Assoc. Off. Anal. Chem. 78, 143A-160A(1995).
2. W. J. Youden and E. H. Steiner 'Statistical Manual of the AOAC' AOAC International, 481 N. Frederick Ave., Gaithersburg MD 20877 USA, 1975.
3. S. L. R. Ellison, 'Uncertainty and Method Validation'. Presented at 113th AOAC International Annual Meeting, September 30, 1999, Houston TX USA.
4. Joint FAO/WHO Codex Alimentarius Commission, Alinorm 97/23A, Appendix II, 1997. FAO, Rome, Italy.
5. FAPAS Secretariat, 'FAPAS Protocol' CSL Food Science Laboratory, Norwich Research Park, Colney, Norwich NR4 7UQ, UK, 5 th ed., 1997.
6. Horwitz, W., Kamps, L. R., and Boyer, K. W., "Quality Assurance in the Analysis of Foods for Trace Constituents" J. Assoc. Off. Anal. Chem. 63, 1344-1354 (1980).
7. Horwitz, W., Britton, P. and Chirtel, S. J., "A Simple Method for Evaluating Data from an Interlaboratory Study" J. AOAC International 81, 1257-1265 (1998).
8. Burke, J. A. 'The Interlaboratory Study in Pesticide Residue Analysis' in H. Frehse and H. Geissbuhler, 'Pesticide Residues' Pergamon Press, Oxford, 1979, pp. 19-28.
9. F. P. Byrne, in 'Accuracy in Trace Analysis' Philip D. LaFleur, Ed., NBS Special Publication 422, 1976. p. 125.
10. U.S.Pharmacopea, U. S. Pharmacopeal Convention Inc., 12601 Twinbrook Parkway, Rockville MD 20852 USA, 1999, p. 1923-4.

Optimization and Evaluation of Multi-residue Methods for Priority Pesticides in Drinking and Related Waters

European Program Standards, Measurements and Testing Project
SMT4-CT96-2142

P. Van. Wiele and F. Van Hoof*†[1], A. Bruchet and I. Schmitz[2], J.L. Guinamant and F. Acobas[3], F. Ventura[4], F. Sacher[5], I. Bobeldijk[6] and M.H. Marecos do Monte[7]

[1] SVW-BELGUIM, [2] CIRSEE-FRANCE, [3] ANJOU RECHERCHE-FRANCE, [4] AGBAR-SPAIN, [5] DVGW/TZW-GERMANY, [6] KIWA-THE NETHERLANDS, [7] LNEC-PORTUGAL

1 INTRODUCTION

This project aims at supporting the implementation of the drinking water directive by providing control laboratories with a limited set of reliable and comparable validated multiresidue methods that allow the monitoring of priority pesticides (of common interest in European countries and of major concern in specific countries) in drinking and related waters.

To reach this objective, seven major private and public European Research centres (see above) established by consensus a list of 38 priority pesticides. They share their analytical capacities and expertise to develop and validate a limited number of multiresidue methods based on off-line Solid Phase Extraction with:
-Gas Chromatography coupled with Mass Spectrometric Detection (GC-MS)
-High Performance Liquid Chromatography coupled with Ultra Violet Diode Array Detection (HPLC-UV)
-and High Performance Liquid Chromatography coupled with Mass Spectrometric Detection (HPLC-MS).

The performance data collected within this project will serve as scientific and technical basis for standardisation work within CEN TC/230 (water quality).

2 PRELIMINARY ACTIONS

2.1 Selection of priority pesticides

The selection of priority pesticides of common interest in European countries has been carried out in various steps and the final list has been reached by consensus between the partners. At first, the compilation of national lists from Germany, Belgium, France (ESU Ecotox List), Austria and U.K., resulted in a first global list of 148 substances. As partners agreed on the procedure used for the establishment of the French list, this one has been considered as a reference. After a first selection based on the presence of the 148

* Author to whom correspondence should be addressed
† Present address: Mechelsesteenweg 64, B-2018 Antwerp, Belgium

pesticides in at least two countries plus France, a second list of 68 pesticides has been obtained.

At this stage, two new lists coming from Spain and The Netherlands and an additional list from Germany have been considered and a second selection has been done. This one was based on the occurrence of the 68 compounds in the different lists and on the exclusion of several substances. Pesticides like aminotriazole, for which extraction and analysis are too particular, chemicals for which derivatization is required before analysis and compounds for which an ISO standard is currently underway, were excluded. Pesticides present in the national lists, but only cited one or two times in the previous list of 68 pesticides were also excluded.

After this second selection, a final list of 38 priority pesticides has been obtained. This list includes 27 herbicides, 7 insecticides, one fungicide and three metabolites of atrazine. Table 1 shows the common list of priority pesticides and the analytical approaches used.

Because of the rapid degradation in water, pyridat and phenmedipham have been dropped from the list of HPLC-MS amenable pesticides. Dicamba and fluroxypyr have been dropped, because they have to be isolated at low pH and need negative ionisation mode for analysis.

The 12 HPLC-MS amenable pesticides were analysed in positive ionisation mode.

2.2 Purity study on the certified pesticide reference component

Each of the reference compounds was supplied with a certificate stating the percentage purity, which has been determined by the supplier, using GC with flame ionisation detection or HPLC with diode array detection. Nuclear Magnetic Resonance was used to confirm the identity of the compound.

GC-MS or HPLC-DAD was used to confirm the purity of the compounds. Stock solutions (approximately 10,00 ng/µl, weighed accurately) of each of the certified pesticide reference compounds were prepared in ethyl acetate for GC-MS or methanol for HPLC-DAD. Each injection of a reference compound solution was followed by a 'blank' solvent injection to ensure that there was no carry-over from one injection to the next. The areas of all peaks in the total ion chromatogram for GC-MS or the areas of all peaks in the specific wavelength chromatogram for HPLC-DAD were normalised and the values of the areas due to any impurity peaks calculated as a percentage of the total area.

The purity checks of the pesticide reference compounds confirmed the information provided by the manufacturer on the purity certificates, however, for pirimicarb, there is a significant difference of 3,5 % found in purity between the manufacturer and the data generated by the work package leader.

2.3 Stability study

Before the start of the intralaboratory study, 40 replicate vials of the three solutions were stored under a variety of conditions: - 21 °C, + 4 °C, + 25 °C and + 40 °C in the dark.

Two replicate vials were transferred to a deep freezer at - 21 °C after 0, 2 and 4 weeks (short time stability study) and after 3 and 6 months (long time stability study), prior to GC-MS and HPLC analysis.

After 4 weeks and 6 months, the test solutions were analysed. The results demonstrated that no substantial degradation (> 25 %) occurred for any compounds at -21 °C and 4 °C over a 6 months period (see figure 1).

Table 2 *Chromatographic analysis of priority pesticides of the common list*

Priority pesticide	GC-MS	HPLC-DAD	HPLC-MS
Alachlor (H)	x		
Aldicarb (I)			x
Atrazine (H)	x	x	
Carbendazim (F)		x	x
Chloridazone (H)		x	x
Chlorpyrifos-ethyl (I)	x		
Chlortoluron (H)		x	x
Cyanazine (H)	x	x	
Desethylatrazine (MA)	x	x	
Desisopropylatrazine (MA)	x	x	
Dicamba (H)			o
Dichlobenil (H)	x		
Dimethoate (I)	x		
Diuron (H)		x	x
α-Endosulfan (I)	x		
β-Endosulfan (I)	x		
Ethofumesate (H)	x		
Fluroxypyr (H)			o
Hydroxyatrazine (MA)		x	x
Isoproturon (H)		x	x
Lindane (I)	x		
Linuron (H)		x	x
Metamitron (H)		x	x
Metazachlor (H)	x		
Metabenzthiazuron (H)		x	
Methomyl I)		x	x
Metobromuron (H)		x	x
Metolachlor (H)	x		
Metoxuron (H)		x	x
Metribuzine (H)	x	x	
Pendimethalin (H)	x		
Phenmedipham (H)	o		
Pirimicarb (I)	x	x	
Propazine (H)	x	x	
Pyridat (H)	.		o
Simazine (H)	x	x	
Terbutylazine (H)	x	x	
Terbutryn (H)	x	x	
Trifluralin (H)	x		
Total	N=23	N=22	N=12

0: skipped from the priority list.
(H): Herbicide (I): Insecticide (F): Fungicide (MA): Metabolite of Atrazine

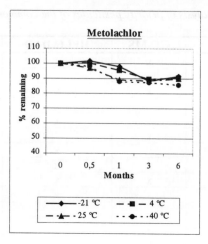

Figure 1 *Stability study for Metolachlor*

2.4 Homogeneity study

Five vials of the 17 HPLC-DAD test solutions and six vials of the 20 GC-MS test solutions were selected at regular intervals within the batch and analysed to assess inter vial homogeneity. Eight replicate injections from a single vial of each method were performed to demonstrate intra vial homogeneity (Figure 2). Both the intra and inter vial relative standard deviation (RSD, in %) didn't exceed 10 % for both methods. Application of an $F_{0.95}$-test[1] demonstrated that in the majority of cases the variance for intra-vial homogeneity was not significantly different when compared with variances associated with inter vial homogeneity. However methomyl (test solution for HPLC-DAD) did exceed the $F_{0.95}$-critical value, which indicates that either the solutions were heterogeneous with respect to this compound, or that the measurement procedure was unsatisfactory. We considered the measurement procedure for methomyl to be unsatisfactory, since methomyl was coeluting with the solvent peak, rendering the quantification of this compound difficult.

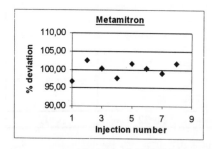
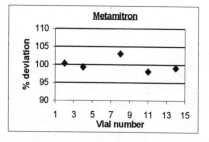

Figure 2 *Intra and inter vial variations for metamitron*

3 INTRALABORATORY STUDY

In this study, the seven partners mentioned as co-authors were involved. Before the start of the study, the work package leaders involved optimised the methods of analyses, which resulted in two detailed multiresidue methods.

3.1 Methods of analysis

3.2.1 HPLC-DAD.   The HPLC-UV/DAD multiresidue method is based on:
1. Solid phase extraction of pesticide residues
    e.g. SDB-1 cartridges (6 ml, 200 mg, J.T. Baker)
    Step 1: Condition cartridge with 10 ml of methanol at 10 ml/min
    Step 2: Condition cartridge with 10 ml of Milli.Q water at 10 ml/min
    Step 3: Condition cartridge with 10 ml of Milli.Q water at 10 ml/min
    Step 4: Load 500 ml of sample onto cartridge at 5-10 ml/min
    Step 5: Dry column with gas for 5 minutes
    Step 6: Soak and collect 2 ml fraction using methanol/acetonitrile (60:40) at 3 ml/min
    Step 7: Collect 8 ml fraction into sample tube using methanol/acetonitrile (60:40) at 3 ml/min
2. Concentration of the organic extracts under a gentle stream of nitrogen till 0,3 ml. The volume is adjusted to 0,5 ml with pure water.
3. Analysis of residues by High Performance Liquid Chromatography coupled with UV Diode Array detection.
    e.g. Supelcosil ABZ+plus column (25 cm x 4,6 mm x 5 μm or 25 cm x 2,1 mm x 3 μm); Supelcosil ABZ+plus guard column (2 cm x 4,6 mm x 5 μm).
    3.2.2 GC-MS.   The GC-MS multiresidue method is based on:
1. Solid phase extraction of pesticide residues
    e.g. SDB-1 cartridges (6 ml, 200 mg, J.T. Baker)
    Step 1: Condition cartridge with 3 ml of ethylacetate at 10 ml/min
    Step 2: Condition cartridge with 3 ml of methanol at 10 ml/min
    Step 3: Condition cartridge with 6 ml of Milli.Q water at 10 ml/min
    Step 4: Load 500 ml of sample onto cartridge at 5-10 ml/min
    Step 5: Dry column with gas for 15 minutes
    Step 6: Soak and collect 2,5 ml fraction using ethylacetate at 0,8 ml/min
    Step 7: Collect 2,5 ml fraction into sample tube using ethylacetate at 0,8 ml/min
2. Concentration of the organic extracts under a gentle stream of nitrogen till 0,3 ml. The volume is adjusted to 0,5 ml with ethylacetate.
3. Analysis of residues by Gas Chromatography coupled with Mass Spectrometric Detection.
    e.g. non polar columns with 5 % phenyl groups:
    - PTE5 (0,25 μm; 30 m x 0,32 mm)
    - BPX5 (0,25 μm; 50 m x 0,32 mm)
    - DB5-ms (0,25 μm; 50 m x 0,25 mm)
    - DB5 (0,33 μm; 12 m x 0,2 mm)
The temperature programme applied, should allow the best separation of the priority pesticides in an acceptable time. The duration time of the analysis should not exceed 45 minutes.

3.2 Statistical treatment of the results

All data received were first screened through the Dixon test for outliers [1].
After screening all data for Dixon outliers, the results were further compared through two tests: 1) the Cochran test, which was used for the removal of laboratories showing significantly greater variability among replicate analyses than other laboratories, was applied as a one tail test at a probability of 2,5 % and 2) the single Grubbs test, which was used as a test for removal of laboratories with extreme averages, was applied as a two tail test at a probability of 2,5 % [2,3].

3.3 Intralaboratory study – results (Table 2 and 3)

An analytical protocol for the intralaboratory study was prepared by the work package leader. It contained instructions for the analyses of the both test solutions and report forms to ensure that results would be reported in standard format.

The protocol allowed participants to select the GC and/or HPLC instrumentation, columns and conditions which they felt were most appropriate. Full details of the chromatographic instrumentation and conditions used, was requested for each test solution, together with examples of typical chromatograms and calibration plots. Finally, each laboratory had to determine basic performance data i.e. linearity, limit of detection, precision (via reproducibility and repeatability) and trueness [2,3].

3.3.1 Linearity. Linearity is controlled via two statistical methods:
1) The calculation of an eight-point calibration curve, based on linear regression and the correlation coefficient 'r' and
2) Each point of the curve is tested on linearity by drawing a linearity plot. For this purpose, the individual response/mass ratios are calculated for each of the eight points. The ratio for each calibration point should not differ more than ± 10 % from the mean of the response/mass ratio of all points of the calibration curve. Only one of the eight calibration points may not satisfy this criterion.

The overall result is that the methods using the HPLC-DAD give linearity for each compound. Problems were encountered receiving linearity for GC-MS. To solve this problem, isooctane will be used as 'keeper' during the interlaboratory study (see Point 4).

3.3.2 Limit of detection. The limit of detection for a specific compound and for each laboratory was calculated as three times the within batch standard deviation using 7 replicate samples spiked with a low concentration of pesticides. The second approach, where the limit of detection is calculated as five times the relative within batch standard deviation of a blank sample, did not yield any results because the blank sample (Evian drinking water) did not give any response at the specific retention time. The average limit of detection for a specific pesticide is calculated as the average of the results of all the laboratories. The maximum laboratory result, which is higher than 5 times the minimum laboratory result, is skipped in the calculation. The same for the minimum laboratory result, which is lower than 5 times the maximum laboratory result, is skipped in the calculation [4,5].
All pesticides have an average limit of detection of 0,01 µg/l except metoxuron and metabenzthiazuron using HPLC and desisopropylatrazine, trifluralin, pirimicarb, cyanazine, metazachlor and α-,β-endosulfan for GC-MS, which have a limit of detection of 0,02 µg/l.

3.3.3 Reproducibility. The reproducibility is expressed as relative standard deviation (in %). The HPLC method is reproducible at the 0,1 µg/l concentration level for all pesticides except for hydroxyatrazine. The GC method is not reproducible for lindane, chlorpyrifos, pendimethalin, α-,β-endosulfan, metamitron and chloridazone

3.3.4 Repeatability. Repeatability is expressed as relative standard deviation (in %). The HPLC method is repeatable for all pesticides at the 0,1 μg/l level except for hydroxyatrazine according to the requirements of Directive 98/83/EG[6].
The GC method is not repeatable for trifluralin, metribuzine, cyanazine, pendimethalin and metamitron.

3.3.5 Trueness. Trueness is expressed as recovery in μg/l and the results are taken from the reproducibility and repeatability experiments. The HPLC method is accurate for all pesticides at the 0,1 μg/l level except for hydroxyatrazine. The GC method is not accurate for trifluralin, lindane, chlorpyrifos, pendimethalin, α-endosulfan and chloridazone.

3.3.6 Conclusion. The results for precision are based on data sets limited to 5 observations or less and thus the values obtained are likely to be estimates of true values.

Concerning HPLC, the degree of intralaboratory agreement was remarkable good, except for hydroxyatrazine, which will be studied in the method using the HPLC-MS-(MS) method.

The method based on GC-MS gave problems for each parameter defined. The basis of these problems is probably due to poor linearity for each of the compounds and of the analytical difficulties of most of the laboratories. Due to this, some results were relatively poor and deleted by the Cochran or Grubbs statistical test.

3.4 Performance in comparison with Drinking Water Directive

The Drinking Water Directive 98/83/EG[6] defines the following performance characteristics for individual pesticides: trueness: 25 % of the parametric value; precision: 25 % of the parametric value and limit of detection: 25 % of the parametric value. The parametric value for individual pesticides is 0,1 μg/l.

In this report, trueness is expressed as % recovery, precision is defined as repeatability and reproducibility, and both expressed as % standard deviation.

The results in bold, shown in Table 2 and Table 3, do not meet the requirements of Directive 98/83/EG (between brackets are the number of results produced by the laboratories after deleting Grubbs and Cochran outliers).

4 RECOMMENDATIONS ON THE METHODOLOGY

Due to the poor results for trueness and precision, hydroxyatrazine will be skipped in the interlaboratory exercise.

After screening of all individual results, obtaining satisfactory results for linearity is one of the major problems in the GC-MS method. After the validation of the results, all the partners decided to do additional tests, namely, the comparison of a 'reference' ethyl acetate with dried ethyl acetate, testing ethyl acetate from different manufacturers and the use of dichloromethane, toluene and isooctane as 'keeper'. After these tests, we concluded that poor linearity results were due to the nature of the solvent. Ethyl acetate is a suitable solvent for the extraction of the pesticides investigated, but is less suitable for the gas chromatographic analysis. For the interlaboratory experiment, isooctane will be used as solvent for the chromatographic analysis. The choice of isooctane is made to avoid the use of chlorinated (dichloromethane) and harmful (toluene) solvents. The mother standard solutions will be diluted in isooctane to obtain better linearity results. Isooctane will be used as a keeper during the extraction and will also be used to dilute the evaporated extract before injection to obtain better precision and accuracy results.

For the preparation of the test solutions for the interlaboratory study, pirimicarb will be bought again and checked for its purity, due to his discrepancy in purity with the manufacturer's results.

For the GC-method, accuracy should be improved for the chlorpesticides lindane, chlorpyrifos and endosulfan, for metribuzine, trifluralin, pendimethalin, metamitron and chloridazone. The last two pesticides can more satisfyingly be analysed with the HPLC-DAD method and will be skipped during the interlaboratory exercise.

Table 2 *HPLC-DAD-UV results*

HPLC-DAD-UV RESULTS AT LEVEL 0,1 µg/l (mean of 5 labs)					
			Precision		
	Linearity		LOD	Repeat.	Reprod.	Trueness
Pesticide	" r "	lin plot	in µg/l	in %	in %	in %
Methomyl	1,000	yes	0,01	14 (n=5)	8,2 (n=5)	90 (n=10)
Desisopropylatrazine	1,000	yes	0,01	4,5 (n=4)	4,7 (n=4)	92 (n=7)
Metamitron	1,000	yes	0,01	13 (n=4)	12 (n=5)	91 (n=9)
Hydroxyatrazine	1,000	yes	0,01	50 (n=5)	50 (n=5)	57 (n=10)
Carbendazim	1,000	yes	0,01	10 (n=5)	10 (n=4)	93 (n=9)
Chloridazone	1,000	yes	0,01	11 (n=4)	14 (n=5)	100 (n=7)
Metoxuron	0,999	yes	0,02	16 (n=5)	9,8 (n=4)	91 (n=9)
Metribuzine	1,000	yes	0,01	7,9 (n=5)	10 (n=5)	92 (n=10)
Simazine	1,000	yes	0,02	7,9 (n=4)	9,8 (n=4)	93 (n=9)
Cyanazine	1,000	yes	0,01	4,8 (n=5)	6,4 (n=4)	91 (n=9)
Pirimicarb	1,000	yes	0,01	15 (n=4)	17 (n=5)	88 (n=10)
Methabenzthiazuron	1,000	yes	0,02	6,0 (n=5)	7,6 (n=4)	88 (n=9)
Atrazine	1,000	yes	0,01	7,3 (n=5)	9,6 (n=4)	90 (n=9)
Chloortoluron	1,000	yes	0,01	4,7 (n=5)	6,2 (n=4)	95 (n=8)
Isoproturon	0,999	yes	0,01	9,2 (n=5)	4,8 (n=4)	90 (n=8)
Metobromuron	1,000	yes	0,01	9,8 (n=4)	9,7 (n=5)	91 (n=10)
Diuron	1,000	yes	0,01	6,3 (n=4)	16 (n=4)	92 (n=6)
Propazine	1,000	yes	0,01	8,8 (n=4)	12 (n=4)	89 (n=8)
Terbutylazine	1,000	yes	0,01	10 (n=5)	16 (n=4)	86 (n=10)
Linuron	0,999	yes	0,01	14 (n=5)	14 (n=5)	88 (n=10)
Terbutryn	0,999	yes	0,02	12 (n=5)	15 (n=5)	84 (n=10)

Table 3 GC-MS results

GC-MS RESULTS AT LEVEL 0,1 μg/l (mean of 5 labs)					
			Precision		
	Linearity		LOD	Repeat.	Reprod.	Trueness
Pesticide	" r "	lin plot	in μg/l	in %	in %	in %
Dichlobenil	0,998	yes	0,01	4,2 (n=4)	18 (n=5)	81 (n=7)
Desisopropylatrazine	0,998	yes	0,02	8,0 (n=4)	11 (n=5)	88 (n=9)
Desethylatrazine	0,999	no	0,01	17 (n=3)	15 (n=5)	91 (n=9)
Trifluralin	0,996	no	0,02	48 (n=3)	23 (n=4)	50 (n=7)
Dimethoate	0,997	no	0,01	5,2 (n=3)	16 (n=4)	80 (n=7)
Simazine	0,999	yes	0,01	20 (n=3)	13 (n=5)	91 (n=9)
Atrazine	0,999	yes	0,01	19 (n=3)	13 (n=5)	94 (n=9)
Propazine	0,999	no	0,01	13 (n=4)	13 (n=5)	92 (n=9)
Lindane	1,000	yes	0,01	24 (n=3)	54 (n=4)	41 (n=8)
Terbutylazine	0,999	no	0,01	19 (n=4)	11 (n=5)	91 (n=9)
Pirimicarb	0,999	no	0,02	21 (n=4)	11 (n=5)	88 (n=9)
Metribuzine	0,998	no	0,01	32 (n=3)	19 (n=5)	73 (n=8)
Alachlor	0,998	no	0,01	16 (n=4)	20 (n=5)	76 (n=8)
Terbutryn	0,997	no	0,01	18 (n=4)	14 (n=5)	87 (n=9)
Ethofumesate	0,999	no	0,01	17 (n=4)	11 (n=4)	92 (n=8)
Metolachlor	0,998	no	0,01	16 (n=4)	17 (n=5)	85 (n=8)
Chlorpyrifos	0,999	no	0,01	18 (n=3)	27 (n=5)	64 (n=9)
Cyanazine	0,997	no	0,02	31 (n=4)	16 (n=4)	82 (n=7)
Metazachlor	0,997	no	0,02	23 (n=4)	26 (n=4)	79 (n=8)
Pendimethalin	0,993	no	0,01	64 (n=3)	31 (n=5)	62 (n=9)
α-Endosulfan	0,999	no	0,02	20 (n=4)	32 (n=4)	66 (n=8)
Metamitron	0,995	no	0,01	52 (n=2)	48 (n=2)	78 (n=4)
β-Endosulfan	0,999	no	0,02	24 (n=3)	28 (n=5)	83 (n=8)
Chloridazone	0,998	no	0,03	80 (n=3)	64 (n=4)	47 (n=7)

5 INTERLABORATORY STUDY

The interlaboratory exercise has been organised with 16 additional European laboratories to validate the methods. An analytical protocol, reporting sheets on hard copy and electronic copy (PC diskette) and five test solutions have been supplied. Two test solutions for GC and HPLC purpose of a unknown concentration were needed to determine the concentration of the different pesticides, two test solutions for GC and HPLC purpose of a known concentration were needed to determine the linearity and the limit of detection and one internal standard solution with anthracene-d10 to control the injection step of the GC device. Two matrices were tested: a commercial drinking water and a tap water. A detailed set-up and methodology of the project was shown on a video produced by the partners involved. This video was supplied to all co-operating laboratories.

Again, the protocol allowed participants to select the GC and/or HPLC instrumentation, columns and conditions which they felt were most appropriate. Full details of the chromatographic instrumentation and conditions used, was requested for each test solution, together with examples of typical chromatograms and calibration plots. This study is still ongoing.

References

1. J.C. Miller and J.N. Miller, 'Statistics for Analytical Chemistry', Chichester, Second edition, 1988.
2. W.J. Youden and E.H. Steiner 'Guidelines for collaborative study procedures', fourth (final) draft, *J. Assoc. Off. Anal.*, 1989, **72**, 694-705.
3. International Organisation for Standardisation, 'Accuracy (trueness and precision) of measurement methods and results', ISO 5725, Geneva, Switzerland, 1994.
4. Th.H.M. Noij and J.H. Knijff 'Organisatie en uitvoering van Methode Evaluerende Ringonderzoeken (MEO's)', Kiwa report SWE 98.002, Nieuwegein, 1998.
5. Th.H.M. Noij and J.H. Knijff 'Validatie van bepalingsmethoden voor organische micro-verontreinigingen', Kiwa report SWE 96.016, Nieuwegein, 1996.
6. Directive 98/83/EC, 1998, PB L 330, 32-54

Validation of Analytical Data in a Research and Development Environment

Ronald Hoogerbugge and Piet van Zoonen

NATIONAL INSTITUTE OF PUBLIC HEALTH AND THE ENVIRONMENT (RIVM), PO BOX 1,
NL-3720 BA BILTHOVEN, THE NETHERLANDS

1 INTRODUCTION

Analytical results are often called validated when standard methods, preferably tested by interlaboratory comparison, are used. The result of any validation procedure is the assessment of a number of validation parameters, like trueness and reproducibility, are determined for a certain scope of application (matrix type concentration etc.). To confirm the validation parameters for a particular laboratory an in-house validation program can suffice, usually performed before the unknown samples are analysed. The real unknown samples will then be analysed in series in combination with batchwise checks and control samples to ensure that the validation parameters as determined for the method in the extensive validation study are maintained.

This validation strategy is less suitable in typical non-routine research and development situations. The interlaboratory validation might be completely impossible, due to the lack of available participating laboratories. Moreover also the within laboratory strategy including a complete validation-application cycle can be very inefficient for a method which might only be used once in a particular study on a limited number of samples. For such incidental use of analytical methods the validation should be much more focused in the primary objective, validation of the results of the study, then on the method itself. Focussing on the results of the study implies that validation of many important variables, such as the concentration of the component, the type of matrices, the operator and all kind of possible environmental parameters are only relevant for the amount of variation which actually occur while performing the study. This requires a tailor-made validation strategy to be performed for the particular study.

To give an example for a standard operating procedure for the validation of research and development results a description of the procure, which is used in our laboratory for several years, is given below. The majority of them is based on choices made in real studies occurring in the last 10 years at our laboratory. Some examples of validations are recently published [1,2,3].

2   OBJECTIVE

This guideline contains a number of strategies for the validation of the results of various types of analytical chemical studies. In special cases the study director can deviate from this guideline.

3   AREA OF APPLICATION

This guideline's primarily purpose is to assure the quality of analysis results as obtained within the analytical chemical laboratory as part of the study. The objective of the study is therefore crucial in the selection of the optimal quality control strategy. This implies that a considerable part of the validation can be integrated in the particular study.

Procedures to quantify performance characteristics for *analytical methods* are described, and defined, in for example [4,5].

4   ORGANISATION OF THE CONTROL REGIME

4.1 Planning

The validation strategy is described in the study plan. The plan may also contain the special requirements as defined by the end-user of the data. The study director is responsible for the exploration of the available capacity and expertise. These are not only the participants who are directly involved in the particular study but also personal working in related fields. Depending on the type of study, examples for that kind of relations can be, the chemometrician, when the study requires special statistical features, facilitating departments, in case of major logistical implications or the quality assurance officer, for special QA/QC demands like the compliance to additional QA/QC requirements like the OECD-GLP.

4.2 During research and reporting

The study director is responsible for the whole set of quality control actions and authorises them by reporting. For practical purpose the following tasks will be used:

1) The operator that conducted a particular (part of the) study authorises these results by signing the research file. The operator checks the validation results against the study plan and contacts the study director when compliance appears to be impossible.

2) A colleague operator checks the research results with the raw data. A selection of the data might be sampled. In this selection samples with deviating results or with particularly important results should be included. Such a check may also be performed by using less accurate but very independent methods of estimation like using a ruler instead of a data system or using a calculator instead of a spreadsheet program. This check should include the possible presence of typical human (incidental) errors like the interchange of samples/components or mistakes by the calculation of correction factors or converting factors and so on. The name of the colleague who has performed these checks should be recorded for example in the logbook of the original operator.

3) The study director checks the consistency of the results in the complete dataset and with available additional information. For large (multivariate) data sets statistical techniques can be used for this check. The study director also checks whether all deviating results, or results with a possible major impact, had all the special attention, which they rationally should deserve. Finally the study director checks whether all deviations from the research plan are properly addressed in the report.

4) The execution of the study (1) and the check by a colleague (2) cannot be done by the same person. Other combinations are possible.

5 SAMPLE IDENTIFICATION/ TREATMENT

Interchange of sample identification, by the operators, can be prevented as much as possible by systematic methods to make extracts dilutions and so on. If feasible the relation between the various stages in the analyses will be checked afterwards. For example for series of samples analysed with an autosampler the relation between the vials and the chromatogram can be ensured by checking after the analysis whether each vial was on its proper position indeed.

For deviating results (from some expectation value) or for results with a major impact the analyses will be repeated and/or confirmed by an independent method.

6 VALIDATION PARAMETERS OF METHODS OF ANALYSES

Although the R&D environment often leads to quality control strategies which are more focused on the validation of the particular study than on the analytical method an adequate history of validation data is considered necessary. Therefore the validation information for the most important analytical techniques are compiled on validation sheets, which are stored in a database. The information can be used by the planning of future research to ensure that the agreements with the end-user, as described in the study plan, are feasible.

7   LIMIT OF DETECTION

{Concentration derived from the smallest response that can be detected with reasonable certainty for a given analytical procedure. In line with [6]}

The limit of detection is generally associated with the signal that equals 3 standard deviations (repeatability conditions) from the background. Assuming a normal distribution this implies that the signal differs from the background with a certainty of more that 99 %.

1) In a chromatogram this is interpreted as three times the noiseband.
2) For additions this is three times the standard deviation obtained at or about the limit.
3) Using an appropriate calibration curve this is the estimated standard deviation at the lowest calibration level.

In the interpretation of the results of the above approaches a number of problems might arise:

The first method might result in a quite different value than the other approaches. The result of the first method can be misleading for extreme (in)stability and or frequency characteristic of the background.

For the latter two approaches it will often be impossible or impractical to perform these experiments for every single sample. Therefore a single sample or a selection of samples is used and assumed to be representative for all samples. The latter two methods also may suffer from the statistical uncertainty in the calculated standard deviation.

For these reasons one preferably does not focus on one calculated number as a claim for the limit of detection. This can be solved by multiplying one the numbers by a factor (>1) which is adjusted to the possible consequences of uncertainties that will probably disturb the statistical assumption but that can not be quantified. Such a factor can also be used to harmonise or round detection limits, which are not significantly different. When reporting detection limits one should also consider the agreements made in the study plan.

Also the results expected in a particular study have their impact on the nature of the measurements to be performed. For studies in which the vast majority of the measurements is to be expected to be reported as below the LOD, a situation often encountered in pesticide monitoring programmes, the control and validation regime should be focussed on assuring exactly this parameter. This is preferably done by standard additions at or about the LOD, and analysing these additions within the series of real samples. Approximately 1% of the additions may be found as "not detected", the RSD values of the additions will be around 30%

The LOD is based on the statistics associated with individual analytical results. In many monitoring studies however the individual results must be considered as intermediate data that are to be used for the best estimate of e.g. intake, trends or contours. The further processing of "non-data" reported as < LOD, is in most cases quite troublesome, it could therefore be necessary to report data which are actually below the previously defined LOD. It should however be emphasised in the report that the uncertainty is large and that the individual results do not comply with the statistical criteria for detectability. These data combined with the other indications on the substance/sample combination, such as ubiquity may give a good estimate of the data to be entered for the LOD (e.g. actual value, half or zero) in the evaluation of the total data-set. The necessity to provide these data and the objective should be properly indicated in the study description because it is often not possible or labour intensive to obtain these data after closure of the project

8   TRUENESS

{The closeness of agreement between the average value obtained from a large series of test results and an accepted reference value. [7]}
Use wherever possible
1) Appropriate reference material
2) Enlistment in a relevant proficiency testing programme

Both approaches have the big advantage that the actual measurement of the trueness is not dependent on the laboratories own resources. An additional advantage of proficiency testing schemes is that the "real" value of the sample is unknown during processing, which means that the situation during the determination of the trueness truly resembles the analysis of real unknown samples.

The disadvantage of proficiency testing as preferred instrument for validation of measurement results can be: a mismatch between the sample composition (matrix and concentration), or a mismatch in timing between the interlaboratory scheme and the study to be validated. Also a corresponding proficiency testing scheme might be entirely non-available.

An important alternative for the approaches mentioned above is the performance of recovery experiments. These must be representative for the real samples with respect to concentration levels, sample type and experimental conditions. The number of experiments needed depends on the variety in these parameters over the total set of samples in relation to the calibration procedure.

Trueness can also be assessed by parallel analysis of (a subset of) the samples by alternative analytical methods, these methods should however be completely independent, at least using alternative isolation techniques.

9   CALIBRATION PROCEDURE

{Calibration function: Mathematical relation between the concentration and the response.}
Note that the calibration should in principle cover the complete analyses procedure.
In our laboratory calibration curves are calculated, or validated using Calwer [8].

The calibration procedure can be more or less extensive as described below for some examples.

9.1 "Comprehensive Calibration"

In this rather extreme example all sources of variance are incorporated in the calibration procedure, important aspect are:
- The calibration points are multi-level additions to the actual matrix
- These calibration samples are analysed within series of the unknown samples

The advantage of this approach is that LOD, recovery, linearity and variance are determined in concurrently with the advancement of the study and are there consistent and as representative as possible. A drawback is that the performance parameters are not known before (at least part of) the samples are analysed.

9.2 "Minimum Calibration"

Only multi-level calibration on the basis of standard solutions is used. All other performance parameters must be determined in additional experiments.

9.3 "Procedural Calibration"

Calibration is based on additions of standards to representative extracts of samples, e.g. soil extracts. This means that the extraction procedure needs a separate validation. This intermediate procedure often is the most practical solution. It is very valuable when the extraction efficiency is strongly dependent on the variety of the samples (as in e.g. soil analysis). In this case it is quite difficult to completely correct for the variance of the extraction procedure in the calibration.

9.4 Internal Standard

The use of appropriate internal standards minimises the variance introduced by the calibration procedure. The key to the proper use of internal standardisation is that in calibration, validation and control samples the internal standard is applied in (as much as possible) the same way as in the analysis of the samples.

The suitability of an internal standard depends on the degree of similarity in behaviour between the internal standard and the target compounds. The highest degree in similarity can be accomplished by the use if isotopically labelled standards. These types of standards however do require a mass spectrometer to distinguish between the internal standard and the target compound. Internal standards can be added at alternative stages of the analytical procedure, obviously the stage at which the addition takes place determines the part of the procedure that is secured by the procedure and the requirements of similarity of the internal standard. Internal standards for the behaviour of a chromatographic system are generally more easy to find than internal standards that mimic the extraction efficiency. The addition of alternative internal standards at different steps in the analytical procedure gives a good indication of the critical steps in the analytical chain.

9.5 Single-point Calibration

In analytical procedures where drift gives a major contribution to the variance in the results than the calibration function itself, the application of a single point calibration procedure is recommended. In this case the concentrations are calculated on the basis of a single external reference point. Linearity and offset of the calibration function must then be checked in separate experiments.

A combination of the advantages of single point calibration and a calibration function can be achieved by treating the external standard in the calculation as an internal standard. This opens the possibility to test whether the variance improves by this procedure.

9.6 Blanks

The analysis of a blank samples (or of a blank without matrix) may provide a value measurable value, thus influencing the LOD. When the blank value is more or less constant it can be incorporated in the calibration procedure thus providing an automatic correction for the blank, including the uncertainty in the blank value in the total uncertainty of the analytical result. This requires the blank samples to be treated in exactly the same way as the calibration samples; it can therefore not be used in the "minimum" calibration approach. Another exception in the standard addition approach because this procedure uses the intercept of the calibration function for the determination of the analytical result.

10 PRECISION

{The closeness of agreement between independent test results obtained under stipulated conditions [7]}. We express all precision measures as the standard deviation of the measurement results.

10.1 Within Laboratory Reproducibility

{Precision under conditions where test results are obtained using the same method on identical test items in the same laboratory with different operators using different equipment. In line with [7] }

If a study is performed over a prolonged period and analysed in multiple series a check is necessary on the comparability of the results obtained in the different series. The best method to assure comparability is the use of a reference sample in combination with a control chart based on the principles of Shewart [9,10]. In multi-component analysis a relevant sum parameter, such as the TEQ in dioxin analysis, or some critical components can be selected on the basis of their occurrence or physico-chemical chromatographic properties. An alternative approach is the use of multivariate control charts [10].

In many instances it is impossible to obtain stable reference samples, less stable reference samples can be used if in the chart a decreasing expected value is used. Another method to control the between series variance is the use of duplicate measurement divided over subsequent series. In comparison to the reference sample method this approach gives a poorer estimate of the time-dependency, however it has a better capability to detect matrix effects. An improvement to this procedure often used in epidemiological studies is the randomisation of the samples in combination with blind duplicates. Blind duplicates have the advantage that they are unknown to the researcher, so that they mimic the actual conditions. Moreover this method may be able to detect incidents like exchange of samples and/or results.

Studies aimed at the detection of trends in the occurrence of pesticides or contaminants are often performed once every 5 or 10 years, this requires a check on the comparability of data obtained several years apart. Basically stable reference materials or the continuous involvement in proficiency testing programmes is necessary to assure comparability on these types of studies. If none of the above is available additional validation studies are needed.

For analytical methods that have a relatively high failure risk or a high variance, with respect to the aim of the study, replicate analysis of all samples may be considered. These measurements should than de be made as independent as possible and should therefore be treated and analysed in different series.

10.2 Repeatability

{Precision under conditions where independent test results are obtained with the same method on identical test items in the same laboratory by the same operator using the same equipment within short intervals of time. [7]}

Many sources of variance can be kept low and/or constant by performing the analysis under the same conditions; this can be assured by performing all the analysis in a short time period. In this case little information becomes available on systematic errors. This is not a viable approach for the assessment of data collected over a prolonged period.

Repeatability conditions however may be preferred if the trueness of the results does not play a role but only the comparison of the relative values within a set of samples. Examples of these studies are homogeneity and case-control studies.

11 CONFIRMATION OF THE IDENTITY OF A COMPOUND

With every measurement there is a probability that the observed response is not entirely or even not al all caused by the target compound. If the specificity of the method used is assessed to be not specific, more certainty concerning the identity of the compound can be achieved by reanalysing the sample with an alternative analytical procedure. The selection of samples to be reanalysed depends on the relative importance of the final result (e.g. non-compliance with an MRL) and the estimated probability of interferences in the analytical method. In large series one can start with the confirmation of a limited number of samples and proceed depending on the result. Another way to economise on the number of confirmatory analysis is the use of composite samples.

The certainty on the identity that can be reached depends on the selectivity of the methods used in combination with their independendness. There are no clear-cut criteria for the assessment of the certainty for identification, therefore, generally, the classification "sufficient certainty" is largely dependent on expert judgement, while for specific areas and or techniques criteria exist. For mass spectrometry such criteria were thoroughly discussed in ref. [11]

The availability of suitable confirmatory method opens the way to the appropriate use of screening methods. In studies where this approach is used one hopes to split the total sample load in a large number of negatives, with a relatively low number of positives, the latter to be confirmed with the alternative technique. The most important validation parameter that has to de determined for the screening method is the percentage of false negatives. This should be acceptable in view of the aim of the study. If the number of false negatives exceeds the acceptable limit one can analyse more samples by the confirmatory technique in order to lower the over-all percentage of false negatives. The confirmatory method should that have a comparable or lower ratio of false positives.

12 UNKNOWN MATRICES

When analysing samples, for which an initial validation was not (sufficiently) representative, re-evaluation of the method is necessary. If only a small number of samples is analysed addition of standard may be sufficient, however when isolation of the analytes from the matrix is considered to be the major problem, this approach is of limited value. In this case variations in the extraction conditions (time, temperature, solvent) may give additional information on the extraction efficiency. For small series extensive labour-intensive extraction procedures may be preferred over more cost-effective methods which have been shown to perform adequately for other sample types. For larger series a mixed approach using a selected number of samples or composite samples may be useful; special attention must be paid to deviant or important results.

13 REFERENCE COMPOUNDS/SOLUTIONS

Standard solution should preferably be prepared using two alternative pathways. The solutions obtained through the alternative pathways are checked. Once validated solution exists they can be used to validate new solutions, by comparison against previously prepared solutions. Mixed standards have the additional feature that the pattern of the different compounds can be compared, this is especially valuable with standards of closely related compounds such as chlorophenols and PCBs.

Solid primary standards for pesticides can be tested by using IR-analysis [12].

In the preparation of mixed standard solutions for large numbers of compounds one could expect that the assignment of peaks to the compounds might lead to errors. The preparation of mixtures containing subsets of the final standard avoids these errors.

References

[1] P. van Zoonen, R. Hoogerbrugge, S.M. Gort, H.J. van de Wiel and H.A. van 't Klooster, TRAC, 18 (1999) 584-593.

[2] E.A. Hogendoorn, R. Hoogerbrugge, R.A. Baumann, H.D. Meiring, A.P.J.M. de Jong and P. van Zoonen, J. of chromatogr., 754 (1996)

[3] R. Hoogerbrugge, 'Validation of Analytical Chemical Results', Chapter 21 in R.C. Schothorst et al (eds.) "Quality Assurance and Quality Control for National Reference Laboratories for Detecting Residues in Biological Samples." Proceedings EU-workshop RIVM Bilthoven March 21-25 (1994).

[4] ISO 9169, Air Quality – Determination of performance characteristics of measurement methods, ISO, Geneva, Switzerland (1994).

[5] ISO 8466 Water quality – Calibration and evaluation of analytical methods and estimation of performance characteristics, ISO, Geneva, Switzerland (1990).

[6] IUPAC, Spectrachim. Aca 33 B (1978) 242.

[7] ISO 3534 Statistics-Vocabulary and Symbols. ISO, Geneva, Switzerland 1993.

[8] S.M. Gort and R. Hoogerbrugge, Chemom. and Intell. Lab. Syst. 28 (1995) 193-199.

[9] ISO 8258, Shewart control charts, Geneva, Switzerland 1991.

[10] D.L. Massart, B.G.M. Vandeginste, S.N. Deming, Y. Michotte and L. Kaufman, "Chemometrics a textbook", Elsevier Amsterdan (1988).

[11] R.A. Bethem and R.K. Boyd, 'J. Soc. Mass Spectrom. 9 (1998) 643-648.

[12] T. Visser, 'Infrared spectra of pesticides', Marcel Dekker, New York (1993).

Performance Validation of a Multi-residue Method for 170 Pesticides in Kiwifruit

P.T. Holland, A.J. Boyd and C.P. Malcolm

HORTICULTURE & FOOD RESEARCH INSTITUTE OF N.Z. LTD, RUAKURA RESEARCH CENTRE, HAMILTON, NEW ZEALAND

1 INTRODUCTION

Multi-residue analysis of fruit and vegetables is carried out for a variety of purposes[1]. One of the most demanding areas is certification of exports where low detection limits for a wide range of pesticides and fast sample turnaround are important requirements. Kiwifruit is now one of New Zealand's main horticultural crops with production exceeding 200,000 tonnes per annum.

The importance of export markets to the industry has led to a strong emphasis on quality systems covering all aspects of the production and marketing of the fruit. Spray schedules and pesticide residues are areas which are carefully managed and monitored. While kiwifruit as a crop is not particularly susceptible to pests or diseases, exports must meet high phytosanitary standards. There are few national maximum residue limits (MRLs) set for kiwifruit in most countries, although there are now several Codex MRLs. Many international purchasers such as large supermarket chains now require fruit to meet limits lower than national MRLs, with 'no detectable residue' being the desired outcome. Thus kiwifruit exporters need testing to be carried on lines of fruit to ensure the various required standards are being met.

A method of residue analysis developed for kiwifruit[2] based on methanol extraction, toluene partion and carbon column cleanup gave excellent results for residues of the pesticides registered for use on kiwifruit and a range of other low-medium polarity pesticides. The method addressed the problems for analysis of kiwifruit presented by the high levels of chlorophylls and organic acids. However certification has become stricter and a method was required which could determine a wider range of pesticides, including more polar materials such as methamidaphos down to low detection limits. The use of ethyl acetate as an extractant and size exclusion chormatography (SEC) as a cleanup for high resolution gas chromatography (HRGC) with selective detection is a versatile approach to multi-residue analysis. It has been adopted as the basis for national methods for residue analysis of foods in many countries. Our laboratory adapted for use on kiwifruit the method developed in The Netherlands[3,4].

The following discussion describes the method as validated for kiwifruit in our laboratory and presents some of the on-going quality assurance data gathered during routine application of the method over several years. The approach adopted to estimation of detection limits and monitoring of method performance is put forward as a practical means of linking in-house method characterisation and validation to the on-going need for measures of method performance during use.

2 METHOD

2.1 Reagents and Materials
Ethyl acetate and cyclohexane (Nanograde, Mallinkrodt)
Sodium sulphate, anhydrous (Mallinkrodt), ashed 600° for 6 hours
Biobeads SX-3, 200-400 mesh (Biorad)
Microfilters, 0.45 μm (LC13, Acro)
FSOT column, 25 m x 0.20 mm i.d. with 0.33 μm 5% phenyl methyl siloxane film (Ultra2, Hewlett Packard)
Deactivated FSOT, 0.53 mm i.d. and 0.20 mm i.d. (Restek)
Capillary splitter assembly (VSOS, SGE).

2.2 Analytical Standards
All pesticide reference materials certified (Prochem, Dr Ehrenstorfer)
Stock standards (100-1000 μg/mL) prepared in ethyl acetate. Held −18°C
Dilutions containing the 170 pesticides distributed between four mixed groups (see results) prepared in ethyl acetate. Held 18°C
Group 1 (45 pesticides): 5 calibration standards 2, 1, 0.5, 0.1, 0.02 μg/mL
Group 2, 3 and 4 (42, 45 and 38 pesticides): 2 calibration standards 0.5, 0.1 μg/mL
Surrogate standard: tetrachlorvinphos 100 μg/mL in ethyl acetate

2.3 Equipment
Food chopper (Hobart)
Homogeniser (T25m Ika)
Solvent pump for SEC (Shimadzu LC6A)
Low pressure LC column for SEC, 500 x 10 mm i.d., with adjustable teflon plunger/bed support (Pharmacia)
Sample injection and fraction collection system for SEC with 1 mL sample loop (Gilson 401/232)
Gas chromatograph with autosampler, programmed temperature vapouriser, electron capture detector and nitrogen-phosphorus detector (Varian 8100/3500 or Varian 8200/3600 with 1078 injector)
Dual channel chromatography data system (Maxima, Waters)
Ion-trap GC-MS system (Saturn 2000, Varian)

2.4 Sample Preparation and Extraction
Fine chop 10 fresh or partially thawed frozen fruit. Homogenise a 50 g sub-sample with 100 mL ethyl acetate, 50 g sodium sulphate, 0.5 g sodium carbonate and 100 μL surrogate standard. Buchner filter through GF/A paper and stand over 20 g sodium sulphate. Remove 25 mL and evaporate to less than 2 mL using a nitrogen stream on a heating block (40°C max.) Make to 2.5 mL with ethyl acetate and add 2.5 mL cyclohexane. Microfilter 1.5 mL into an autosampler vial.

2.5 Size Exclusion Chromatography (SEC)

Prepare a 450 x 10 mm column of Biobeads SX3 in SEC solvent (ethyl acetate/cyclohexane 1:1 v/v) and stabilise with 1.0 mL eluant flow. Calibrate with group 1 mix so fraction collection starts and finishes to collect 50 ± 5% of fluvalinate and 100 ± 5% of azinphosmethyl. Inject 1 mL of filtered extract and collect pesticide fraction (ca. 23-43 minutes). Evaporate fraction to less than 1 mL using a nitrogen stream on heating block (40° max). Make to 1.0 mL with ethyl acetate.

2.6 HRGC Determination

Set up GC with a 1 m x 0.53 mm i.d. FSOT retention gap, 25 m x 0.20 i.d. column and outlet splitter using lengths of 0.20 mm i.d. FSOT to achieve a split ratio of 4:1 nitrogen-phosphorus detector (NPD)/electron capture detector (ECD). Column pressure 48 psi. Detector temperature 300°C. Make 1 μL injections into PTV (75° 0.2 min, 160°C/min to 250°C, hold). Column oven temperature programme: 60° 2 min, 40°/min to 150° hold 2 min, 1.8/min to 200°C, 40°/min to 260° hold 16.48 min.

Calibrate GC with five group 1 standards and two standard each for groups 2, 3 and 4. Repeat after every 10-12 samples. Every 60 sample prepare control and spike recovery samples using unsprayed kiwifruit as a control. Fortify sub-samples of control with group 1 at 0.04 and 0.2 mg/kg and groups 2, 3 and 4 at 0.1 or 0.2 mg/kg each. Carry these QA samples through the complete method.

2.7 Calculations

Construct calibration curves (linear or quadratic) based on peak areas for both NPD and ECD channels. Identify pesticide peaks in samples from relative retention to surrogate standard and relative response on the two detectors.

Concentration in mg/kg = μg/mL x 0.40

2.8 Confirmation

Confirmation of identity and concentration of pesticides identified in HRGC is carried out using ion-trap GC-MS using similar GC conditions to screening and acquiring full-scan EI spectra. Concentrate the extracts further for samples with low level residues (0.01-0.05 mg/kg). Prepare two standards with concentrations bracketing those in the sample extract. Use the Saturn 2000 target compound software to search the acquired spectra and calculate concentrations for identified peaks.

2.9 Quality Assurance (QA)

The following QA criteria are applied.

Recovery of surrogate standard: 70-120%

Control kiwifruit: no interfering peaks

Spike recoveries,　　group 1　0.04 and 0.2 mg/kg: 70-120%

　　　　　　　　　　groups 2, 3, 4　0.1-0.2 mg/kg: 70-150%

Limits of detection (LODs):

For group 1 calculate the LOD for each pesticide using the US-EPA formula on the last 6 sets of 0.04 mg/kg spike recovery data (in mg/kg).

　　　　LOD = 3.36 x (std. expand standard deviation of recovery)

Compare the calculated LOD to the signal/noise ratio 3:1 equivalent concentration and use the greater as the method LOD.

For groups 2, 3, 4 calculate approximate LODs using data for the lowest spike level to estimate 3:1 signal/noise ratio equivalent concentrations.

Prepare control charts (Shewhart plots) for the recovery and LOD data of the most important and the most sensitive pesticides over at least a 12 month period. Check that current data remains within the $\pm 1\sigma$ and $\pm 2\sigma$ bands.

2.10 Reporting

Set the limit of quantitation (LOQ) for each analyte as 3.3 x LOD.

For confirmed residues above the LOQ, report to significant figures.

For confirmed residues between LOD and LOQ, report to 1 significant figure and mark as 'concentration in a region of uncertainty'.

For residues below the LOD that have been confirmed by GC-MS and are absent from blanks or control samples, report as 'trace'.

3 RESULTS AND DISCUSSION

The method has proved stable and robust. Adding sodium carbonate during extraction reduces levels of acidic interferences and improves recoveries of some basic pesticides such as prochloraz. The SEC cleanup effectively and reproducibly removes lipids, chlorophylls and other high molecular weight co-extractives that otherwise rapidly lead to reduced HRGC performance.

The PTV injection system offers improved responses over splitless injection for less stable pesticides such as acephate and for higher molecular weight pesticides such as synthetic pyrethroids. Regular changing of the injector, liner and pre-column minimises priming and enhancement effects. The post-column split to two detectors allows screening for a wide range of pesticides with many giving useful responses on both. Figures 1 and 2 show parts of the dual chromatograms for a calibration standard and a kiwifruit sample positive for low levels of pirimiphos-methyl.

Table 1 presents an overall summary of the recovery and LOD data for the 170 pesticides. Tables 2-5 summarise key performance data for the four groups of pesticides. The mean recovery and LOD data were calculated from 6 sets of spike data gathered during the analysis of over 350 samples of kiwifruit from the 1999 harvest. The absence of data for a pesticide on one of the detectors does not indicate a total lack of response but rather that it was too weak or poorly resolved for good quantitation. For the group 1 pesticides (Table 2) the LOD data was calculated using the US-EPA formula[5]. The LODs with very low values (below 0.01 mg/kg) would generally be set at 0.01 mg/kg for reporting purposes. The LODs for Group 2, 3 and 4 are extrapolated estimates based on the signal:noise ratios of fortified samples and therefore are on a less sound basis. However the relatively linear calibrations and stable recoveries for the majority of pesticides in these groups gives confidence that the method can reliably screen kiwifruit for absence of residues of these pesticides to low levels.

Table 1 *Overall method performance data – Mean pesticide recoveries (%) and LODs (mg/kg) based on data for fortified kiwifruit samples.*

		ECD		NPD	
	No.	Rec.	LOD	Rec.	LOD
Group 1	45	87	0.009	91	0.012
Group 2	42	102	0.012	106	0.054
Group 3	45	87	0.010	82	0.025
Group 4	38	114	0.019	105	0.047

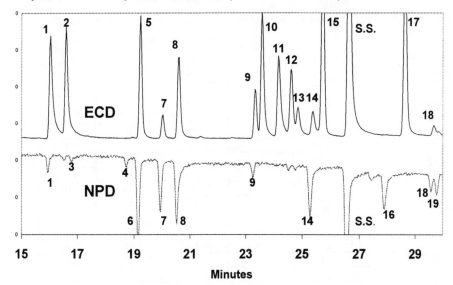

Figure 1 *HRGC chromatogram (part) of Group 1 mixed standard 0.1 µg/mL. 1 metribuzin, 2 vinclozolin, 3 carbaryl, 4 methiocarb, 5 dichlofluanid, 6 pirimiphos-methyl, 7 malathion, 7 chlorpyrifos, 9 pendimethalin, 10 captan, 11 folpet, 12 fluazinam, 13 procymidone, 14 methidathion, 15 endosulfan-alpha, 16 fenamiphos, 17 ppDDE, 18 myclobutanil, 19 buprofezin, S.S. tetrachlorvinphos (0.5 µg/mL).*

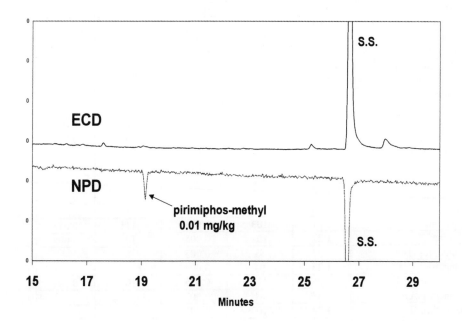

Figure 2 *HRGC chromatogram (part) of kiwifruit extract with incurred residue.*

Table 2 *Recovery and LOD data. Group 1 pesticides, 0.04 mg/kg spike in kiwifruit.*

Pesticide	RRT	Electron Capture		Nitrogen-Phosphorus	
		Recovery %	LOD mg/kg	Recovery %	LOD mg/kg
dichlorvos	0.18	77	0.028	60	0.023
methomyl	0.32			70	0.062
trifluralin	0.42	80	0.013	77	0.019
dimethoate	0.47	95	0.013	88	0.005
simazine	0.48			104	0.008
atrazine	0.49			99	0.006
propazine	0.50			82	0.010
terbuthylazine	0.52			96	0.009
diazinon	0.56	88	0.021	90	0.013
chlorthalonil	0.57	59	0.015	85	0.016
etrimphos	0.59			89	0.014
metribuzin	0.65	72	0.028	100	0.006
vinclozolin	0.67	94	0.009	91	0.015
carbaryl	0.67			87	0.020
methiocarb	0.74			95	0.007
dichlofluanid	0.75	88	0.007		
pirimiphosmethyl	0.75			84	0.006
malathion	0.78	101	0.005	87	0.004
chlorpyrifos	0.80	99	0.005	85	0.003
pendimethalin	0.89	98	0.005	86	0.007
captan	0.90	88	0.007		
folpet	0.92	95	0.018		
fluazinam	0.93	38	0.008		
procymidone	0.94	99	0.004		
methidathion	0.96	114	0.008	99	0.007
endosulfan-A	0.97	78	0.006		
tetrachlorvinphos-SS	1.00	103		97	
fenamiphos	1.04			97	0.010
ppDDE	1.07	84	0.008		
myclobutanil	1.10			93	0.004
buprofezin	1.11			91	0.004
bupirimate	1.13	98	0.011	100	0.016
endosulfan-B	1.13	86	0.010		
endosulfan-sulphate	1.18	86	0.010		
ppDDT	1.18	84	0.007		
tebuconazole	1.19			101	0.010
captafol	1.20	91	0.012		
iprodione	1.21	103	0.006	111	0.018
phosmet	1.22	102	0.005	100	0.008
bromopropylate	1.23	83	0.005		
bifenthrin	1.23	94	0.004		
azinphosmethyl	1.26	118	0.015	101	0.009
permethrinA	1.33	81	0.003		
permethrinB	1.34	100	0.005		
alphacypermethrin	1.43	90	0.007		
FluvalinateA & B	1.55/156	47	0.006		
deltamethrin	1.63	94	0.010		

Table 3 *Recovery and LOD data. Group 2 pesticides, 0.1 mg/kg spike in kiwifruit.*

Pesticide	RRT	Electron Capture Recovery %	LOD mg/kg	Nitrogen-Phosphorus Recovery %	LOD mg/kg
methamidaphos	0.17	77	0.03	99	0.02
dichlobenil	0.21	73	0.01	74	0.02
propham	0.25			101	0.04
omethoate	0.35			109	0.03
propachlor	0.36	105	0.01	98	0.05
cymoxanil	0.36			119	0.05
methabenzthiazuron	0.42			106	0.29
bromoxynil	0.42	105	0.01		
thiometon	0.45			81	0.02
carbofuran	0.49			109	0.04
lindane	0.51	90	0.01		
pyrimethanil	0.54			105	0.03
triallate	0.58	107	0.01	97	0.07
dimethenamid	0.64	153	0.02	135	0.05
parathion-methyl	0.66	122	0.01	109	0.01
tolclofos-me	0.67	109	0.01	99	0.02
prometyn	0.70			102	0.02
linuron	0.75	112	0.03	116	0.06
aldrin	0.76	85	0.01		
metolachlor	0.78	118	0.02	105	0.18
parathion-ethyl	0.80	117	0.01	107	0.02
bentazone	0.83	101	0.01	171	0.13
cyprodinil	0.87			102	0.03
pyrifenox-A	0.90	111	0.01	101	0.05
quinalphos	0.92			106	0.02
phenthoate	0.93	119	0.01	104	0.02
pyrifenox-B	0.98	112	0.01	103	0.06
tetrachlorvinphos-SS	1.00	85		89	
chlorfluazuron	1.04	25	0.03		
imazilil	1.05	65	0.01	102	0.10
oxadiazon	1.10	114	0.01		
oxyfluorfen	1.12	114	0.01		
kresoxim-methyl	1.13	110	0.01	107	0.07
ppDDD	1.15	96	0.01		
benodanil	1.16	135	0.01		
benalaxyl	1.18			108	0.04
propiconazole-A	1.18	105	0.01	106	0.02
propiconazole-B	1.19	105	0.01	104	0.01
diflufenican	1.20	117	0.01	109	0.03
dicofol	1.23	102	0.01		
fenpropathrin	1.24	117	0.01	107	0.04
cyhalothrin	1.29	72	0.01		
bitertanol	1.33			132	0.09
cyfluthrin-A	1.38	115	0.01		
cyfluthrin-B	1.39	109	0.01		
cyfluthrin-C	1.40	98	0.01		
cyfluthrin-D	1.40	96	0.01		

Table 4 *Recovery and LOD data. Group 3 pesticides, 0.1 mg/kg spike in kiwifruit.*

Pesticide	RRT	Electron Capture		Nitrogen-Phosphorus	
		Recovery %	LOD mg/kg	Recovery %	LOD mg/kg
diuron	0.20	90	0.02	69	0.09
acephate	0.24			45	0.02
methacrifos	0.28	86	0.01	65	0.01
tecnazene	0.36	51	0.01		
demeton-s-me	0.37			98	0.01
naled	0.40	94	0.01	89	0.03
dicrotophos	0.41			72	0.01
monocrotophos	0.43			141	0.01
HCB	0.45	50	0.01		
clomazone	0.50			75	0.04
quintozene	0.52	54	0.01		
disulfoton	0.56			55	0.01
isazophos	0.59			76	0.01
phosphamidon	0.65	105	0.02	85	0.02
chlorpyrifos-me	0.66	86	0.01	75	0.01
alachlor	0.69	88	0.01	79	0.04
ioxynil	0.73	93	0.02		
fenitrothion	0.74	99	0.01	82	0.01
fenthion	0.79			76	0.01
cyanazine	0.81	97	0.01	76	0.01
nitrothal-iso	0.84	99	0.01	81	0.15
heptchlor-epoxide	0.87	66	0.01		
penconazole	0.89	98	0.01	85	0.04
chlorfenvinphos	0.92	101	0.01	84	0.02
furalaxyl	0.94			82	0.05
trans-chlordane	0.94	68	0.01		
opDDE	0.97	75	0.01		
cis-chlordane	0.98	67	0.01		
tetrachlorvinphos-SS	1.00	87		79	
dieldrin	1.05	72	0.01		
opDDD	1.09	80	0.01		
azaconazole	1.10	97	0.01	85	0.02
cyproconazole	1.13			88	0.03
fluazifop-p-but	1.15			80	0.03
opDDT	1.15	70	0.01		
ethion	1.16	96	0.01	89	0.01
carbophenthion	1.17	78	0.01	88	0.01
hexazinone	1.19			88	0.01
dinocapA	1.21	95	0.01		
dinocapB	1.22	99	0.01		
dinocapC	1.24	100	0.01		
tetradifon	1.25	79	0.01		
fenarimol	1.29	136	0.01	85	0.01
prochloraz	1.36	106	0.01	98	0.02
cypermethrinA to D	1.41-1.44	101	0.01		
fenvalerateA & B	1.52/1.55	107	0.01		

Table 5 *Recovery and LOD data. Group 4 pesticides, 0.1 mg/kg spike in kiwifruit.*

Pesticide	RRT	Electron Capture Recovery %	LOD mg/kg	Nitrogen-Phosphorus Recovery %	LOD mg/kg
EPTC	0.21			70	0.01
mevinphos	0.24	62	0.02	101	0.02
heptenophos	0.33	107	0.03	91	0.01
propoxur	0.36			99	0.02
diphenyl-amine	0.37			92	0.04
chlorpropham	0.39			96	0.05
bendiocarb	0.42			100	0.03
phorate	0.43	106	0.03	84	0.02
dichloran	0.47	141	0.01		
terbumeton	0.51			94	0.02
terbufos	0.52	109	0.03	90	0.02
terbacil	0.57	119	0.01	100	0.05
pirimicarb	0.62			85	0.02
acetochlor	0.66	110	0.01	103	0.07
heptachlor	0.67	81	0.01		
metalaxyl	0.70			98	0.08
terbutryn	0.74			92	0.02
bromacil	0.75	102	0.01	96	0.06
phorate-sulfoxide	0.76	144	0.04	116	0.05
phorate-sulfone	0.78	126	0.02	103	0.03
fenpropimorph	0.80			83	0.06
triadimefon	0.81	120	0.01	98	0.06
bromophos-me	0.84	101	0.01	93	0.03
thiabendazole	0.89			159	0.09
chlozolinate	0.91	111	0.01	96	0.29
triadimenol	0.93			135	0.10
bromophos-ethyl	0.97	102	0.01	103	0.02
tetrachlorvinphos-SS	1.00	100		86	
iodofenphos	1.03	106	0.01	95	0.04
prothiophos	1.05	109	0.01	94	0.02
metamitron	1.10			246	0.10
endrin	1.11	116	0.01		
flusilazole	1.11			90	0.04
fensulfothion	1.15	149	0.02	123	0.02
oxadixyl	1.16			101	0.02
triazophos	1.17			115	0.01
norflurazon	1.18	175	0.02	151	0.14
EPN	1.23	103	0.01	107	0.01
tebufenpyrad	1.24			80	0.02
phosalone	1.26	115	0.01	110	0.02
pyrazophos	1.30	134	0.01	113	0.02
coumaphos	1.35	143	0.01	125	0.02
quizalfop-methyl	1.43	152	0.01	123	0.05
difenconazole-A	1.57	168	0.02		
difenconazole-B	1.59	161	0.02		

There are some points to note regarding particular pesticides. In group 1 dichlorvos and methomyl barely meet the acceptance criteria and the LODs are elevated. These early eluting pesticides are subject to losses that are not easily controlled. Recoveries for fluvalinate (two isomers) are deliberately low due to the SEC fraction cut. Increasing the recovery of fluvalinate results in increased levels of chlorophyll and other co-extractives in the pesticide fraction. The SEC cutoff is very reproducible, as evidenced by the low LODs for fluvalinate, and levels measured in samples can be confidently corrected for recovery.

In group 2 the urea pesticides methabenzthiazuron and chlorfluazuron were not reliably determined by the method due to GC instability. Residues are extremely unlikely in kiwifruit. In group 3 there were unsatisfactory recoveries for tecnazene, HCB and quintozene although the LODs are adequate for screening purposes and again residues are extremely unlikely in kiwifruit.

In group 4 several pesticides show enhanced recoveries (>120%), probably due to protective effects of coextractives during GC relative to standards in solvent. These effects will reduce the quantitative reliability of any measured residues but do not invalidate the screening ability of the method, apart from metamitron (recovery 246%).

Control charting of recovery data over a more extended period has also shown the method to be capable of providing reliable results on an ongoing basis. Figure 3 is the Shewhart plot for low level spike recoveries of chlorpyrifos in kiwifruit gathered over a 3 year period during analysis of over 1000 samples. Figure 4 is a similar plot for iprodione, in this case using NPD data. Even for this much less stable pesticide on the less sensitive detector, very acceptable trends in recovery were obtained.

The LOD data for group 1 pesticides calculated using the US-EPA formula can similarly be charted for a three year period by using a moving band of 6 data points for each pesticide across the 19 sets of low level spike recovery data. Figures 5 and 6 give such charts for the LODs calculated from the chlorpyrifos and iprodione data in Figures 5 and 6. Given the direction connection of LOD to recovery variability, these are essentially control charts of control charts. The stable nature and strong detector responses for chlorpyrifos lead to the low mean LOD of 7 μg/kg which varied by less than a factor of two over the 3 year period (Figure 3). The variation in LOD for iprodione on the NPD was also low over the period (mean 18 μg/kg, see Figure 4).

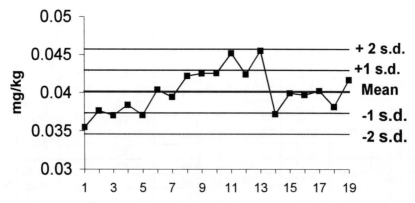

Figure 3 *Control chart of chlorpyrifos recoveries 1997-99, 0.04 mg/kg spike (ECD)*

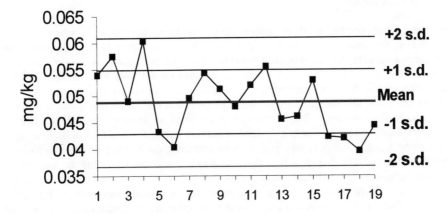

Figure 4 *Control chart of iprodione recoveries 1997-99, 0.04 mg/kg spike (NPD*

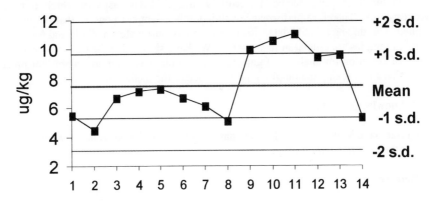

Figure 5 *Control chart for chlorpyrifos limits of detection 1997-1999*

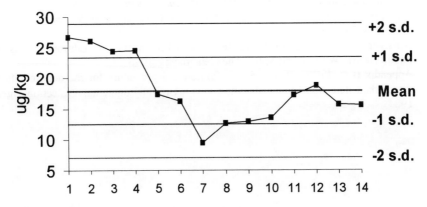

Figure 6 *Control chart for iprodione (NPD) limits of detection 1997-1999*

4 CONCLUSIONS

The basic multi-residue method studied here has been employed in many residue laboratories for analysis of foods, particularly fruit and vegetables. However the comprehensive data set for kiwifruit gathered over several years of routine use has provided further validation of the method as adapted for use in that crop. This performance validation has confirmed the reliability and robustness of the method and demonstrated that low detection limits and high precision can be routinely obtained for a wide range of pesticides. The method is a compromise between providing accurate quantitation and detection limits (US-EPA statistical basis) for a set of pesticides most likely to be found (Group 1) and less rigorous screening for a range of other pesticides. This provides a practical approach to rigorous performance validation for the analytes of greatest significance. Although mass-spectrometry is now the preferred determination step for many multi-methods, selective detectors in the post-column split configuration have been found to be economical and versatile for screening of samples such as kiwifruit where there is a low frequency of positives.

Multi-residue methods remain very dependant on rigid quality control and skilled analysts for their performance, particularly in the HRGC aspects which remain the weakest links. Therefore performance validation is essential to ensure that the method as applied is producing reliable data. Recovery data for surrogate standards and for low level fortified samples are crucial in this respect. We have shown that the data on precision and limits of detection for analytes, if gathered on a regular basis, are invaluable for objective validation and tracking the on-going performance of the method.

Acknowledgements

Ray Isaac, Jane Veitch and Jay Dudson have provided dedicated technical assistance during the kiwifruit testing programme.

References

1. P.T. Holland and C.P. Malcolm, in 'Emerging Strategies for Pesticide Analysis', ed. J. Sherma and T. Cairns, CRC Press, Boca Raton, 1992, p71.
2. P.T. Holland and T.K. McGhie, *J. Assoc. Off. Anal. Chem.*, 1983, **66**, 1003.
3. A.H. Roos, A.J. Van Munsteren, F.M. Nab and L.G.M.Th. Tuinstra, *Anal. Chim. Acta*, 1987, **196**, 95.
4. 'Analytical Methods for Pesticide Residues in Foodstuffs. 6th Edition. Ministry of Public Health, Welfare and Sport, The Netherlands, 1996.
5. Appendix B to 40CFR Part 136 – Definition and procedure for the determination of the method detection limit – Revision 1.11. US Environmental Protection Agency, 1984.

Effects of Sample Processing on Pesticide Residues in Fruit and Vegetables

A.R.C. Hill[1], C.A. Harris[2] and A.G. Warburton[2]

[1] CENTRAL SCIENCE LABORATORY, SAND HUTTON, YORK YO41 1LZ, UK
[2] PESTICIDES SAFETY DIRECTORATE, MALLARD HOUSE, 3 PEASHOLME GREEN, YORK YO1 7PX, UK

1 INTRODUCTION

A sample taken and sent to the laboratory for determination of pesticide residues is generally known as a laboratory sample. Depending on the nature of the sample, it may require preparation and processing before analysis. Such preparation includes procedures such as removal of adhering soil from potatoes, withered leaves from lettuce, "tops" from carrots, removal of seeds from stone fruit, and so on. Sample preparation is not readily amenable to validation and is not considered further in this paper. Sample processing is intended to make the sample homogeneous and to convert it to a suitable state for analysis. Sample processing usually involves a comminution process (chopping, blending, grinding, etc.) and/or mixing. Sample processing is amenable to validation and, since it can influence results significantly, should be validated.   In practice, laboratory sample processing is rarely checked because validation is often perceived as being difficult or unnecessary.

Sample preparation and processing in the laboratory are not the same as domestic or industrial preparation and processing of food, although there are some similarities. Domestic and industrial procedures frequently employ physical, chemical and/or microbiological processes that are unlikely to be used in laboratory sample processing. The effects of domestic and industrial preparation and processing on pesticide residues have been, and continue to be, widely studied for pesticide registration and consumer risk assessment purposes[1-5] and also by food manufacturers.[6]

In contrast, laboratory sample processing has received little attention. This lack of attention is potentially serious, because there tends to be a major difference between the processing procedures adopted in residues monitoring laboratories and those adopted for residues trials for registration purposes. In general, residues monitoring laboratories process samples at room temperature whereas, in general, for pesticide registration samples are deep frozen and processed in the presence of dry ice (a process sometimes termed "cryogenic milling").

In most cases, but perhaps not invariably, it may be reasonable to assume that any losses of pesticide that occur during laboratory sample processing are likely to reflect similar losses that occur during food preparation and eating by consumers. Thus, in practice, there is unlikely to be an under-estimation of consumer risk as a consequence of the analytical under-estimate. Consideration of the validity of such assumptions is beyond the scope of this paper.

If a pesticide is likely to be lost as a result of sample processing, it is certain that the proportion of loss is likely to differ according to the nature, duration, temperature and (in certain cases) acidity of the process(es) used. Maximum residue limits (MRLs) for pesticides define the critical (i.e. worst-case limit of) good agricultural practice (GAP) in the use of pesticides. If the processing carried out in generating the results for setting an MRL was much less likely to lead to pesticide losses than the processing adopted for monitoring food for compliance with that MRL, not only will residues arising from GAP appear to be comfortably within the MRL but even gross misuse of the pesticide may go undetected. Trading disputes between producers and purchasers of raw agricultural materials may arise similarly.

Determination of pesticide residues for compliance with MRLs requires that the test portion analysed is representative of the laboratory sample. The test portion is unlikely to be representative if the analyte is not more or less homogeneously distributed within the laboratory sample before the test portion is withdrawn. Residues incurred as a consequence of pesticide use in food production are rarely, if ever, distributed uniformly between, or within, the individual units of fruit and vegetables.[7,8] Thus a laboratory sample of such commodities must be disintegrated and mixed, to achieve an acceptable degree of homogeneity.

The physical processes of disintegration and mixing are usually simple. In some cases, the potential for loss of pesticides is obvious, for example fumigant residues may be lost readily by volatilisation. Similarly, losses of pesticides that are prone to hydrolysis might be expected if homogenisation involves lengthy exposure to water and extremes of pH.[2] Enzymes and other reactive chemicals can be liberated from damaged plant cells and may react with the analytes of interest.[2]

In this paper we do not address loss of water because, except where extreme conditions are used, it is unlikely to lead to significant changes of residue concentration. Neither do we consider losses of fumigants and other highly volatile pesticides since sample disintegration usually leads to a loss of such residues. In this short paper, we consider a few examples which show that sample processing can and should be checked for potential adverse effects on the accuracy of residue determination. In the UK, openness about the results of such checks has led to public concern about the accuracy of the national monitoring programme,[9] although the procedures used in the UK are similar to those used elsewhere in the world.

2 EFFECTS OF SAMPLE PROCESSING IN PRACTICE

2.1 Dithiocarbamates

The problem of losses of dithiocarbamate fungicides during sample processing of fresh plant materials has been widely recognised for many years.[10] An early report showed that up to 30% of maneb was lost after addition to pre-chopped chilled or frozen kale.[11] Much greater losses may occur if dithiocarbamates are present during the disintegration process, especially if the process leads to a fine purée.

Most analysts adopt one of three strategies to avoid the problem. The simplest is to scale up the analysis so that the whole, or a large proportion, of the laboratory sample is analysed in a single determination, without prior disintegration, although this requires large volume glassware and large quantities of reagents. Alternatively, segments may be cut (with minimum tissue disruption) from the fruit or vegetables, for smaller-scale analyses, but the variability of residue levels within the laboratory sample necessitates the analysis of a number of replicate portions. Lastly, samples may be frozen quickly, disintegrated to a powder in the presence of dry ice and mixed, and the analysis of sub-samples commenced without allowing the material to thaw. This approach is sometimes referred to as "cryogenic milling" and has been widely adopted for the generation of data intended for registration of pesticides.

Cryogenic milling can be less convenient and somewhat more costly than disintegration at room temperature and it may not be suitable for all pesticides. Nonetheless, the temperature of cryogenic milling is well below the freezing point of most fruit and vegetables and hence the rates of ordinary chemical reactions are drastically reduced, so the process can be expected to be widely applicable.

2.2 Chlorothalonil

Losses of chlorothalonil during sample processing have been acknowledged by the UK Working Party on Pesticide Residues (WPPR),[12] although the supporting experimental data are unpublished. Intermittent poor recovery of chlorothalonil had been experienced by laboratories contributing to the WPPR over some years but the causes were studied at the Central Science Laboratory only after an experiment, intended for other purposes, produced unexpected results.

Five pesticides at equal concentrations in the same solution were sprayed onto whole butterhead lettuce. The lettuce were chopped finely using a Hobart Food Cutter and the levels of the pesticides were determined (Table 1).

Table 1 *Residue Levels of Five Pesticides, Applied to Whole Lettuce in Equal Amounts and Determined Immediately After Comminution (n=34)*

	bifenthrin	dimethoate	chlorothalonil	procymidone	vinclozolin
mean mg/kg	1.34	1.25	0.18	1.06	1.00
s.d.	0.10	0.28	0.04	0.12	0.13

The concentrations of bifenthrin, dimethoate, procymidone and vinclozolin were more or less equal, as expected. In contrast, the level of chlorothalonil (tetrachloro*iso*phthalonitrile) was about one-fifth of the levels of the other compounds. Concurrent determinations of recovery provided excellent quantitative data for all five compounds. Effectively, internal standardisation with the other four pesticides had shown that chlorothalonil was lost as a consequence of the sample processing.

Although chlorothalonil has a higher vapour pressure than the other pesticides, its volatility is nonetheless low. Chlorothalonil is not a highly reactive compound: it is thermally stable at ambient temperatures and in gas chromatography; it is stable to UV light in solution and the solid state; and it is stable in acidic or moderately alkaline solutions, although it does hydrolyse slowly at pH > 9.[13] Codex and EU MRLs for approved uses on fruit and vegetables are generally in the range 0.5 to 10 mg/kg, so residues of this non-systemic fungicide are relatively stable in crops and rapid losses during sample processing were not expected.

Comparable experiments (using HCB as internal standard) conducted on fresh broccoli, celery, lemons and lettuce, showed that losses of chlorothalonil differed between the commodities but were more or less time-dependent in all. Losses from lettuce did not occur if the plants were "killed" by brief heating in a microwave oven, prior to application of the chlorothalonil. Processing of onions led to complete loss of chlorothalonil but, unlike the other commodities, the pesticide was also lost if it was added to ethyl acetate extracts of onions. Broadly comparable results were subsequently obtained by colleagues working independently, who investigated the effects of processing leeks, garlic, lettuce and tomatoes, although, in this case, little or no loss of chlorothalonil occurred with the "Little Gem" lettuce used.

The fate of the pesticide in finely chopped lettuce and in onion extracts was investigated further, using both unlabelled and radiolabelled [U-ring ^{14}C]-chlorothalonil, TLC, HPLC, GC-MS and LC-MS.

Addition of chlorothalonil to pre-chopped lettuce led to the loss of about half the pesticide within one day and most of the radioactivity associated with this loss could not be extracted from the lettuce with ethyl acetate or water. The degradation products were not identified but were apparently "bound" to insoluble materials. There was no indication that 4-hydroxy-2,5,6-trichloro*iso*phthalonitrile, a known plant metabolite of chlorothalonil, was produced in a free state.

Chlorothalonil added at 0.5 µg/ml to ethyl acetate extracts of onion (equivalent to 1 mg/kg onion), was degraded completely within one day. Three-quarters of the radioactivity was recovered as three breakdown products, one accounting for half of the total radioactivity and all being more polar than chlorothalonil itself. LC-MS with negative-ion electrospray showed a major degradation product, more polar than chlorothalonil itself (which was detected as the anion of 4-hydroxy-2,5,6-trichloro*iso*phthalonitrile), was probably a derivative involving the analogous 4-thiol. The degradation product was evidently not the free 4-thiol but further characterisation was hindered because the corresponding anion was the largest fragment that could be generated with electrospray ionisation. Probable reaction with sulfur compounds in onion extracts was supported by observation that chlorothalonil was completely stable in extracts that had been passed through alumina impregnated with silver nitrate. In retrospect the occurrence of such reactions is not completely unexpected, as one of the mechanisms of chlorothalonil fungicidal activity involves reaction with glutathione. However, reaction rates are difficult, if not impossible, to predict.

"Cryogenic milling" of lettuce avoided losses of chlorothalonil but the technique was ineffective with onions, presumably because sufficient quantities of reactive sulfur compounds were generated during the thawing that occurred immediately prior to the solvent extraction. Heating onions in a microwave oven to "kill" them, or the addition of zinc acetate, prior to comminution, was similarly ineffective. However, comminution of onions in the presence of a strong acid (e.g. *ortho*-phosphoric added at 1 to 2 M) can inhibit losses of chlorothalonil. A combination of acid addition and cryogenic milling may be the most effective approach and is under investigation.

Chlorothalonil losses during sample processing are clearly a significant problem if the analyst does not take appropriate action. Determination of recovery is not normally intended to identify problems at sample processing and it is unlikely to measure the true accuracy of determination of chlorothalonil residues. In retrospect, occasional poor recovery was an indication of the underlying problem but it was infrequent and difficult to repeat, so the cause was not investigated. The unexpected behaviour of this compound provides a good example of why it is important to demonstrate the validity of laboratory sample processing procedures.

2.3 Captan, captafol, folpet and dicofol

Captan, captafol, folpet and dicofol present a more complicated picture, although their initial degradation routes are well known. They are rather like chlorothalonil in that occasional low recoveries may be experienced but dicofol has also been observed to degrade in solution.[14] They tend to be pyrolysed very readily during injection for gas chromatography: captan and captafol forming tetrahydrophthalimide; folpet forming phthalimide; and *p,p'*- and *o,p'*-dicofol forming *p,p'*- and *o,p'*-dichlorobenzophenone. These degradation products can also be formed metabolically, and such degradation might also occur during sample processing. As the metabolites are not currently incorporated into MRLs for these pesticides, it is not acceptable to use a method that involves complete conversion of residues to the degradation products. Thus analysts must be sure that they do not unknowingly degrade the pesticides at any stage of the analysis.

Experiments carried out at the Central Science Laboratory showed significant losses of captan, captafol and folpet occurred during processing of apples, lettuce and tomatoes at room temperature (Table 2). Corresponding losses from grapes were relatively small at room temperature and losses were effectively prevented in all four crops by "cryogenic milling". Recovery of the pesticides added after processing at either temperature, and immediately before extraction, was quantitative with good repeatability. It was shown that solutions of dicofol must be acidified (e.g. with acetic acid), to prevent degradation. Addition of acid (e.g. *ortho*-phosphoric acid) at extraction helps to ensure good recovery of added dicofol. No degradation of dicofol was found during sample processing at room temperature. Although the dicofol solution added before processing contained a small quantity of acetic acid (20 µl in 1 kg sample), and thus the pesticide was initially stabilised, fresh fruit and vegetables tend to be slightly to strongly acidic, so it was reasonable to concluded that dicofol is more likely to degrade in solution than during sample processing.

2.4 Other pesticides

A similar experiment to that described above, involved other pesticides and vegetables (Table 3) but was "internally standardised" in a manner equivalent to that adopted for the data provided in Table 1.

The results confirmed the adverse effects of sample processing on chlorothalonil and showed clearly that dichlofluanid is also unstable when lettuce samples are processed at room temperature. The apparent loss of iprodione from onions has not yet been investigated further.

The data in Table 3 derived from two different sample sizes (7 replicates each of 30 g and 210 g). Propham was also added to the potatoes but the results from the two sample sizes conflicted and they were omitted from the Table. Our experience indicates that interpretation of data from sample processing experiments may become more problematic, the more complex the experiment carried out.

Table 2 *Effects of sample processing temperature on captan, captafol, folpet and dicofol, added to fruit and vegetables at 1 mg/kg (n=7)*

	percentage surviving, (s.d.)			
	captan	captafol	folpet	dicofol
Room temperature				
apples	41 (7)	62 (4)	50 (6)	110 (22)
grapes	69 (9)	81 (6)	86 (4)	103 (9)
lettuce	4 (2)	64 (7)	18 (6)	not determined
tomatoes	40 (11)	57 (9)	57 (11)	103 (6)
"Cryogenic milling"				
apples	81 (10)	95 (8)	93 (10)	114 (15)
grapes	98 (2)	107 (6)	98 (2)	92 (12)
lettuce	91 (5)	96 (6)	97 (5)	not determined
tomatoes	89 (6)	95 (11)	96 (4)	100 (3)

Table 3 *Effects of sample processing temperature on various pesticides, added to vegetables at 1 to 2 mg/kg (n=14)*

percentage of the pesticide surviving, (s.d.)

Lettuce	chlorothalonil	dichlofluanid	iprodione*	tolclofos-methyl	vinclozolin
room temperature	79 (2)	23 (14)	100 (-)	95 (6)	96 (4)
"cryogenic milling"	97 (2)	92 (6)	100 (-)	101 (2)	99 (3)

Onions	chlorothalonil	chlorpyrifos	fenpropimorph*	iprodione	tolclofos-methyl
room temperature	0 (-)	100 (6)	100 (-)	79 (4)	103 (6)
"cryogenic milling"	0 (-)	96 (5)	100 (-)	75 (10)	102 (7)

Potatoes	chlorpropham	tolclofos-methyl	thiabendazole*
room temperature	87 (6)	91 (9)	100 (-)
"cryogenic milling"	91 (5)	94 (8)	100 (-)

* pesticide used as the internal standard

For example, the effects of sample processing were investigated further, using the "internal standardisation" method but adding a large number of pesticides simultaneously. The expectation that this would produce data on many compounds simultaneously was partially fulfilled, in that major losses could be distinguished, but the vagaries of multi-residue determination made it more difficult ascertain whether minor losses had occurred.

In a preliminary experiment with 90 pesticides (all amenable to multi-residue GC-MS determination but representing many different groups of compound) in apples and lettuce, it was concluded that only apparent losses greater than about 30% were likely to represent true losses. Pesticides wholly or apparently partially lost according to this criterion were biphenyl (lettuce only), bitertanol (apples only), chlorothalonil, chlozolinate (lettuce only), dichlofluanid, dichlorvos (lettuce only), ethoxyquin, etridiazole, heptenophos (apples only), isofenphos (apples only), prochloraz (apples only), quinomethionate (lettuce only), tebuconazole (apples only) and tolylfluanid.

Another experiment was intended to assess the effects of sample processing in three UK laboratories contributing to the national monitoring programme for residues in fruit and vegetables. Two laboratories processed samples at room temperature, whereas the other processed samples in a part-thawed state (following storage in a freezer). In this case, 105 pesticides, which may be used on various fruit and vegetables, were added concurrently to apples, lettuces, oranges or tomatoes at 1 mg/kg. Seven replicate samples were processed by each of the 3 procedures. In addition to the tests of sample processing, recovery of the pesticides was also determined (3 replicates) in parallel.

With few exceptions, recovery results were within the range 80 to 120% but results from processing were generally much more variable (most relative standard deviations in the range 20 to 35%). This made interpretation difficult, although general trends were apparent. The commodities could be ranked broadly in order of adverse effects as: lettuce>apples>oranges>tomatoes. As might be expected, the adverse effects involved greater losses and more pesticides at room temperature than in processing partially thawed material, at least in the cases of lettuce, oranges and tomatoes . It is not clear why apples did not follow the same trend. Losses of chlorothalonil, dichlofluanid, etridiazole and tolylfluanid were evident from all four commodities, with greater losses at room temperature, although losses of etridiazole and tolylfluanid from processed oranges were small. Among the more surprising observations was that more than half of the thiabendazole was apparently lost from oranges at room temperature. Although less thiabendazole was lost from the semi-thawed oranges, this is one of several adverse effects that require further investigation.

The pesticides so far shown to be lost during sample processing have few or no features in common and a single chemical mechanism does not appear to be involved. Hydrolysis is widely considered to be the principle mechanism for loss during domestic and industrial food processing but volatilisation, oxidation and reduction reactions have also been reported.[2] The reaction of thiols with chlorothalonil is an interesting substitution reaction. Hydrolysis may predominate where enzymes and other active chemicals have effectively been destroyed by boiling, etc. Laboratory sample processing does not normally involve high temperatures but the procedure should be designed avoid all chemical changes. Comminution or blending of fruit and vegetables can generate "soups" with a potentially wide range of chemical activity and, in addition under such conditions, losses of volatile pesticides may be inevitable.

3 CONCLUSIONS

Losses of pesticide residues have been shown to occur during comminution of fruit and vegetables. Depending on the pesticide and sample matrix involved, the effects may range from undetectable to highly significant. A single, simple mechanism of loss is not involved and it is reasonable to assume that trace levels of certain other organic compounds, besides pesticides, may be similarly affected. Except for compounds of high volatility, the types of compound that will be prone to losses are not predictable with certainty. Fortunately, only a few pesticides appear to be affected seriously.

Sample processing procedures are rarely validated at present. Measurement of losses during processing is relatively simple and it should be done, to ensure that the accuracy of results is not seriously compromised. Sample processing procedures generally evolve much more slowly than other analytical procedures, so the validation may hold good for many years.

However, adverse effects during processing may be difficult to reproduce. They may also vary between laboratories using the same procedure and they may differ between different varieties of the same commodity. For these reasons, it may be necessary to adopt procedures, such as "cryogenic milling", that are inherently less likely to lead to losses by chemical reaction or volatilisation. At present, we have few data to guide us and much work remains to be done to optimise processing procedures, to avoid chemical degradation of organic analytes.

Validation of the sample processing should be planned cautiously and conducted carefully, as an over-ambitious approach may generate data that are difficult to interpret.

Acknowledgements

Unpublished data on the effects of laboratory sample processing, referred to in the text, were generated by S. L. Reynolds, S. Nawaz, M. J. Duff, P. Harrington and A. R. C. Hill at the Central Science Laboratory. The work was funded by the Pesticides Safety Directorate.

References

1   Food and Agriculture Organization of the United Nations, "FAO manual on the submission and evaluation of pesticide residues data for the estimation of maximum residue levels in food and feed", FAO, Rome, 1997, p. 27.
2   R Mackenzie, HNC Thesis, University of Hull, 1997.
3   Advisory Committee on Pesticides, 'Consumer risk assessment of insecticide residues in carrots', Pesticides Safety Directorate, York, 1995.
4   Advisory Committee on Pesticides, 'Evaluation of aldicarb', report 109, Pesticides Safety Directorate, York, 1994.
5   Advisory Committee on Pesticides, 'Evaluation of carbaryl', report 155, Pesticides Safety Directorate, York, 1996.
6   H. B. Chin, *in* "Pesticide Residues and Food Safety", B. G. Tweedy, H. J. Dishburger, L. J. Ballantine and J. McCarthy, American Chemical Society, 1991, chapter 19.
7   Advisory Committee on Pesticides, "Unit to unit variation of pesticide residues in fruit and vegetables", Pesticides Safety Directorate, York, 1997,
8   Pesticides Safety Directorate, "Report on the International Conference on Pesticide Residues Variability and Acute Dietary Risk Assessment" PSD, York, http://www.gov.uk/aboutmaf/agency/psd/news/finalrep.pdf, 1999.
9   F. Lawrence, J. Blythman and J. Wilson, The Guardian, September 16, 1999, London, p. 1.
10 A. R. C. Hill, *in* 'Emerging Strategies for Pesticide Analysis', T. Cairns and J. Sherma, CRC Press, Boca Raton, 1992, Chapter 9, p. 217.
11 S.F. Howard and G. Yip, *J. Assoc. Off. Anal. Chem.*, 1971, **54**, 1371.
12 Ministry of Agriculture, Fisheries and Food, 'Annual Report of the Working Party on Pesticide Residues: 1994', *Suppl. Pesticides Register*, p. 3, HMSO, London, 1995.
13 C. D. S. Tomlin, 'The Pesticide Manual', British Crop Protection Council, Bracknell, 11th edition 1997, p. 227.
14 A. Andersson, 'Report on European Commission Proficiency Tests on Pesticide Residues in Fruit and Vegetables 1996/1997, Proficiency Tests I and II', National Food Administration, Uppsala, Sweden, 1997.

Testing the Efficiency and Uncertainty of Sample Processing Using ^{14}C-Labelled Chlorpyrifos: Part 1. Description of the Methodology

B. Maestroni, A. Ghods, M. El-Bidaoui, N. Rathor, T. Ton and Á. Ambrus

FAO/IAEA TRAINING AND REFERENCE CENTRE FOR FOOD AND PESTICIDE CONTROL, FAO/IAEA AGRICULTURE AND BIOTECHNOLOGY LABORATORY, A-2444 SEIBERSDORF, AUSTRIA
e-mail: B.maestroni@iaea.org, http://www.iaea.org/trc

ABSTRACT

In order to obtain accurate and repeatable results the whole laboratory sample of 1.5 - 3 kg should generally be cut into pieces and homogenised to obtain statistically well mixed matrix before the representative analytical portions are withdrawn for analysis. The method described for the determination of sample processing uncertainty is making use of radiolabelled compounds, but it can also be applied with unlabelled pesticides, although its application takes much longer and the estimated uncertainty of sample processing is less precise.

The method is based on surface treatment of plant materials with radiolabelled Chlorphyrifos, determination by LSC of the activity of ^{14}C-Chlorphyrifos in ethyl-acetate extracts from 5 replicate analytical portions of different sizes, calculation of the recoveries, statistical elaboration of data, estimation of the sampling constant values and determination of the uncertainty of sample processing.

Key Words: *uncertainty, sample preparation, sample processing, sampling constant, radiolabelled compounds.*

1 INTRODUCTION:

In 1967 Youden (1) defined the overall random error of an analysis (S_R) as a function of the random errors in each stage of the analysis by

$$S_R = \sqrt{(S_S)^2 + (S_{SP})^2 + (S_A)^2} \qquad \text{eq.(1)}$$

Equation 1 gives the overall error in terms of the variances contributed by the sampling (S_S), sample processing (S_{SP}) and analysis (S_A). The expression can be modified to incorporate additional stages. Wallace and Kratochvil (2) elaborated a method for the estimation of sampling uncertainty and studied the sampling component by adopting the concepts of sampling constants developed by Ingamells and Visman (3).

The preparation of the analytical sample from which the analytical portion is withdrawn may consist of two distinct procedures. According to the definitions introduced by Hill and Reynolds (4), sample preparation is the procedure used, if required, to convert the laboratory sample into the analytical sample by removal of parts (soil, stones, bones,...) not to be analysed. Sample processing is the procedure (e.g. cutting, grinding, mixing) used to make the analytical sample acceptably homogeneous with respect to the analyte distribution, prior to removal of the analytical portion. Sample preparation and processing must be carried out according to the aim of analysis. For instance, for assessing the dietary intake of consumers the edible parts of the commodities shall be analysed, or for providing data for estimation of maximum residue limits, CAC (5) specifies the portion of commodities to which MRLs apply.

The method of sample preparation may be the source of substantial systematic and random errors, which cannot be estimated. To overcome this problem each laboratory must strictly follow the appropriate instructions preferably given in the form of SOPs. Sample processing and not sample preparation is considered in this paper.

The method adopted to determine the efficiency and the uncertainty of sample processing is taken from Ambrus (6), who applied the sampling constants concept for the analysis of pesticide residues, to calculate the size of analytical portion that will hold the variability of sample processing to a specified level.

Since the variation of sample processing cannot be determined directly as a self standing uncertainty component, the best approach is to analyse several replicate test portions of a homogenised sample to obtain the combined uncertainty of the sample processing (S_{SP}) and analysis (S_A). This paper describes the methodology to determine the efficiency and uncertainty of sample processing.

1.1 Principle of the method:

Ingamells and Visman defined a sampling constant as the weight of a single increment that must be withdrawn from a well-mixed material to hold the relative sampling (withdrawing and processing) uncertainty to 1% with 68% level of confidence. K_S can be determined from the relation:

$$K_S = W \times CV_{SP}^2 \qquad \text{eq. (2)}$$

where
W = weight of a single increment (in our case it is the analytical portion)
CV_{SP}^2 = uncertainty of sample processing

According to Wallace and Kratochvil it is possible to determine with a high level of confidence whether a material is well-mixed by analyzing two sets of increments of widely differing weight ($W_{Lg}/W_{Sm} \geq 10$), where W_{Lg} is the large portion size and W_{Sm} is the small portion size. If the matrix is well mixed then:

$$K_{S(Sm)} = K_{S(Lg)} \qquad \text{eq.(3)}$$

Since the average residue concentration of the small and the large analytical portion is the same (R), the CV^2 (S^2/R^2) can be substituted with S^2.

$$S_{Lg}^2 x W_{Lg} = S_{Sm}^2 x W_{Sm} \qquad \text{eq. (4)}$$

$$S_{Lg}^2 = S_{Sm}^2 \times \frac{W_{Sm}}{W_{Lg}} \qquad \text{eq. (5)}$$

or in terms of variance

$$V_{SP\,Lg} = V_{SP\,Sm} \times \frac{W_{Sm}}{W_{Lg}} \qquad \text{eq. (6)}$$

In order to check the homogeneity of the chopped sample the F-test is applied to compare the variance of sample processing of large versus small test portions (eq. 5). We use two tail F-test because either of the two values ($V_{SP\,Lg}$ or $V_{SP\,Sm} \times \dfrac{W_{Sm}}{W_{Lg}}$) can be larger. The F-test should be applied at 90% or lower level of confidence since the consequences might be severe if one decides a sample is well-mixed when it's not! If the difference is not significant the processed sample is statistically well mixed and the K_S can be calculated from the W_{Lg} and CV_{SPLg} (which is more precise than an estimate based on small sample increments as shown in part II (7)). From the Ks value (eq. 2) we can then calculate the uncertainty of sample processing. Prior to this we have to separate the analytical contributions (S_A) and the effect of sample processing (S_{SP}) (described in 3.6.).

2 EQUIPMENT AND MATERIALS:

2.1 Equipment

The following equipment are required for the study: Laboratory Chopper up to 5 kg capacity: Stephan UM12 (S-ch); commercial food processor up to 2 kg capacity (P-ch); Waring blender, 1 litre steel containers (WB) New Hartford-USA; high-speed Ultra-Turrax T 25, IKA, Germany; Refrigerated Centrifuge (Sigma 4K15); Centrifuge tubes 250 ml; Centrifuge tubes 25 ml; Thick wall glass beakers, 1 litre; Cylinders 50, 100 and 500 ml; Pipette, 1 ml; Hamilton micro syringe 500 µl; Top load balance 0.01 g; Freezer, -20 °C;

Beckman Liquid scintillation counter LS 6000TA (LSC); LSC polyethylene vials, Camberra-Packard.

2.2 Chemicals

The following chemicals are required for the study: Chlorpyrifos analytical standard (Dr. Ehrenstorfer), treating solution: acetone solution of ^{14}C-Chlorphyrifos (Internationale Isotope, München), Sodium hydrogen carbonate, residue grade ; Sodium sulphate anhydrous, residue grade; Ethyl acetate, residue grade; Milli-Q water (dionex); dry-ice Linde, Ultima Gold XR liquid scintillation cocktail, Packard bioscience.

3 METHOD:

3.1 Objectives of the study

The objectives of the study are to set up a robust method for the determination of the uncertainty of sample processing, to test the efficiency of sample processing with different choppers, and to test different procedures that could decrease the uncertainty of the sample processing.

3.2 Experimental design

The method is based on surface treatment of plant materials with radiolabelled Chlorphyrifos, determination by LSC of ^{14}C-Chlorphyrifos activity in ethyl-acetate extracts of 5 replicate analytical portions of different sizes, calculation of recoveries, statistical elaboration of data, estimation of the sampling constant values and determination of the uncertainty of sample processing.

The scheme of withdrawal of analytical portions from the processed analytical sample is essential to ensure the statistical design for the calculation of the sampling constant **(Figure 1)**. 5 replicate analytical portions of 250g (or 150 g) and 25g (or 15 g) are withdrawn after the homogenization carried out in a chopper, extracted and counted for ^{14}C-activity.

According to the objective of the study, a subsample of 400-800 g can be further homogenised in a Waring blender and 5 replicate analytical portions of 50 g and 5 g are withdrawn, extracted and analysed.

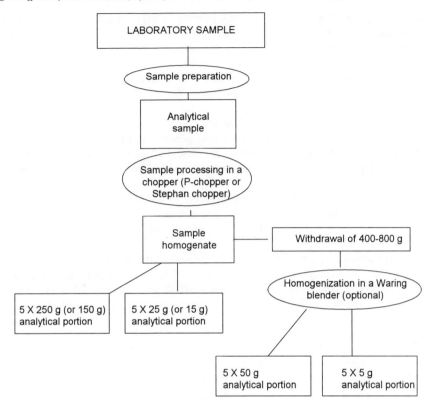

Figure 1 *Scheme of withdrawal of analytical portions*

3.3 Treatment and processing of the samples

Prepare the treating solution, a mixture of labelled and unlabelled pesticide so that the commodity is treated to get 0.5 mg/kg residue level and ca 3000 dpm/g specific activity.
Count by liquid scintillation counting the radioactivity of 3 X 10 μl treating solution. Determine the actual specific activity (s.a.) in dpm/μl; calculate the volume of treating solution ($V_{μl} = 3000_{dpm/g}$ * weight$_g$ of matrix to be treated (M) / s.a.$_{dpm/μl}$) and enter the value of the specific activity expressed in dpm/g (= s.a $_{dpm/μl}$.× $V_{μl}$ / M_g) in cell H4 as shown in **appendix 1.**

Weigh approximately 2 - 4 kg of plant material, cut natural units of vegetable samples into half in the longitudinal direction if applicable. Place the units with their cut surface on clean aluminium foil and carefully apply with a Hamilton syringe the treating solution on their upper surface in small portions. *Make sure that the applied material doesn't run off the surface.*

Wait 15 minutes to dry the residues onto the surface by letting the acetone evaporate. Carefully avoid touching the treated part place the units in the bowl of the chopper. Close the lid and chop the material for 1 minute. Transfer the material from the lid and the wall with a large spatula and continue chopping for another minute. Visually examine the consistency of the matrix and estimate the time required to get a homogeneous material by continued chopping.

Withdraw 5 replicate analytical portions of each size (see fig. 1) from the chopper (the material should be taken as a single increment from a confined position from the bowl) into beakers and centrifuge tubes respectively. Record the exact weight of the analytical portion to 0.01 g accuracy. Insert the data in cell B7 (appendix 1). First withdraw slightly more material than required, and adjust the correct weights after all portions have been withdrawn to avoid segregation of processed sample. Be careful that the quick operation should not result in a spill of material or inaccuracy! If segregation of the minced sample is observed repeat the homogenisation between withdrawals. It is obligatory in case of tomato.

If the study also aims at testing the effect of a double processing procedure on the uncertainty of sample processing, then a representative subsample of the chopped material is further subjected to homogenization in a blender according to the procedure described below.

Transfer the material from the lid and from the wall of the chopper, and homogenize for 20 seconds the remaining material. Weigh 400-800 g homogenised material in a 1 litre container (WB). Add 20% (~80 ml) water (Milli-Q grade) except for tomato and lettuce (8), and withdraw 5 replicate analytical portions of 50 g and 5 g as described above.

3.4 Extraction

Add NaHCO$_3$ in 1:6 weight ratio to the analytical portions. Add ethyl acetate in 2 ml/g sample ratio (9). Record the weight and insert the data in cell (B9) (appendix 1). Add sodium sulphate anhydrous (fine powder) in 1:1 w/w to the sample. Mix the sample. Extract the sample with 25 T Ultra Turrax at a setting of 11000 rpm, for 2-4 min. Adjust the rpm, if necessary, to assure that the whole mass inside the extraction vessel is moving and vortexing.

Weigh, just after extraction, the centrifuge tubes and the 1 l beaker (which contains the 250 or 150 g sample extract). Leave the 250 g (150 g) sample extract standing for 30 minutes in a fume hood, well covered with aluminium foil, while centrifuging the other extracts for 10 min at 2500 rpm speed. Weigh again all vessels and keep record of the weight differences. The weight measurement is carried out as an internal quality control measure to check that the recovery is not affected by the evaporation of the solvent.

3.5 Measurement of radioactivity

Pipette 5 times (at least 5 replicates should be analysed to increase the accuracy and precision of the estimated residue content) 1 ml organic supernatant into polyethylene vials and record the mass of the extract to 0.01 g accuracy. Insert the values into cells (D7:H7) as shown in appendix 1. Add 10 ml liquid scintillation cocktail, close tightly the vials for the determination of the ^{14}C-activity. Determine the radioactivity (dpm) of each vial by liquid scintillation counting. Set the counter to allow 3 times 5 minutes reading for each vial. Include a ^{14}C commercial standard <1µCi (10×10^4 dpm) in each batch to monitor the counting precision and accuracy. Although the extracts are generally coloured, the results are not affected by the colour quenching since the LSC allows an automatic quench compensation to overcome the matrix effects. The recovery is calculated as the ratio of the measured and applied radioactivity.

3.6 Calculation of the variability of sample processing

To analyze the results, create a matrix for each portion size in an Excel spreadsheet as shown in **Table 1**. The replicate analytical portions (at least 5) are the rows (h). The replicate measurements of 1 ml of the same analytical portion are the columns (n). Make sure the recovery is calculated taking into account the corrections, as shown in appendix 1, which are meant to improve the precision of the estimation of the uncertainty components. The stepwise calculation is shown in details in table 1. Calculate the variance of all recoveries (V_T) by taking all data as a single sample with mean **R** and $[(h*n)-1]$ degrees of freedom. Calculate the variance of analysis (V_A) as the average of the variances of each analytical portion with $[h*(n-1)]$ degrees of freedom. Apply one-tailed F test at 95 % confidence level to test if V_T is significantly larger than V_A. If $V_T > V_A$, then

$$S^2{}_{SP} = S^2{}_T - S^2{}_A \qquad \text{eq. (7)}$$

Calculate the estimated variance of sample processing $S^2{}_{SP}(V_{SP})$ for each portion size. Then apply a two tails F test, at 90% confidence level, to check that $V_{SP\,LG}$ and $V_{SP\,SM} \times W_{SM}/W_{LG}$ are not significantly different (refer to *eq. (6)*). If $F_{calculated} < F_{tabulated}$ the sample is homogeneous. Then you can calculate the sampling constant from the large portion size. The uncertainty of sample processing can now be estimated for any analytical portion size above W_{sm} from the K_S value.

3.7 Comments on the method

The criteria we followed in all experiments was to homogenise for a minimum of 2 minutes and to further process the sample until we reached a visually good homogenised sample. **Table 2** shows the chopping time that was needed to visually achieve a good consistency for different materials-either processed fresh/frozen or minced in the presence of dry-ice- using a Stephan chopper, a commercial food processor or a Warring Blender. The chopping time should be optimised in every laboratory depending on the equipment used and the type of samples analysed.

Table 1: *Example of the matrix for the determination of* CV_{SP}

ANALYTICAL PORTION, from 1…h	REPLICATE of the extracts of the same analytical portion, from 1…n				Calculations
1	Recovery $R_{1,1}$	Recovery $R_{1,2}$	Recovery $R_{1,…}$	Recovery $R_{1,n}$	= variance $(R_{1,1} : R_{1,n}),= V_{A,1}$
2					

ANALYTICAL PORTION, from 1…h	REPLICATE of the extracts of the same analytical portion, from 1…n				Calculations
3					
…					
h	Recovery Rh$_1$	Recovery Rh$_2$		Recovery R$_{h,n}$	= variance (R$_{h,1}$:$_{Rh,n}$)= V$_{A,h}$

R =average of all the recoveries
V_A = average ($V_{A,1}$…. $V_{A,h}$….) and [h*(n-1)] degrees of freedom
V_T =variance of all recoveries and [(h*n)-1] degrees of freedom
Apply one tail F test$_{0.05}$: $F_{calculated} = V_T / V_A$
If $F_{calculated} > F_{tab} \Rightarrow V_T \gg V_A$. Then $V_{SP} = V_T - V_A$ and $CV_{SP} = \sqrt{(V_{SP})}/ R$
IF all the portion sizes (small and large) passed the F test, then verify the homogeneity of the chopped sample. Apply two tail $F_{0.1}$ test to $V_{SP\,LG}$ and $V_{SP\,SM} \times W_{SM}/W_{LG}$, where v_{num} =h_{LG}-1 and v_{den} =h_{SM}-1
If $F_{calculated} < F_{tab} \Rightarrow$ the analytical sample is well mixed. NOW calculate K_S using the estimated CV_{SP} of the large portion.
$K_S = W * CV_{SP}^2$
The uncertainty of sample processing for any analytical portion size (W),bigger than W_{sm}can be estimated from the sampling constant value:
$CV_{SP} = \sqrt{((K_S)/W)}$

The suggested method is actually estimating the combined uncertainty of sample processing and extraction. In a separate set of experiments (the results will be published elsewhere) the precision of extraction was determined by fortifying the analytical portions with ^{14}C-Chlorphyrifos just before extraction. The CV_{EXTR} was less than 1.5 % and this value is much smaller than the CV_{SP} obtained for the 25 g and 50 g analytical portion. Therefore the effect of the extraction uncertainty is negligible compared to the sample processing component (7).

The efficiency of sample processing was tested in our laboratory with apples, oranges and lettuce to demonstrate the application of sampling constant for the estimation of the uncertainty of sample processing. The results will be discussed in part II (7).

Table 2 *Chopping time for different matrices*

	fresh sample S-chopper (min)	fresh sample P-chopper (min)	fresh+dry-ice-P-chopper (min)	frozen- P-chopper (min)	frozen+dry-ice-P-chopper (min)	Warring Blender (min)
apple	2	3-5	5-8	6-20	3-5	3
tomato	2	2-3	2-4	-	2-6	2-3
carrot	2	-	-	-	-	
orange	2-3	4	-	-	-	2
lettuce	2-6	-	-	-	-	2

4 CONCLUSIONS:

The use of radio labelled compounds gives a great advantage in residue analysis because it allows one to quantify the analyte directly in the extract, without clean-up, and therefore eliminates one source of uncertainty while maintaining the uncertainty of analysis around ≤2%. The method is also very quick and robust. In fact, once the extracts are sealed in LSC vials, there is no change in the ^{14}C-Chlorphyrifos activity and it is possible to measure the radioactivity of the samples even after some time.

The suggested methodology for the determination of sample processing uncertainty can also be applied with unlabelled compounds (6), but its application takes much longer and the estimated uncertainty of sample processing is less precise because of the larger V_A.

The method described is based on the surface treatment of selected fruits and vegetables with radio labelled compounds. The residue is deposited on the surface of the peel amounting to a small portion of the total mass to imitate the worst scenario regarding the initial inhomogeneity of residues in the sampled units. Since field treated crops have more uniform surface residues distribution than the laboratory surface treatment on a small portion of the whole surface, we can assume that during the analysis of field incurred residues the efficiency of sample processing would be equal to or better than that determined by our tests.

References

(1). Youden,, W., J. Assoc. off. Anal. Chem., 50, 1007-1013, (1967).
(2). Wallace, D. and Kratochvil, B., Analytical Chemistry, 59, 226-232, (1987).
(3). Ingamells, C.O.; Switzer, P. Talanta 20, 547-568, (1973).
(4). Hill, A.R.C. and Reynolds, S.L., Guidelines for in-house validation of analytical methods for pesticide residues in food and animal feeds, Analyst,124, 953-958, (1999).
(5). FAO Codex Alimentarius Commission, Vol.2/A "Portion of commodities to which MRLs apply".
(6). Ambrus, A., Solymosné, E.. and Korsós, I.; Estimation of uncertainty of sample preparation for the analysis of pesticide residues, J. Environ. Sci. Health, B31, (1996).
(7). Maestroni, B., Ghods, A., El-Bidaoui, M., Rathor, N., Jarju O., T.Ton. and Ambrus, A.; Testing the efficiency and uncertainty of sample processing using ^{14}C-labelled Chlorpyrifos; Part II. FAO/IAEA/IUPAC/AOAC International workshop on Method validation, Budapest, Nov 1999.
(8). Kadenczki, L., Arpad, Z., Gardi, I., Ambrus, A. and Gyorfi, L., "Column extraction of residues of several pesticides from fruits and vegetables: a simple multiresidue analysis method", J. Assoc.Off.Anal.Chem., 75, 53, (1992).
(9). Non-Fatty Foodstuffs - Multiresidue methods for the gas chromatographic determination of pesticide residues, Part 2: Methods for extraction and clean-up, Method "P", CEN/TC 275 N 245, (1997).

Next Page: **Appendix 1** Excel spreadsheet for the corrections of recovery values

	A	B	C	D	E	F	G	H	I	J
1	STUDY		Sample matrix:					Date:		
2	Chopper:		Chopping time/speed:					Operator:		
3	Blender:		Amount:				[g]	Operator:		
4	Activity of solution:		Added vol.		Activity dpm/g sample:			3980		
5	Sample Id.		Replicate measurements of the activity of approx. 1 ml EtAc extracts							
6	*Analytical portion 25 /1*			1	2	3	4	5	Ave dpm/g	
7	Sample [g]	25.04		0.92	0.91	0.91	0.9	0.9	0.893235	
8				0.514329	0.508739	0.508739	0.503148	0.503148		
9	Tot.EtAc g	44.79								
10	1min		Count,	1839.03	1797.1	1809.5	1776.45	1801.18		
11			Recovery	0.898389	0.887553	0.893678	0.8871032	0.899453	3.38E-05	

Callout boxes:

- weight sample equivalent =D7*$B$7/$B$9
- weight of the 1 ml extract
- radioactive count in dpm
- Average recovery =average(D11:H11)
- Recovery =D10/D8/$H$4
- V_{ai}= Variance of recoveries= var(D11:H11)

Testing the Efficiency and Uncertainty of Sample Processing Using ^{14}C-Labelled Chlorpyrifos: Part II

B. Maestroni, A. Ghods, M. El-Bidaoui, N. Rathor, O.P. Jarju, T. Ton and Á. Ambrus

FAO/IAEA TRAINING AND REFERENCE CENTRE FOR FOOD AND PESTICIDE CONTROL, FAO/IAEA AGRICULTURE AND BIOTECHNOLOGY LABORATORY, A-2444 SEIBERSDORF, AUSTRIA
e-mail: B.maestroni@iaea.org, http://www.iaea.org/trc

ABSTRACT

The methodology presented in part I (1) was applied to fruit and vegetable samples, representative of different classes of commodities to estimate the uncertainty of sample processing.

Different choppers and different procedures that could decrease the uncertainty of the sample processing were investigated. It was demonstrated that the efficiency of sample processing depends on the equipment used, the type of processed matrix and the degree of inhomogeneity of the pesticide residue in the sample. It was observed that the synergistic effect of a double processing procedure and the homogenisation in the presence of dry-ice improves the uncertainty of sample processing of apples and tomatoes more than 6 fold.

Uncertainty of sample processing, stability of pesticides residues, analytical portion size, processing fresh or frozen samples in the presence of dry-ice, double processing procedures, are all factors to be taken into account in the optimisation of the sample processing procedure.

Key words: *sample preparation, sample processing, homogeneity, uncertainty*

1 INTRODUCTION

The estimation and reporting of the uncertainty of the analytical results is a relatively new requirement (2). In the past, laboratories concentrated only on the precise performance of the core analytical procedure from the point of extraction to the instrumental determination. The effects of sample processing on the accuracy and uncertainty of the results, and also on the stability of residues drew very little attention. Analysts took for granted that a test portion of chopped and minced sample was sufficiently homogeneous for the purpose of analysis. In this paper we present some results that highlight the importance of estimating the uncertainty of sample processing. The stability of residues during processing was described by El-Bidaoui et al. (3).

2 METHOD

As described in part I (1), ^{14}C- labelled chlorpyrifos was used as a model substance. This was very advantageous since it allowed a precise (with a typical relative standard deviation ≤ 2 %) and quick determination of the analyte directly in the extract, without clean-up. By eliminating the effects of the rest of the analytical procedure, the precision of the final results was significantly improved.

In order to get realistic information regarding the distribution of residues in a chopped sample, a relatively small portion of the surface of selected fruits and vegetables (apples, tomatoes, carrots, lettuce and oranges) was treated with radio-labelled chlorpyrifos to imitate the worst scenario regarding the initial inhomogeneity of residues in the sampled units.

The efficiency of sample processing was compared using different equipment -a Stephan chopper **(S-ch)** and a commercial food processor **(P-ch)**.

Different procedures that could decrease the uncertainty of sample processing were studied. Two consecutive processing steps were considered: in the first step the whole laboratory sample was processed in a chopper, while a representative portion of 400-800 g was withdrawn for a second homogenisation step in a Warring blender. Except for tomato and lettuce, the second step of homogenisation of vegetable samples was always carried out in the presence of 20% water, as suggested by Kadenczki et al. (4), in order to improve the homogenisation and successfully blend the sample and to avoid the formation of an aggregate rolling on the top of the mixing blades, as in the case of carrot.

The processing of fresh and frozen apples and tomatoes in the presence of dry-ice was also studied. By adding small pieces of dry-ice up to 27% w/w, the fresh sample was gradually getting deep-frozen and comminuted until free flowing powdery material was obtained. The chopped material was then left at room temperature and stirred occasionally to sublime the CO_2.

3 RESULTS AND DISCUSSION

The goal of any sample processing procedure is to deliver uniform residue distribution in analytical portions. If the starting sample unit does not have a uniform composition (e.g. meat products), it is unlikely that any pesticide residue will be uniformly distributed in that unit. Hence sample processing is vital to ensure that high residue pockets will not exist in the chopped sample (5). Although the procedure looks simple, we observed that it is not always easy to obtain homogeneous samples. The percentage of sample processing experiments which resulted in a statistically well-mixed (6) material is shown in **Table 1.** In 13% and 12% of the cases, the homogenisation did not produce well mixed tomato and orange samples respectively. The reason for non-homogeneous samples could be attributed to the limits of the chopping devices, the inevitable random error and operator handling. This shows the importance of regularly checking the homogeneity of samples as an internal quality control measure. Moreover, it emphasises the importance of having proper skills and training to perform even easy operations such as sample processing.

Table 1: *Statistics for homogenisation experiments*

	% of experiments where the sample was statistically well-mixed (chopper)	% of experiments where the sample was not statistically well-mixed
lettuce	100	
tomato	87	13
apple	93	7
carrot	100	
orange	88	12

3.1 Reproducibility of analysis

As described in part I (1), the ^{14}C activity was determined in a minimum of 5 x 1 ml portion of organic extract from one analytical portion. This allowed the calculation, with relatively high precision, of the uncertainty of analysis which takes into account not only the error of the liquid scintillation counting, but also the operator error in processing the analytical portions.

Figure 1 shows the reproducibility of analysis (CV_A%) for different commodities and different test portion sizes. In most of the experiments we were able to achieve a CV_A smaller than 2% for all test portions. Some higher values were found for the 5 g and 250 g test portions of lettuce and frozen apple respectively, possibly due to the difficulty in handling such sample sizes or the segregation of water, especially for the 5g sample size. It is therefore advisable to regularly mix the matrix between the withdrawals.

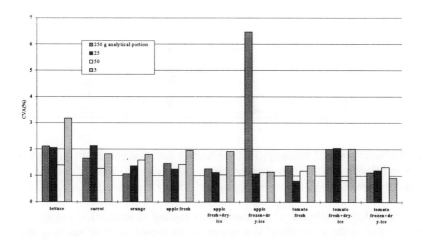

Figure 1 *Reproducibility of analysis for 250 g, 25 g, 50 g and 5 g test portions.*

3.2 Reproducibility of sample processing step

Figure 2 presents the reproducibility of sample processing, where 250 g and 25 g analytical portions were withdrawn just after the homogenisation in a chopper and 50 g and 5 g analytical portions were withdrawn after a consecutive homogenisation in a blender. The variances of sample processing were pooled after elimination of outliers with the Cochran test and the typical reproducibility of sample processing was calculated as

$$CV_{typ} = \frac{\sqrt{V_{ave}}}{Q_a} \qquad \text{eq. (1)}$$

where V_{ave} is the average (pooled) variance of sample processing and Q_a the average recovery obtained in a commodity. $CV_{typ\,sp}$ was calculated only from those experiments that gave a statistically well-mixed sample. Fig.2 shows that the estimation of the sampling constant, based on the variance of sample processing of the larger analytical portion (250 g or 50 g), is very well justified since their reproducibility is significantly higher. Lettuce, carrot and orange showed very good reproducibility compared to tomato samples. Tomato is a difficult matrix and the variety and the ripening stage may be key factors affecting the uncertainty of sample processing.

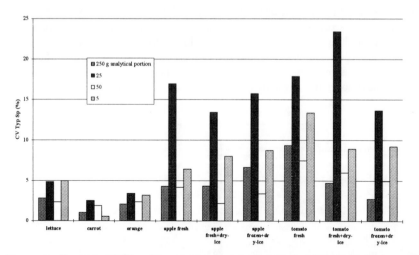

Figure 2   *Reproducibility of sample processing (except for lettuce and carrot that were homogenised in a Stephan-chopper, all the other matrices were homogenised in a P-chopper).*

As with every statistical estimation, the estimated sampling constant (Ks) has its own uncertainty. Therefore the efficiency of sample processing cannot be described by a single sampling constant, but should be considered as a range. The Ks ranges were calculated as follows: the sample processing variances, categorised by matrix and processing device, obtained from repeated experiments, were tested with the Cochran test ($\alpha=0.05$) for the presence of outliers (7). Then the V_{Sp} were pooled and a confidence interval based on the Chi-square distribution (χ^2) at 95% probability was obtained for the average variance.

$$vs^2/\chi^2_{0.025} \leq \sigma^2 \leq vs^2/\chi^2_{0.975} \qquad \text{eq.(2)}$$

The true variance of sample processing (σ_{SP}^2) has a 95% probability of being within the calculated range. The number of degrees of freedom was $v = n^* (h-1)$ where, $n =$ number of pooled variances, each characterised with a number of degrees of freedom equal to $[(h^*n-1)-(h^*(n-1))] = (h-1)$.

Equation (2) can be used to calculate the minimum and maximum value of Ks from the typical variance for each matrix and each chopping devise used. The typical sampling constant ($K_{S\ TYP}$) values are calculated from $K_{S\ TYP} = W *CV_{SP}^2$, where CV_{SP} is obtained from the pooled variances of sample processing as described above, and W is the weight of the large analytical portion.

3.3 Utilisation of the sampling constant

Figure 3 shows the estimated uncertainty of sample processing as a function of the test portion size. The CV_{SP} is calculated for a medium (2.5 kg), high (6.6 kg) and very high (17.4 kg) Ks value, for fresh apples processed with dry-ice, frozen apples processed with dry-ice and fresh apples respectively, homogenised in a commercial food processor. A relationship is also shown for apple samples processed in a Stephan chopper and having a lower Ks (1.6 kg). The estimated uncertainty of sample processing varies according to the sampling constant value (Ks) and to the analytical portion size. It is very clear from the figure that this principle can be used to optimise the test portion size for an analytical procedure. A 30 g analytical portion represents a good compromise, especially if the Ks is in the low-medium range. On the other hand analysing a 100 g or 150 g analytical portion does not improve the situation in terms of CV_{Sp}, but certainly increases the cost of analysis.

A method of analysis that uses a 5 g analytical portion may have great advantages in terms of solvent consumption and analysis time, but definitely is not the best choice in terms of uncertainty of sample processing.

3.4 Efficiency of sample processing

Figures 4 and 5 show the range and typical values of the sampling constants for samples chopped in a P-chopper or an S-chopper and further homogenised in a Warring Blender. K_S1 and K_S2 refer to the sampling constant of the first and second processing step respectively. Low Ks values are synonymous with low CV_{SP} or in other terms in a "good efficiency of sample processing", which depends on the chopping and mincing equipment used and the type of sample material (matrix)

Taking the example of the fresh apple homogenised in the presence of dry-ice, its typical K_S1 was 2.5 kg and K_S2 was 0.27 kg. This means that by analysing a 30 g sample, taken after the first processing step, the uncertainty of sample processing is already about

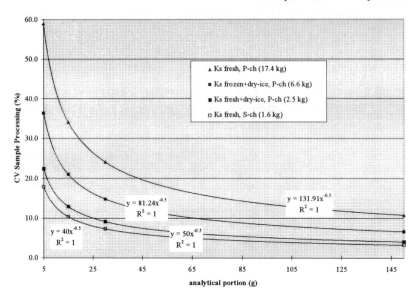

Figure 3 *The estimated uncerainty of sample processing for apple samples processed in a commecial food processor and in a Stephan chopper for different sampling constant values (Ks) and different analytical portion sizes.*

9.1%; alternately, it is possible to further blend a 400 g subsample and have 3.0% combined uncertainty when processing a 30 g analytical portion (eq. 3).

Orange is an example of a commodity that was very easy to homogenise with the P-chopper (Ks1 $_{orange}$= 0.32 kg), followed by fresh and frozen apples and tomatoes in the presence of dry-ice. Figure 5 shows that the efficiency of sample processing increased with the processing of samples in the presence of dry-ice, which made the samples more fragile, produced a powdery consistency and hence the sample was more likely to become well mixed during the homogenisation. The processing of frozen samples in the presence of dry-ice was also more efficient. Fresh apples and tomatoes gave high Ks1 values and this is in line with previously reported results (6).

In general, the efficiency of further homogenisation in a Warring blender resulted in low typical KS_2 values ranging from 0.12 to 1.29 kg, with the exception of fresh tomato which gave 3.94 kg. If the laboratory sample is already chopped to a well-mixed condition further blending of 400-800 g, in the presence of 20% water, substantially improves the homogeneity of the sample, resulting in smaller Ks values.

A visual observation of the chopped and homogenised samples was also carried out. 5 g of homogenate - chopped in a commercial food processor and in a Warring blender-were diluted in 1 l water. The peels contained in the water were then collected on filter papers. The dried filters are shown in **Figures 6-7-8** for tomato and apple samples. The different comminution of treated peels influenced the distribution of residues among the analytical portions and hence the Ks ranges. It is interesting to note that the size of the peels changed according to the status of the sample (fresh or frozen) and homogenisation procedure (processing carried out in the presence or absence of dry-ice).

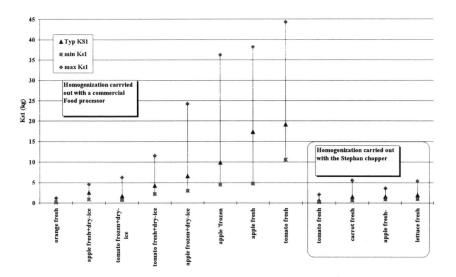

Figure 4 *Calculated ranges of the sampling constant (K_{SI}) for fresh, frozen and/or with dry -ice apples, oranges and tomatoes when the homogenisation is carried out in a chopper (P or S-ch)*

The density and the size of the peels were bigger for the fresh samples as shown in Figures 6a and 6c. The visual observation was in perfect agreement with the calculated typical Ks values shown in fig. 5.

Tomato is a very difficult commodity to homogenise because of the fragility of the peel which breaks easily and because of the elasticity shown during processing, which causes the peel to fragment only along the shearing direction of the blades, thus maintaining a relatively high area/weight ratio (fig. 6a, 7a and 8a). Furthermore, due to the high water content the segregation of water phase and pulp and peels is instantaneous. These observations are in line with the values of reproducibility of sample processing presented in fig. 4. Processing fresh tomato samples is less reproducible than fresh or frozen samples processed in the presence of dry-ice.

Apples, on the contrary, were easier to homogenise since the peels remains attached to the pulp and they were more easily cut into small pieces.

The reason for very low Ks ranges for orange (fig. 5) can also be visually explained from Figure 9; it is evident that the first chopping step in a common food processor was already very efficient in the comminution of orange peels and hence in the homogenisation of the sample. The size of the peels was very small, compared to fresh tomato or apple peels (fig. 6a and 6c). In most of the orange experiments the uncertainty of sample processing was very low, and in some cases the homogenisation of the sample was so good that there was no significant difference between V_T and V_A. This means that the uncertainty of sample processing was included in the 2% of $CV_A\%$.

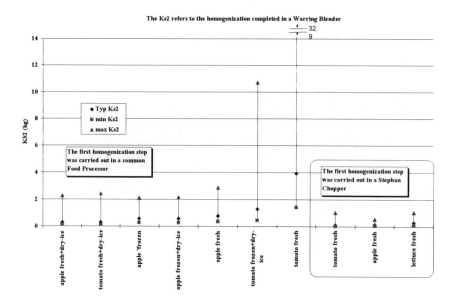

Figure 5 *Calculated ranges of sampling constant values (K_{S2}) for apple , tomato, carrot and lettuce when the chopped samples are further homogenised in a Warring blender .*

3.5. Comparison of choppers

Ambrus et al. (6) stated that the homogeneity of the processed sample, and consequently the sampling constant, depends on the chopping and mincing equipment used, but is independent from the actual pesticide residue present.

In our experiments we compared the use of two different choppers: a Stephan chopper (S-ch) and a semi-industrial Food processor (P-ch). Ambrus et al. (6) reported the efficiency of sample processing for another commercial food processor (K_{sapple}= 11-21 kg). As expected we found that the S-chopper gave smaller ranges of Ks and even smaller absolute values (**Fig. 4**). Lettuce and carrot chopped in a S-chopper gave very good reproducibility of sample processing compared to apple and tomato chopped in a P-chopper. These results show the importance of checking, in each laboratory, the uncertainty of sample processing since it depends strongly on the type of equipment used.

If the sample processing uncertainty is carried out with unlabelled pesticides, easily and reproducibly extractable stable compounds (e.g. Chlorpyrifos, Lindane, OPs, OCs, pyrethroids) should be used for determining the K_S value for a given sample processing method. The K_S will be applicable for all pesticides determined under the given condition.

3.6. Options to reduce the uncertainty of sample processing

In order to reduce the uncertainty of sample processing we examined three options.

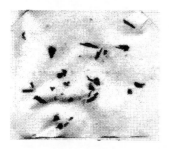

Figure 6 (a): *FRESH TOMATO: homogenization carried out in a P-chopper*

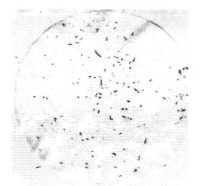

Figure 6 (b): *FRESH TOMATO: further homogenization carried out in Waring blender*

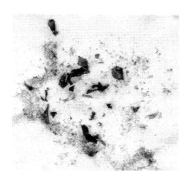

Figure 6 (c): *FRESH APPLE : homogenisation carried out in a P-chopper*

Figure 6 (d): *FRESH APPLE: further homogenisation carried out in Waring blender*

Figure 7 (a): *FRESH TOMATO*
processed with DRY ICE:
homogenisation carried out in a P-
choppe)

Figure 7 (b): *FRESH TOMATO*
processed with DRY ICE: further
homogenisation carried out in a Waring
blender

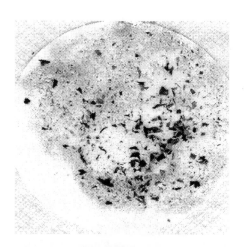

Figure 7(c): *FRESH APPLE processed*
with DRY ICE: homogenisation carried
out in a P-choppe)

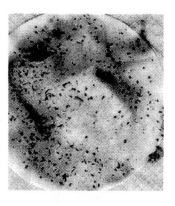

Figure 7 (d): *FRESH APPLE processed*
with DRY ICE: further homogenisation
carried out in aWaring blender

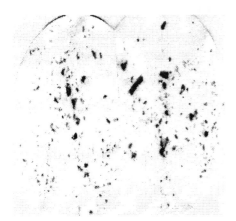

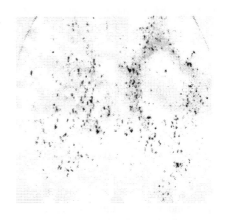

Figure 8 (a): *FROZEN TOMATO processed with DRY ICE: homogenisation carried out in a P-chopper*

Figure 8 (b): *FROZEN TOMATO processed with DRY ICE: further homogenisation carried out in a Waring blender*

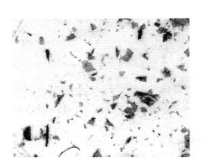

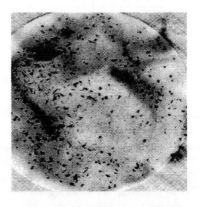

Figure 8 (c): *FROZEN APPLE processes with DRY ICE: homogenisation carried out in a P-chopper*

Figure 8 (d): *FROZEN APPLE processed with DRY ICE: further homogenisation carried out in a Waring blender*

3.6.1. Improvement through double processing. First, we tested the applicability of a two phase processing procedure including the initial mincing with large capacity choppers, which can process the whole laboratory sample, and further chopping of a smaller portion in a Warring Blender. The combined uncertainty of sample processing, (CV_{Sp2}) will depend on the sampling constant K_{S1} of the minced sample, the size of the portion further homogenised (W_2), the sampling constant of the homogenised portion K_{S2} and the test portion, W_A, which is withdrawn for analysis.

$$CV_{SP2} = \sqrt{\frac{K_{S1}}{W_2} + \frac{K_{S2}}{W_A}} \qquad \text{eq. (3)}$$

The relation of the sampling constants, the portion of processed samples withdrawn and the corresponding uncertainties of sample preparation [$CV_{1\ SP}$ (first step) and $CV_{2\ ST}$ (second step)] is illustrated in **Table 2**.

Table 2 *The effect of 2-step sample processing on the uncertainty of sample processing ($CV_{SP2,\ 400\text{-}30\,g}$ (%) means $W_2=400\,g$ and $W_A= 30\,g$)*

Matrix	$K_{S1\,(kg)}$	$CV_{Sp1},$ 30 g (%)	$CV_{Sp1},$ 400 g (%)	$CV_{Sp1},$ 800 g (%)	$K_{S2\,(kg)}$	$CV_{Sp2},$ 400-30 g (%)	$CV_{Sp2,\ 800\,-}$ 30g (%)
tomato fresh S-ch	0.6	4.4	1.2	0.8	0.12	2.3	2.2
apple fresh S-ch	1.6	7.2	2.0	1.4	0.15	3.0	2.6
tomato frozen+dry-ice	1.7	7.5	2.1	1.5	1.29	6.9	6.7
tomato fresh+dry-ice	4.2	11.9	3.3	2.3	0.29	4.5	3.9
tomato fresh	19.1	25.3	6.9	4.9	3.94	13.4	12.5

From the table it is clear that if the analytical portion (30 g) is withdrawn just after the first chopping step the error may be very high and definitely a double processing step reduces the uncertainty about 2-3 times. For comparison, the CV_1 for 400g and 800 g are reported. The mass of subsample that is further homogenised in a Warring blender and the corresponding CV_{SP1} represent a limiting factor.

The second chopping step (WB) reduced the size of the peels, as noted in figures 6b, 6d, 7b, 7d, 8b, 8d, 9b, 9d, giving rise to a far more homogeneous sample. The advantage of the 2-step procedure in reducing uncertainty of sample preparation is evident, however further studies are required to check the stability of residues under these conditions.

3.6.2. Improvements through dry-ice. Second, we used dry-ice in the processing of the samples. This is a procedure adopted sometimes by residue monitoring laboratories in countries where the cost of analysis is not a limiting factor. The use of dry-ice definitely improves the homogeneity of the matrix, and simultaneously reduces the chance of

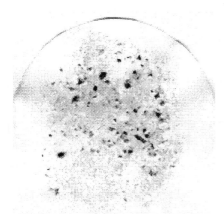

Figure 9 (a): *FRESH ORANGE: homogenization carried out in a P-chopper*

Figure 9(b): *FRESH ORANGE: further blending carried out in a Warring blender*

decomposition of pesticide residues during sample processing. The results obtained for apple are shown in **Table 3**.

Table 3 *Comparison of the CV_1 and CV_2 for the processing of fresh, frozen or with dry-ice, apples (m_2= 400 g and m_A= 30 g).*

	$K_{S1 (kg)}$	$CV_{Sp1,30g}$ (%)	$K_{S2 (kg)}$	$CV_{Sp2, 400-30 g}$ (%)
apple fresh+dry-ice	2.5	9.1	0.27	3.9
apple fresh	17.4	24.0	0.78	8.3
apple frozen+dry-ice	6.6	14.8	0.58	6.0
apple frozen	9.9	18.1	0.57	6.6

The addition of dry-ice to the processing of fresh apples gave 2.6 and 2.1 fold improvement to the uncertainty of sample processing, whether the analytical portion was withdrawn after the first or the second processing step. In the case of tomato samples (table 2) the addition of dry-ice reduced 2-3 times the uncertainty of sample processing.
The synergistic effect of the double sample processing procedure and the addition of dry-ice improved 6.2 and 6.5 fold the uncertainty of sample processing, for apples and tomatoes respectively (see also table 2).

3.6.3. Improvement through frozen status. Third, we processed deep-frozen apples with the option of adding dry-ice during mincing. This last option is the regular practice of pesticide manufacturing companies, but it requires special and expensive equipment which are usually not available in pesticide residue monitoring laboratories. The results are shown in table 3. The "frozen status" compared to the "fresh status" did not significantly improve CV_1 and CV_2 (1.3 folds improvement). The addition of dry-ice to the processing of frozen samples further reduced the uncertainty, but to a smaller extent, from 18.1 to 14.8 (CV_1) and from 6.6 to 6.0 (CV_2). If the samples are processed frozen, in the presence of dry-ice and subjected to a further homogenisation in a Warring blender the uncertainty of sample processing of apples can be reduced 4 fold, compared to a 6.2 fold reduction discussed above. The same results were observed for tomato samples. Further studies are required to check the stability of residues under this conditions.

4 CONCLUSIONS:

The uncertainty of sample processing (CV_{SP}) depends on the degree of inhomogeneity of the pesticide residue in the sample. Therefore, in our experiments the worst scenario, in which the residue was not homogeneously distributed on the surface of the commodities, was simulated. Since field treated crops have more uniform surface residues than the surface treatment on a small portion of the whole surface, it is possible to assume that, during the analysis of field incurred residues, the efficiency of sample processing will be equal to or better than that determined by our tests.

When the processed sample was statistically well mixed (homogeneous) the efficiency of sample processing was characterised by estimating the typical sampling constant, and the uncertainty of sample processing, within the tested analytical portion mass range, was calculated from $K_S = W * CV^2$.

The sampling constant concept (8) is an excellent tool to predict the uncertainty of sample processing (CV_{SP}) for different test portion sizes, or for selecting the test portion size that assures a target level of CV_{SP} which fits the purpose of the analysis.

The efficiency of sample processing depends on the equipment used and the type of processed matrix. It may also depend on the variety and maturity of the commodity. The lowest KS_{typ} was obtained for orange sample and the highest for fresh tomato samples when chopped in a commercial food processor. On the contrary, tomato gave the lowest KS1 when chopped in a Stephan chopper followed by carrot, apple and lettuce samples. We strongly recommend that each laboratory checks the homogeneity and the efficiency of sample processing, which cannot be derived from the literature or from other laboratories. Eventually the efficiency of sample processing should become a routine internal quality control check.

We also suggest that analysts observe the consistency of the processed matrix, to check whether it is still carrying large pieces of peel (0.5 mm) that would largely contribute to the inhomogeneity of the sample.

We observed that a double processing procedure was a great improvement in reducing the uncertainty of sample processing (2-3 fold improvement), as well as the processing of samples in the presence of dry-ice which increased the efficiency of sample processing by 2-3 times. However the stability of residues under this condition still have to be confirmed.

We observed that the synergistic effect of the double sample processing procedure and the addition of dry-ice reduced the uncertainty of sample processing by 6.2 and 6.5 fold for apples and tomatoes respectively.

An optimisation process is always involved when deciding on which sample processing procedure to adopt. Uncertainty of sample processing, stability of pesticides residues, analytical portion size, processing fresh or frozen samples in the presence of dry-ice, double processing procedures, are all factors to be taken into account.

According to Ambrus (9) there is little point in trying to reduce the combined uncertainty of sample processing and analysis below 12%. Since the uncertainty of sample processing can be more easily kept at a lower level, the optimum would be a 5% uncertainty of sample processing. A CV_{SP2} equal to 5% was obtained in our laboratory for fresh apples and tomatoes by a double processing step in the presence of dry-ice (tables 3 and 4).

Although the analysis is longer and less precise, the suggested method to check the uncertainty of sample processing can also be applied with unlabelled compounds (6). Since the uncertainty of sample processing depends on the type of sample material (matrix), on the inhomogeneity of the pesticide residues in the sample, and on the equipment and procedure used for sample processing, but is independent from the actual residues present, easily and reproducibly extractable stable compounds (i.e. Chlorpyrifos, Lindane, OPs, OCs, pyrethroids) should be used for determining the K_S value for a given sample processing method. Once the K_S is determined for typical commodities and surface residues, the results can be applied for all pesticides and there is no need to test it for all commodities-pesticide residues combinations.

References

(1) B. Maestroni, A. Ghods, M.El-Bidaoui, N. Rathor, T.Tam and A.Ambrus "Testing the efficiency and uncertainty of sample preparation using ^{14}C-labelled chlorpyrifos: PART I. Description of the methodology",. FAO/IAEA/IUPAC/AOAC International workshop on Method validation, Budapest, Nov. 1999.

(2). FAO/IAEA/AOAC/IUPAC Guidelines for single-laboratory validation of analytical methods for trace-level concentrations of organic chemicals, International workshop on principles and practices of method validation, Nov. 4-6, 1999- Budapest.

(3) M. El-Bidaoui, O.P.Jarju, B.Maestroni, Y.Phaikaew and A.Ambrus "Stability of pesticide residues during sample processing and storage" FAO/IAEA/IUPAC/AOAC International workshop on Method validation, Budapest, Nov. 1999.

(4) L. Kadenczki, Z.Arpad, I.Gardi, A.Ambrus and L.Gyorfi, "Column extraction of residues of several pesticides from fruits and vegetables: a simple multiresidue analysis method", J. Assoc.Off.Anal.Chem. 75, 1992, 53.

(5) E.A. Bunch, D.M. Altwein, L.E. Johnson, J.R. Farley and A.A.Hammersmith, "Homogeneous sample preparation of raw shrimp using dry-ice", J. AOAC Int., 78, 1995, 3.

(6). A. Ambrus, E. Solymosne and I. Korsos., Estimation of uncertainty of sample preparation for the analysis of pesticide residues, J.Environ.Sci. Health, B 31, (3) (1996).

(7) G.W. Snedecor and W. G. Cochran, Statistical methods, 6th ed. The Iowa State University Press, Ames (1967), pp.480-483.

(8) D. Wallace and B. Kratochvil, Analytical Chemistry, 59, 226-232 (1987).

(9). A. Ambrus, Quality of residue data, In "Pesticide Chemistry and Bioscience" by Brooks, G.T. and Roberts, T.R. RSC,1999

Testing the Effect of Sample Processing and Storage on the Stability of Residues

M. El-Bidaoui[1], O.P. Jarju[1], B. Maestroni[1], Y. Phakaeiw[2] and
Á. Ambrus[1]

[1] FAO/IAEA TRAINING AND REFERENCE CENTRE FOR FOOD AND PESTICIDE CONTROL.
FAO/IAEA AGRICULTURE AND BIOTECHNOLOGY LABORATORY, A-2444 SEIBERSDORF,
AUSTRIA
e-mail: M.Elbidaoui@iaea.org, http://www.iaea.org/trc
[2] DEPARTMENT OF AGRICULTURE, DIVISION OF AGRICULTURAL TOXIC SUBSTANCES,
BANGKOK, THAILAND 109000

1 INTRODUCTION

Storage temperature, enzymatic activity, water content and pH of the sample matrix are
the major factors contributing to pesticide residue degradation. Hydrolytic resistance of a
pesticide gives an initial idea on its expected residue stability during storage and
processing [1].

A laboratory sample is generally subject to processing in order to obtain a
representative analytical portion. This treatment may affect the stability of the analytes
present and lead to biased results.

Pesticide manufacturers provide information on the stability of pesticide residues
during storage and sample processing (a requirement for pesticide registration) some of
which are published by FAO/WHO [2]. Nevertheless, the conditions of sample processing
may be entirely different in food monitoring laboratories, and their effect should be
studied on the stability of residues.

Present guidelines for accreditation of Chemical Laboratories do not set specific
requirements for testing the stability of analytes [3, 4 and 5]. Hill proposed the inclusion
of testing the effect of sample processing in the validation scheme for analytical methods
[6] because of two major concerns regarding: a) the eventual potential loss or degradation
of the analytes, and b) the degree of homogeneity reached in the analytical sample during
this step i.e. the variability of average analyte concentration in the analytical portions.
Others [7, 8 and 9] pointed out the importance and the consequence of the uncertainty in
sample processing on pesticide residue data, and tested the efficiency of different
equipment and conditions. Sample homogeneity is crucial for repeatable and reproducible
analysis, however the processes and the conditions applied to attain homogeneity were not
studied in relation to stability.

Our study aimed primarily to assess the effect of sample processing, using
different equipment and process, therefore the commodity was treated on the surface to set

a known initial residue level; secondly to estimate the effect of storage on the stability of pesticide residues. Considering the primary goal of the study, analytical standard mixtures were applied on the surface of tomato fruits, instead of incorporating the analytes in a pre-processed sample matrix which is the recommended method for studying the deep-frozen storage stability [1, 10], in order to focus on the processing effect. The fruits were processed fresh, shortly after surface treatment, or after storage at $-20^{\circ}C$ overnight, or for one, two, three and four weeks. The pesticide mixtures contained ^{14}C-radiolabelled chlorpyrifos, since the quick and accurate measurement of the ^{14}C activity made it ideal for use as an internal standard (I.S) to correct for the effect of different sources of variability in the analytical procedure based on the ethyl acetate extraction [11].

2 EXPERIMENTAL

2.1 Equipment

The following specialised equipment was used in the present study:

Food processor, Mod. PSP-500 6C, for maximum 2 kg of sample (referred to as P-chopper); Warring Blender with 1 litre stainless-steel container; Ultra-Turrax T25 (IKA); Rotary vacuum evaporator, (Büchi R114R); Refrigerated centrifuge (Sigma 4K 15); semi-automatic Gel Permeation Chromatography (GPC) apparatus with Pyrex glass column (1 x 25 cm) operated under constant nitrogen pressure; liquid scintillation counter (LSC) Beckman LS 6000TA; HP 6890 Gas chromatograph equipped with ECD and NPD and an auto-injector (HP 6890 series, HP G15).

2.2 Chemicals and Reagents

Pesticides analytical standards >95% a.i. (Dr. Ehrenstorfer); ethanol solution of ^{14}C chlorpyrifos with a specific activity of 1.1MBq/mg (Internationale Isotope, München); ethyl acetate; cyclo-hexane; isooctane; dichloromethane, acetone, all pesticide residue analytical grade (Merck); Bio-Beads SX-3; $NaHCO_3$; Na_2SO_4, anhydrous; LSC cocktail (Ultima gold, XR Packard); Milli-Q Water (Millipore), commercial dry ice slabs.

2.3 Surface treatment and fortification mixtures

The mixtures contained stable and unstable pesticides, presenting different physico-chemical properties i.e. resistant or not to acidic and/or alkaline media.

2.3.1. ^{14}C chlorpyrifos solution. 7.5μg/mL solution of chlorpyrifos in ethyl acetate with ~2500000 dpm/mL (dpm: disintegration/min.) specific activity, was used to prepare the radiolabelled surface treatment mixtures.

2.3.2. Fortification and calibration mixtures. The composition and concentration of the stock solutions are given in Table 1.

Intermediate solutions of approximately 50 μg/mL concentration (diluted in isooctane from the stock solutions) were used to fortify individual analytical portions of 25 g, at 0.5 mg/kg level, in order to perform the recovery studies (2.4.2).

The same intermediate solutions were used to prepare calibration solutions in isooctane at 25, 50, 75, 100, 150 and 200 ng/mL level for the ECD and 50, 150, 250, 350, 450 and 550 ng/mL level for the NPD.

Table 1 *Composition of stock solutions of analytical standard mixtures*

ECD mixture in isooctane	mg/mL	NPD mixture in dichloromethane	mg/mL
Lindane	1.53	metamidophos	1.15
Diazinon	1.69	dimethoate	1.55
Chlorothalonil	1.51	pirimicarb	1.52
Parathion-methyl	1.57	vinclozolin	1.48
Chlorpyrifos	1.57	pirimiphos-methyl	1.65
Folpet	1.52	chlorpyrifos	1.51
α endosulfan	1.53	iprodione	1.49
β endosulfan	1.51	azinphos methyl	1.49
p,p-DDT	1.51	azinphos ethyl	1.65

2.3.3 Surface treatment mixtures. Mixtures in Table 1 were diluted with the ^{14}C chlorpyrifos solution (2.3.1) to obtain the treatment solutions shown in Table 2. 1mL of the ECD and NPD mixtures were used to treat the surface of the 400 g tomato samples and 2mL for the 2kg sample, at about 0.5 mg/kg level. 5µl of the treating solutions was counted in triplicate before each application to determine their specific activities.

Table 2

ECD mixture	µg/mL (400g)	µg/mL (2kg)	NPD mixture	µg/mL (400g)
lindane	192	537	metamidophos	144
diazinon	211	590	dimethoate	194
chlorothalonil	189	529	pirimicarb	190
parathion-methyl	196	549	vinclozolin	185
chlorpyrifos	196	549	pirimiphos-methyl	207
^{14}C chlorpyrifos	6.46	4.84	^{14}C chlorpyrifos	6.46
folpet	190	531	chlorpyrifos	189
α endosulfan	191	535	iprodione	187
β endosulfan	189	529	azinphos methyl	187
p,p-DDT	189	529	azinphos ethyl	207

This surface treatment resulted in an activity of approximately 1000 and 2500 dpm/mL in the final cleaned tomato extracts for the storage stability and the processing studies, respectively.

2.4 Sample Preparation

2.4.1. Blank samples. The tomatoes were washed and soaked in normal water, then washed three times with Milli-Q water to remove contamination from the surface and then dried. Blank samples were handled in the same way as the surface treated samples i.e.

extracted immediately, frozen overnight or stored at -20°C for one or several weeks. One blank sample was included in each analytical batch processed by an analyst.

2.4.2. Fortified analytical portions. To study the reproducibility of the analytical procedure, the NPD and the ECD mixtures were used to fortify 25g analytical portions from the blank samples at 0.5 mg/kg level for recovery tests. The procedure was then followed as in 2.4.5. Two spiked analytical portions were included in each analytical batch.

2.4.3. Surface treatment. Tomato fruits weighing approximately 400g were cut into halves and the units placed with the cut surface on a clean aluminium foil. The surface treatment mixture (2.3.3) was carefully applied in equal amounts on the top part of the fruits using a Hamilton syringe to obtain approximately 0.5 mg/kg residue concentration.

Thirty minutes waiting time was allowed for the adsorption of the residues onto the surface and evaporation of the solvent. The sample was then either processed immediately or packed in hermetically sealed polyethylene bags and placed into a deep-freezer at -20°C. Each study was performed twice.

When the samples were processed, the aluminium foil, syringe, volumetric flask containing the treatment solution were rinsed with 2-3mL of acetone and added to the sample during the processing step. These materials were washed again and the activity of washing solution was checked. The remaining ^{14}C activity, taken into account in the calculation, was in average 2.5 % of the total activity applied.

2.4.4. Sample processing. The surface treated sample (~400g) was homogenised in a Waring blender for one minute. The material from the walls were transferred into the blender with a spatula and homogenised at high speed for another 2 minutes.
The frozen tomato samples were allowed to thaw completely before homogenisation.

2.4.5. Extraction and clean-up. 2 x 5 x 25g analytical portions were withdrawn from the Waring blender and weighed into 250mL centrifuge tubes, and analysed by two analysts.
The analytical portion was extracted using the Ultra-Turrax at 11000 rpm/min for one minute, after adding 4.17g of sodium hydrogen carbonate, 50.0mL of ethyl acetate and 25g of anhydrous sodium sulphate. The material remaining on the Ultra-Turrax' head was transferred back as much as possible to the centrifuge tube. The crude extract was weighed before and after centrifugation for 10 minutes at 2500 rpm speed. 3 x 1mL of the supernatant was pipetted into LSC vials and weighed, 10mL of LSC cocktail was added and the vials tightly closed for the determination of the ^{14}C activity referred to as the $^{14}C_{ext}$.

Further, 20mL of the extract was weighed and filtered through 60g anhydrous sodium sulphate (10-60 mesh) into a 100mL round-bottom flask. The filter cake was washed with 3 x 20mL ethyl acetate and the combined filtrate was evaporated using the Büchi rotary evaporator to approximately 1-2mL.

The concentrated organic extract was transferred with a Pasteur pipette to a 15mL test tube and the flask washed with 5 x 1mL ethyl acetate. The combined extract was then evaporated under a gentle nitrogen stream to 1mL, about 1mL of cyclo-hexane was then added, mixed and evaporated again to approximately 0.8-1.0mL. The volume was finally adjusted to exactly 1mL with a (1:1 v/v) cyclo-hexane:ethyl-acetate mixture. The gross mass of the test tube was also noted.

500 µl of the solution was injected into the GPC and eluted with a cyclo-hexane:ethyl acetate (1:1, v/v) mixture under constant nitrogen pressure with 1-2mL/min flow rate.

The first 8mL were discarded (the LSC revealed no activity in this fraction). The next 30mL fraction was collected in a 50mL round bottom flask and concentrated to 0.5-1mL using a vacuum rotary evaporator. The concentrated eluent was transferred to a test tube and the round bottom flask washed 5 times with 1mL isooctane. The final volume was adjusted to 10mL with isooctane and 2mL (after concentration under nitrogen stream) for the determination with ECD and NPD, respectively. The gross weights were recorded.

2 x 1mL (ECD) and 2 x 0.5mL (NPD) were withdrawn from the organic extract and weighed for LS counting ($^{14}C_{ana}$) and 250µl was transferred and diluted to 1mL into the auto-sampler vial for GC quantification.

The weighing was part of the internal quality control to reduce the uncertainty of the volumetric measurements.

2.5 Samples for the storage stability

The storage stability study of residues was performed only with the radiolabelled ECD mixture. About 2 kilograms of tomato were surface treated and divided in four parts of known weights and stored for one, two, three and four weeks at -20°C in hermetically sealed polyethylene bags. The frozen samples were homogenised in the P-chopper in the presence of about 200g of dry ice in order to increase the homogenisation efficiency [9] and to process the sample in frozen condition to minimise the effect of this step. After one minute homogenisation, the material from the lid and the walls were transferred into the chopper with a spatula and homogenised for another 2 minutes. The analytical portions (25g) were withdrawn five minutes after the processing in order to allow the sublimation of the CO_2. The extraction procedure was followed as in 2.4.5.

2.6 Gas chromatographic analysis

Operating conditions:

Oven temperature programme: 70°C (1 min.), 20°C/min to 160°C, 4°C/min to 275°C (5 min.).

Carrier gas: helium 1.5mL/min, constant flow. Injection volume: 1µl.

ECD: 300°C:

Column: CP-sil 5CB fused silica capillary column (25m x 0.32mm, film thickness 0.25µm) with CP fused silica retention gap (2.5m x 0.32mm).

Splitless injector: 250°C. Purge vent open after 1 minute.

NPD: 300°C

Column: HP-5 (30m x 0.32mm, film thickness 0.25µm) with CP fused silica retention gaps (2.5m x 0.32mm)

PTV injection (splitless mode): 70°C (0 min.), 600°C/min to 250°C (2 min.), 100°C/min to 70°C.

2.7 Liquid scintillation counting

Beckman LSC was used to determine the activity of the ^{14}C in the extracts prepared according to section 2.5.4. Each sample was counted for 3 times 5 min. A ^{14}C commercial standard <1μCi (10×10^4 dpm) was included in each batch in order to monitor the counting precision and accuracy.

The instrument provided also an automatic quench compensation to overcome the matrix effect.

The measurement of radioactivity could be performed with a relative standard deviation ≤ 1 %.

2.8 Statistical evaluation

One way ANOVA was used to compare the average recoveries of a pesticide obtained by the two analysts, in the case of surface treated samples extracted immediately (day 0) and samples stored frozen overnight (day 1). Followed by a *t*-test to compare results of day 0 and day 1.

Two-way ANOVA was used to assess the effect of the different storage periods on the stability of pesticide residues and to separate the effects of this controlled factor and the replicate analysis performed by two analysts, considered as the random factor.

The Dixon test was used to determine the outliers.

A confidence interval of 95% was applied for all the statistical evaluations.

3 RESULTS AND DISCUSSION

3.1 Blank samples

The chromatograms of the blank samples showed no significant interference.

3.2 Recovery studies

Quantification was done by external standard comparison using 6-point matrix matched calibration with correlation coefficients of the regressions ranging from 0.994 to 0.9999.

All pesticides were well separated from each other, allowing for good quantification.

Table 3 shows the mean recoveries from the fortified analytical portions, the number of replicates (n) and the relative standard deviation (RSD %) indicating the within laboratory reproducibility of the analysis.

Table 3

Pesticide	Detector	analytical recovery %	n	R.S.D %
Azinphos-ethyl	NPD	91.4	8	12.0
Azinphos-methyl	NPD	106.6	8	11.8
Chlorothalonil	ECD	83.2	30	17.3
Chlorpyrifos	ECD	84.4	33	18.8
Chlorpyrifos	NPD	95.2	12	8.9
Chlorpyrifos (average NPD-ECD)		**87.2**	**45**	**17.16**
pp-DDT	ECD	86.9	33	15.8
Diazinon	ECD	84.6	34	14.4
Dimethoate	NPD	101.4	7	15.8

Testing the Effect of Sample Processing and Storage on the Stability of Residues

Pesticide	Detector	analytical recovery %	n	R.S.D %
α-Endosulfan	ECD	87.5	34	16.3
β-Endosulfan	ECD	87.2	34	16.6
Folpet	ECD	79.2	28	21.8
Iprodione	NPD	97.5	12	10.9
Lindane	ECD	85.6	33	16.5
Metamidophos	NPD	85.5	6	21.5
Parathion-methyl	ECD	88.1	32	22.0
Pirimicarb	NPD	94.5	12	9.4
Pirimiphos-methyl	NPD	89.3	12	8.53
Vinclozolin	NPD	95.5	12	8.94
Qa = 90.2			**CV_{typ} = 15.9**	

All average recoveries were within the acceptable range (70-110%) with an RSD < 23%, the limit for the within laboratory reproducibility [12].

From these recovery values, we estimated the typical performance characteristics of the method i.e. a typical CV_{typ} = 15.9 % and recovery Qa = 90.2 %.

The typical CV was calculated as $CV_{typ} = \sqrt{V_{ave}}/Qa$ (1)

Where V_{ave} is the average variance of recoveries and Qa is the average recovery of all analytes [13].

Although the analytical recoveries of three analytes (folpet, metamidophos and parathion-methyl) were close to the limit of the within laboratory reproducibility (23%), the CV_{typ} indicated that the analytical conditions and the procedure were acceptable for carrying out the stability study.

3.3 Analysis of ^{14}C residues

The average proportion of the initial ^{14}C chlorpyrifos activity, applied on the surface of the fruits, $^{14}\overline{Cext}$ detected in all crude extracts was 88.6% (with a CV_{14Cext} of 5.5%). In the final cleaned extract its average proportion, $^{14}\overline{Cana}$ was 87.3% with a CV_{14Cana} of 7.9%. The difference was not significant.

The average analytical recovery determined from spiked analytical portions for chlorpyrifos (87.2%, Table 3) is not significantly different from the average proportion (88.6%) measured from surface treated fruits.

These findings indicate that
- the chlorpyrifos is not lost during the sample processing, but rather during the extraction step and
- from the filtration and clean-up phases to the determination, there is no loss of chlorpyrifos.

Consequently, the ^{14}C chlopyrifos could be used as an internal standard.

The average proportion of the $^{14}\overline{Cext}$ obtained from the totality of the studies (88.6%) is in full agreement with the 88% calculated from a large number of measurements made for studying the efficiency of sample processing [8]. It is the best estimate of the extraction efficiency and can be used as an indicator of the initial concentration of the analytes on the treated commodity.

3.4 Correction with the I.S.

The correction with the I.S. aimed to overcome any random or systematic variability due to the analytical procedure and the probable inhomogeneity of the processed sample.

The sources of variability in the determination of pesticide residues in surface treated samples can be divided into three parts: The effect of sample processing and extraction (I); the uncertainty of filtration, evaporation and GPC procedures (II); and the determination of the analyte in the final extract (III). Part I is characterised by the CV_{14Cext}, the combined uncertainty of I, II and III can be quantified based on the LSC ($^{14}CV_{ana}$) and the GC (CV_{GCana}) determinations. We know the typical CV of the LSC counting ($CV_{LSCtyp} = 1\%$). From these data we could calculate the uncertainty of II and the gas chromatographic detection (CV_{GC}).

$$CVana = \sqrt{CV^2{}_I + CV^2{}_{II} + CV^2{}_{III}} \qquad (2),$$

Therefore the CV_{II} calculated from the LSC measurement is: $(7.9^2 - 5.5^2 - 1^2)^{0.5} = 5.58\%$. The contribution of the GC analysis (CV_{III} or CV_{GC}) to the uncertainty of the chlorpyrifos determination was calculated as:

$$CV_{GC}ana = \sqrt{CV^2{}_I + CV^2{}_{II} + CV^2{}_{III}} \qquad (3),$$

Inserting the appropriate values into the equation (3) we obtain for:
CV_{III} or $CV_{GC} = (17.16^2 - 5.58^2 - 5.5^2)^{0.5} = 15.26\%$.

The use of the ^{14}C as I.S is justified as the variability of the results obtained with the GC analysis (15.2%) is much higher then with the LSC determination (1 %). The large difference in the precision of the detection method is also reflected by the variability of results obtained for the chlorpyrifos with the GC (17.16%) and the LSC (7.9%).

The correction with the IS ^{14}C was done in two steps:
• The first step aimed to compensate for the variability of the initial residues in the analytical portions (ap).

We assumed that for an analyte A: CA_{iap}/CA_t, (where CA_{iap} is the theoretical initial concentration of analyte **A** in the ith analytical portion to be extracted and CA_t is its initial concentration in the treating solution) is equal to $^{14}Ciext/^{14}\overline{Cext}$ (i. = 1.... n),
Hence, $CA_{iap} = CA_t \times {}^{14}Ciext /^{14}\overline{Cext}$ \qquad (4)

• The second correction step took into consideration the variability due to the analytical procedure (filtration, GPC clean-up, evaporation, dilution, etc.). It consisted of multiplying the residues measured with the GC analysis R_{GC} with a factor (f) equal to the ratio of $^{14}Ciext/^{14}Ciana$ (i. = 1...n) to give an R-c (c for corrected). R-c would correspond to the recovery obtained after processing, so consequently:
$Cc = C_{GC} \times fi$ \qquad (5)

Dividing by the theoretical initial concentration C_t, we get for chlorpyrifos:

$$Rc\text{-}c = R_{GC} \times {}^{14}Ciext/{}^{14}Cana \tag{6}$$

For the rest of the analytes, we are interested in the ratio CA_c/CA_{ap}, which is actually the corrected proportion of the initial residue recovered Ra-c:

$$CA_c/CA_{ap} = CA_{GC} \times fi \, / \, CA_t \times {}^{14}Ciext \, /{}^{14}\overline{Cext} \tag{7}$$
$$= R_{GC} \times {}^{14}Ciext/{}^{14}Ciana \, / \, {}^{14}Ciext \, /{}^{14}\overline{Cext}$$

So, for the other analytes:

$$Ra\text{-}c = R_{GC} \times {}^{14}\overline{Cext}/{}^{14}Cana \tag{8}$$

This principle was very useful as indicated below, but it has its limitations.
For instance, it may not reflect the systematic errors occurring during the GPC clean-up. (The GPC calibration using a ${}^{14}C$ chlorpyrifos showed that this compound elutes between the 9-12mL fraction of the collected volume (30mL total). Therefore a correction for pesticides eluting earlier than the chlorpyrifos improves the accuracy, while the same correction for the late ones i.e. folpet, azinphos-methyl, etc. [11] might introduce a bias when the first fraction contains more then 8mL. The relatively low uncertainty ($CV_{II} = 5.58\%$) calculated for the intermediate steps of the procedure indicates that such error did not occur in our experiments.

Table 4 and 5 show the effect of the correction on the harmonisation of the results with regard to the two-way ANOVA test, the example below is taken from the storage stability study.

Without correction, the ANOVA test failed for six compounds out of nine indicating that the different storage period have significant effect on the result, while after correction the test failed for only three compounds.

Table 4 *Recoveries not corrected for p,p-DDT and the corresponding 2-way ANOVA test*

(n)	Week 1		Week 2		Week 3		Week 4	
	analyst 1	analyst 2	analyst 1	analyst 2	analyst 1	analyst 2	Analyst 1	analyst 2
1	79.7	76.8	82.3	81.8	80.4	90.1	63.7	100.9
2	80.5	79.9	70.9	80.05	73.8	96.4	58.4	67.9
3	74.8	77.2	70.1	90.4	74.9	99.7	66.5	88.9
4	83.5	84.3	74.4	78.7	70.2	100.7	60.4	81.7
5	73.8	85.0	82.1	81.2	75.7	94.9	71.7	78.4

Source of Variation	F	P-value	F critical
weeks	**3.9116**	0.019287	2.960
analysts	**3.3133**	0.007544	2.250

All calculated F values were higher than the critical values, showing clearly a significant difference between weeks and between analysts. This indicates that the storage periods have an effect on the residue stability.

Table 5 *Corrected recoveries for p,p'-DDT and the corresponding 2-way ANOVA test*

(n)	Week 1		Week 2		Week 3		Week 4	
	analyst 1	analyst 2	analyst 1	analyst 2	analyst 1	analyst 2	analyst 1	analyst 2
1	69.2	78.1	65.8	76.0	72.1	94.5	96.7	76.0
2	69.0	93.5	69.4	75.9	95.5	91.8	83.8	71.8
3	69.2	112.0	71.5	98.9	76.3	106.9	73.6	82.1
4	68.1	76.9	74.6	101.1	79.8	104.8	79.5	73.9
5	73.7	*	72.9	96.8	91.7	86.5	75.5	84.2

* outlier.

Source of Variation	F	P-value	F critical
weeks	**2.1507**	0.117126	2.960
replicates	**1.2068**	0.331162	2.250

The correction of the results allowed the p,p'-DDT to pass the 2-way ANOVA test, revealing a non-significant difference between weeks and between analysts. Therefore, this indicates no storage effect on the residue stability.

3.5 Combined uncertainty

The corrected results carry a combined uncertainty deriving firstly from the uncertainty of the chromatographic measurement, and secondly from the uncertainty of the LSC count.

Ra-c being a result of a multiplication and division (Ra-c = R_{GC} x $^{14}\overline{Cext}/^{14}Cana$ (8)), its uncertainty would be expressed as:

$$CV_{comb} = \sqrt{CV^2{}_{typ} + CV^2{}_{ext} + CV^2{}_{ana}}$$ (9)

Since the best estimate of the uncertainty of analysis is the CV_{typ} (15.9%) and the uncertainty of sample processing is incorporated in CV_{ext}, the calculation of the combined uncertainty may be performed as: CV_{comb} = $(15.9^2+5.5^2+7.9^2)^{0.5}$ = 18.6%. For the uncertainty of the uncorrected results, we omit the 7.9% and take into account only the best estimate for the uncertainty of sample processing (5.5%); this would give 16.8%. The difference between the two estimates (18.6 and 16.8 %) of uncertainty of measured residues is negligible compared to the improvement in the accuracy of the results.

3.6 Sample processing effect

The typical sample processing uncertainty expressed as $^{14}CV_{ext}$ of 5.5% was in line with previous findings [9, 14].

Table 6 shows the average analytical recoveries (taken from table 3) and the corrected (Ra-c) proportions of the initial residues measured after processing of fresh samples (day 0) and samples kept deep-frozen overnight (day 1).

Table 6 *Recoveries from the spiked analytical portions and the proportions (%) of initial residues measured and after processing surface treated samples*

Pesticide	Analytical recovery %	Day 0	CV $_{day\ 0}$	Day 1	CV $_{day\ 1}$
Azinphos-ethyl	91.37	88.99	14.54	88.6	12.60
Azinphos-methyl	106.61	98.2	11.48	99.03	14.38
Chlorothalonil	83.2	43.9	11.07	52.3	21.09
Chlorpyrifos (ECD)	84.4	101.0	10.61	104.0	15.2
Chloripyrifos (NPD)	95.2	113.9	7.26	107.5	12.27
pp-DDT	86.9	100.0	5.89	95.4	15.3
Diazinon	84.6	77.3	9.80	70.4	13.3
Dimethoate	101.4	107.9	11.07	96.2	12.39
α-Endosulfan	87.5	97.1	8.63	100.4	17.4
β-Endosulfan	87.2	93.9	8.09	98.1	14.5
Folpet	79.2	50.6	11.89	43.5	13.8
Iprodione	97.5	92.6	9.59	101.1	13.01
Lindane	85.6	97.0	10.29	102.0	8.07
Metamidophos	85.5	93.96	30.68	77.39	20.86
Parathion-methyl	88.1	98.1	9.69	99.3	15.3
Pirimicarb	94.5	102.0	8.96	105.6	12.1
Pirimiphos-methyl	89.3	94.0	7.37	84.1	9.25
Vinclozolin	95.5	102.0	9.75	108.66	8.71
Qa		**91.8**	**12.12**	**90.75**	**13.99**

The average proportion of residues measured on day 0 and 1 in surface treated samples were compared to the average analytical recoveries. Using the *t*-test, no significant difference was revealed, indicating that the concentration of the compounds was not affected by the processing except for chlorotalonil, folpet and diazinon.

Chlorothalonil and folpet degraded approximately at the same rate during processing, 37-47% and 36-45% of their initial amounts were lost, respectively. The degradation of diazinon was slower leading to a loss of 9-17%. (Table 6 and Table 7). The loss due to volatilisation can be excluded as the vapour pressure of these compounds [chlorothalonil: 0.076 mPa (25°C); diazinon 12 mPa (25°C); folpet 0.021 mPa (25°C)] is lower than or comparable to the vapour pressure of lindane [5.6 mPa (20°C)] [15].

Diazinon is hydrolysed in acidic media [DT_{50} (20°C) is 11.77h at pH 3.1], and the pH of tomato ranged between 4.2-4.9, therefore chemical hydrolysis could not be the only source of degradation. Also, the physical properties of the other compounds do not explain their degradation either.

Considering the effect of the overnight freezing, the *t*-test showed a significant difference from day 0 to day 1 for diazinon, dimethoate and folpet. Yet the proportion of the initial residues at day 1 for dimethoate was not different from its analytical recovery, indicating no significant loss.

The losses for diazinon and folpet are probably due to the thawing time observed before sample processing and not to the overnight storage at – 20°C as shown in 3.7.
The Rc decreased also for methamidophos from day 0 to day 1, but the *t*-test revealed no significant difference between the two values, indicating no degradation.

As explained in 3.5, we can calculate the combined CV for the corrected proportions of initial residues also (eq. 9) using the CV_{typ} at day 0 and day 1, this would give 15.5 and 16.9 % respectively. These CV-s are more realistic in fact, as the correction was applied on the results obtained for the surface treated samples. Despite the degradation of the folpet and chlorothalonil and the inclusion of the sample processing step, the typical recoveries Qa for day 0 and day 1 are again in line with the previous value obtained with the spiked analytical portions (90.2 %). This indicates a good reproducibility for the method.

3.7 Storage effect:

The storage stability at -20°C was studied only for a relatively short period of one to four weeks to obtain additional supportive evidence to the findings on the effect of sample processing.
The trend with time was expressed with the first order equation and as shown in Figure 1:

$$R_t = R_0 \; x \; e^{-kt} \tag{10}$$

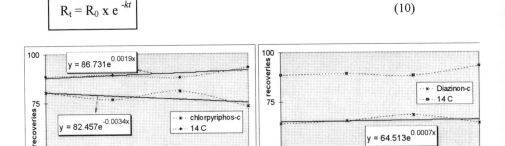

Figure 1 *Radiolabelled, cold chlorpyriphos and diazinon, showing no degradation*

The two-way ANOVA showed a significant difference between the average residues measured at different times. But these differences do not indicate degradation considering the absence of a trend.

The two-way ANOVA indicated significant difference between the average residues measured at week 1 and week 2 also for α-endosulfan, which was attributed to the high variability between analysts. The average residues were comparable during the last 3 weeks, indicating no significant degradation.

The large variability of residues may be partly explained by the fact, that the present study design is not ideal for testing storage stability, which can be more properly tested with the fortification of pre-homogenised samples [1, 10]. Fortifying individually analytical portions excludes the uncertainty of processing and a potential error introduced by the non-uniform surface treatment. However, the study of the processing effect on the residue stability requires the surface treatment of the commodity.

Table 7 *Proportion of the initial residues recovered during the whole period of the study*

day	^{14}C	lindane	Diazinon	chlorothalonil	parathion-methyl	chlorpyriphos	folpet	α-endosulfan	β-endosulfan	p,p'-DDT
0	85.4	97.0	77.3	43.9	98.1	101.6	50.6	97.1	93.9	100.3
1	89.1	102.3	70.4	52.2	99.9	103.8	43.5	99.9	97.8	95.4
7	88.6	74.5	63.9	58.4	104.3	80.4	51.1	60.5	72.1	78.9
14	89.2	75.2	65.3	35.3	89.8	76.7	40.0	72.1	71.7	80.3
21	88.1	83.5	68.1	61.0	101.	81.1	64.8	79.2	85.4	90.0
28	93.1	81.2	64.1	52.7	88.4	72.9	38.5	71.3	76.4	79.7

4 CONCLUSION

For testing the effect of processing on the stability of analytes, the surface treatment of the commodity is a practical way to fix the initial residue levels. The drawbacks are the time needed for the application, the heterogeneity in the applied amounts and the possibility of running off some portion of the fortifying solution. All these points will have consequences on the calculated correction factor (fi) of the IS and therefore the evaluation of the data.

The results obtained during this study revealed that the sample processing - characterised with a typical uncertainty of 5.5%- might be a very critical part of any residue analytical method. Around 40% of the 0.5 mg/kg initial residues of chlorothalonil and folpet degraded within less than five minutes of processing at room temperature.

The varying effect of different sample processing procedures cannot be observed in inter-laboratory comparisons or proficiency testing schemes. Loss or degradation during sample processing may be a significant source of systematic error leading to incomparable analytical results obtained in different laboratories. Therefore, it requires careful checking in each laboratory under its particular condition.

Despite the fact that longer time would be needed for the storage stability study in order to determine and quantify accurately any degradation, this study has proven that no degradation occurred for the tested pesticides during one month. The stability of pesticide residues should be tested in specifically designed studies covering the likely maximum storage period.

The quantification with ^{14}C is very quick, accurate and precise. The use of radiolabelled pesticides is a convenient tool to assess the effect of every step in the analytical method on the residues measured. It helps to identify the critical points that require special care and attention from the analyst; hence, its possible uses as an I.S. Nevertheless the correction with the ^{14}C is by no means a solution to overcome the variability deriving from the lack of precision.

The concept of correction with a I.S. does not necessarily require the use of radiolabelled compound. Analytes of proven stability may be used as a cheaper alternative. However, the precision of the determination of the I.S. is critical as it may significantly affect the uncertainty of the corrected results.

In addition to its stability during processing, the I.S-s should be chosen taking into account their elution properties on clean-up columns and their suitability as I.S. for the final chromatographic detection method.

References

1. H. Egli, Storage stability of pesticide residues. *J. Agric. Food Chem.* **1982**, *30*, 861-866.
2. FAO/WHO publications, Pesticides residues in food - Evaluations, Part I, Residues.
3. EURACHEM Guidance document No. 1, WELAC Guidance document No. 2, Accreditation for chemical laboratories.
4. Criteria for laboratories. Document number SC 00, Ed. 1996, Dutch Council for Accreditation.
5. NF V03-110, December 1998. Inter-laboratory validation procedure for an alternative method compared to a reference method.
6. A. R. C. Hill and S. L. Reynolds, Guidelines for in-house validation of analytical methods for pesticide residues in food and animal feeds. Analyst, 1999, **124**, 953-958.
7. A. Ambrus, Etelka M. Solymosné and Ibolya Korsós, Estimation of uncertainty of sample preparation for the analysis of pesticide residues. J. Environ. sci. health, B31(3), 443-450 (1996).
8. B. Maestroni, A. Ghods, M. El-Bidaoui, N. Rathor, T. Tam and A. Ambrus, Testing the efficiency and uncertainty of sample processing using ^{14}C-labelled chlorpyrifos: PART I: Description of the methodology, FAO/IAEA/IUPAC/AOAC International workshop on principles and practices of method validation, Nov. 4-6, 1999-Budapest.
9. B. Maestroni, A. Ghods, M. El-Bidaoui, N. Rathor, O. Jarju and A. Ambrus, Testing the efficiency and uncertainty of sample processing using 14C-labelled chlorpyrifos: PART II, FAO/IAEA/IUPAC/AOAC International workshop on principles and practices of method validation, Nov. 4-6, 1999- Budapest.
10. IVA guidelines, Residues studies. Part II Storage stability of residue samples - iva industrieverband Agar e. V. (German pesticides manufacturers' Association) 1992.
11. prEN 12393-2 (final draft), Non-Fatty Foodstuffs - Multiresidue methods for the gas chromatographic determination of pesticide residues, Part 2: Methods for extraction and clean-up, Method "P", CEN/TC 275 N **245**, May 1997.
12. FAO/IAEA/AOAC/IUPAC Guidelines for single-laboratory validation of analytical methods for trace-level concentrations of organic chemicals, International workshop on principles and practices of method validation, Nov. 4-6, 1999- Budapest.
13. A. Ambrus, Worked example for validation of multi-residue methods. FAO/IAEA/IUPAC/AOAC International workshop on principles and practices of method validation, Nov. 4-6, 1999- Budapest.
14. A. Ambrus, Quality of residue data, in "Pesticide chemistry and Bioscience", by Brooks, G. T. and Roberts, T. R., RSC, 1999.
15. The pesticide manual, 11th edition, C D S Tomlin ed., British crop protection council, 1997.

AOAC International Collaborative Study on the Determination of Pesticide Residues in Nonfatty Foods by Supercritical Fluid Extraction and Gas Chromatography/Mass Spectrometry

Steven J. Lehotay

US DEPARTMENT OF AGRICULTURE, AGRICULTURAL RESEARCH SERVICE, EASTERN REGIONAL RESEARCH CENTER, 600 EAST MERMAID LANE, WYNDMOOR, PENNSYLVANIA 19038, USA

1 INTRODUCTION

AOAC International is a volunteer-based scientific organization that has developed an Official Methods Program designed to provide an inter-laboratory validation process for analytical methods. The collaborative study process and acceptance criteria of Official Methods have changed over time, and the most recent guidelines for the collaborative study process have been published in detail.[1-2] For quantitative methods, a minimum of 8 laboratories reporting valid data in the analysis of at least 5 materials are needed. Statistical criteria, essentially defined by the Horwitz equation,[3] should be met for a method that has undergone a collaborative study to become an AOAC Official Method.

AOAC Int. is structured to contain a number of methods committees categorized by topics. The committees review collaborative study protocols and reports and vote on whether a method should be adopted as a first action Official Method. These committees report to the Official Methods Board for additional review of collaborative study reports before methods achieve final action Official Method status. AOAC Int. membership also votes on whether a method should become an Official Method. First action and final action Official Methods are published in *J. AOAC Int.* and *Official Methods of Analysis* by AOAC Int.

General Referees (GRs), who report to the methods committees, are assigned to particular sub-topics. GRs oversee collaborative studies that fall within the scope of the sub-topic. One of the roles of the GR is to ensure that methods have been demonstrated to attain acceptable in-house results prior to the initiation of the collaborative study.

Associate Referees (ARs) are those who plan, coordinate, and report the findings of the collaborative study. Collaborators are those who analyze the test samples using the analytical method exactly as it is written in the protocol. The AR is often the individual or group who originally developed the method and has volunteered to conduct a collaborative study. The functions and responsibilities of the AR include: 1) perform in-house method validation, 2) design the collaborative study, 3) write the method protocol, 4) recruit collaborators, 5) prepare and ship the test samples, 6) monitor progress of the collaborators, 7) compile and statistically analyze the results, and 8) write the collaborative study reports and publications. Statistics and Safety Advisors are assigned to each methods committee (and collaborative studies within them) to provide their expertise to the methods committee, GRs, and ARs.

The author of this paper served as the AR in the AOAC Int. collaborative study, "Determination of Pesticide Residues in Nonfatty Foods by Supercritical Fluid Extraction (SFE) and Gas Chromatography/Mass Spectrometry (GC/MS)." This project is a rare example of a multiclass, multiresidue collaborative study, and furthermore, it entails the use of relatively new instrumentation that does not necessarily conform to the types of conditions used in more traditional analytical methods. The object of this paper is to share observations and experiences from this unique AOAC Int. collaborative study.

1.1 SFE and GC/MS

SFE is an instrumental approach that utilizes a supercritical fluid rather than a liquid solvent for the extraction of analytes from sample matrices. The most common extraction fluid in SFE is CO_2 due to the relatively low pressure (73 atm) and temperature (31 °C) at which it becomes a supercritical fluid. Major advantages of SFE over liquid-based methods include the greater degree of selectivity offered by controlling the temperature and pressure of extraction, and the avoidance of a solvent evaporation step to concentrate the sample extract (the CO_2 reverts to the gaseous state after extraction). Other advantages include the potential cost, labor, and time savings by using an automated instrument, and the reduction of hazardous waste generation and glassware needs. Some of the disadvantages include a small sample size (2-5 g), potentially long method development times, and possible matrix effects.

SFE instruments have been commercially available since approximately 1990. In 1992, companies started offering automated SFE instruments, and at the time, there were at least 6 companies marketing analytical SFE instruments. Currently, only 2 companies continue to sell these type of instruments. There are several reasons for this decline, one of which pertains to the lack of inter-laboratory validated methods using SFE.[4-5] An inherent difficulty with the adoption of new analytical technologies is that many laboratories are not willing to invest in a new approach until the approach has undergone inter-laboratory validation, but new technologies cannot undergo this type of validation until enough laboratories have already invested in them.

Traditionally, pesticide residues are detected in GC systems using element selective detectors such as the flame photometric detector (FPD) or electron-capture detector (ECD).[6] GC/MS has been used for many years mainly as a structure elucidation and confirmation tool of unknown and/or targeted chemicals. Until the mid-1980's, however, GC/MS had the reputation as a sophisticated technique requiring expensive instrumentation for use by mass spectroscopists only. As computers improved and costs reduced, GC/MS instruments became more compact and user friendly. By the early 1990's, commercial bench-top instruments were available that achieved acceptably low detection limits and reliable operation to permit their application to multiresidue pesticide analysis by analysts in the laboratory. GC/MS has several advantages in the multiclass, multiresidue analysis of pesticides which include: 1) simultaneous quantitation and identification of analytes, 2) universal detection independent of elemental composition, 3) high degree of selectivity, and 4) reasonably low detection limits in complex matrices. Disadvantages still include the high cost and complexity of analysis, but the differences versus GC with selective detectors are often small.

In 1993, the author's laboratory obtained SFE and GC/MS instrumentation and began method development studies. The combination of SFE and GC/MS led to a more powerful means to quickly and effectively analyze a diverse range of pesticide residues in nonfatty samples. In 1995, Lehotay and Eller[7] and Lehotay, *et al.*[8] published their SFE and GC/MS approach for the multiclass, multiresidue analysis of pesticide residues in fruits and vegetables. The following years were spent optimizing and evaluating the approach and gaining a better understanding of its benefits and limitations. In-house method validation studies were conducted and an SFE pilot study was initiated in 4 states in the U.S. that had access to the instruments. A few laboratories began implementing the SFE and GC/MS method for pesticide analysis in routine monitoring, and the number of labs acquiring the instruments for a variety of applications grew.

In 1997, the author was appointed AR in the AOAC Int. collaborative study, "Determination of Pesticide Residues in Nonfatty Foods by SFE and GC/MS." By 1998, the author had identified approximately 40 laboratories that had SFE and GC/MS instruments and began to write the collaborative study protocol.

2 DISCUSSION

There are several reasons not to conduct inter-laboratory collaborative studies. First of all, they are a great deal of work, especially in the case of a multiclass, multiresidue study applicable for a variety of sample matrices. The AR must be prepared to spend months of time to perform the assigned role. Also, the AR must learn daunting statistics and follow detailed format and administrative requirements imposed by AOAC Int. The sponsoring laboratory often bears significant costs in terms of preparing and shipping samples and purchasing supplies. Because of these and other factors, frequent solicitations by AOAC Int. for ARs in collaborative studies for in-house validated methods are often unheeded. Furthermore, collaborators must also purchase supplies and set aside time, instrumentation, and other facilities to conduct the analyses. In the case of a new or unfamiliar method, the analysts must be trained using the approach, and have access to the necessary instruments in good working condition. As a result, many laboratories cannot, or choose not to participate in collaborative studies. Other reasons not to conduct collaborative studies include legitimate questions on the validity of the results. If the study fails, reasons other than the quality of the method may be to blame (*e.g.* sample inhomogeneity, poor analyst performance, analyte degradation).

In light of all these negative factors, why did the author choose to conduct an AOAC Int. collaborative study? Despite the criticisms that can be made about the validity of the results in an inter-laboratory validation, it remains the most extensive validation procedure of a method that is typically performed. Most governmental and industrial organizations recognize analytical results from AOAC Offical Methods as acceptable for their needs, and contract laboratories are often required (or prefer to) use Official Methods when available.

There were also a few professional and personal reasons that the author chose to become an AR in the study, but the main reason to conduct the SFE and GC/MS collaborative study pertained to the dissemination of the approach among laboratories that were not regularly taking advantage of these techniques. The collaborative study served as a training exercise for labs that did not have the opportunity to realize the potential benefits of the instruments.. This was the main benefit to collaborating labs interested in learning more about the approach. The risk of failure was potentially greater by including inexperienced labs, but the goal to provide training and experience was felt to be more important.

2.1 Matrix-Matched Standards

The traditional approach of using calibration standards in pure solvents has been demonstrated to give high-biased results in the GC analysis of certain organophosphorus (OP) and other pesticides. One issue that needed to be resolved in the collaborative study protocol was the use of "matrix-matched" calibration standards for analysis. Matrix matching requires blank matrix extracts to be used as the solvent for calibration standards. Thus, matrix blanks must be available and more time and work is required to conduct extra extractions for making standards. The "matrix enhancement effect" occurs due to the adsorption or degradation of certain analytes on the glass surfaces in the GC liner and column.[9] The presence of matrix components reduces the degree of analyte losses in comparison to a standard in pure solvent, thus leading to a high bias in the result for affected pesticides unless precautions are taken. The use of matrix-matched calibration standards is the most common way to overcome this effect, and in fact, European Union guidelines require matrix-matched standards in pesticide residue analysis unless the matrix enhancement effect has been demonstrated not to be a factor.[10]

However, the policies of U.S. agencies that regulate pesticides do not permit the use of matrix-matched standards. The AOAC Int. Methods Committee on Residues and Related Topics had to make a decision on whether to permit the use of matrix-matched standards in the SFE and GC/MS collaborative study. The AR presented evidence to the committee from the scientific literature demonstrating the extent of the matrix enhancement effect and asked the committee for approval to use matrix matched standards. No members or GRs voiced opposition to the concept despite regulatory agency policies in the U.S. and past objections from former members of the committee.

2.2 Choice of Pesticides, Concentrations, and Matrices

Of the more than 800 registered pesticides worldwide, approximately 350 are amenable for GC analysis.[6] The number of nonfatty food commodities in which pesticide residues may occur also number in the hundreds. The concentrations at which the residues should be detected for monitoring or regulatory enforcement purposes also vary from approximately 1 to 50,000 ng/g. The permutations of pesticide combinations, commodities, and concentrations (low, mid, high) numbers in the millions. The complete validation of any method for the multiclass, multiresidue analysis of pesticides in foods is practically impossible, and surely any method would fail to achieve acceptable results for certain permutations. Therefore, the choice of pesticides, concentrations, and matrices in the study were made judiciously to provide a diverse range of combinations to demonstrate the analytical capabilities of the method.

The validation of a method for a certain pesticide in one commodity does not mean that a different pesticide in another matrix will give acceptable results with the method, but the likelihood is greater if the pesticides and commodities are similar. This is why the choice of pesticides were made from different major classes and quite different fruits and vegetables were chosen in the collaborative study. Due to time constraints of the volunteer collaborators, the analysis of the samples was limited to 2 weeks for a single analyst. A chemist trained and experienced in SFE and GC/MS and equipped with automated instruments could analyze 12 or more samples per day, but few collaborators were expected to be able to perform at this level. The AR limited the number of total extractions and analyses to 42 samples (half of which were calibration and QC samples). This would take an inexperienced analyst with manual instruments 1 week if all went well, and 2 weeks if the analyses needed to be repeated. In this situation, 3 commodities were as many sample types as time allowed, and 5 fortified pesticides per commodity was felt to be appropriate. Any incurred pesticide residues in the chosen commodities would also be analyzed as part of the study.

The choice of commodities was made by studying the results of monitoring programs, proficiency test samples, and the precedence of previous inter-laboratory studies.[11,12] The fruit representative, apples, was chosen due to its high consumption, moderate homogeneity, relatively high number of incurred residues detected, and high sugar and water content. The green vegetable selected was string beans, and the root crop selected was carrots which was a less moist and more homogeneous matrix.

The only previous multiclass, multiresidue collaborative study had chosen the same 6 pesticides in each of 3 commodities for analysis.[12] In the SFE and GC/MS study, rather than fortify the same pesticides in each commodity, the number of pesticide analytes was increased by adding different pesticides to each commodity, except for chlorpyrifos which was spiked in all 3 commodities. A total of 13 commonly detected pesticides from the major classes in fruits and vegetables were selected and divided equally among the commodities so that a relatively wide range of volatility, polarity, elemental composition, and uses were covered. Furthermore, the AR had previous experiences with each pesticide and commodity and felt that the choices were sufficiently challenging.

In apple, the chosen fortified pesticides (and their classes) consisted of the insecticides chlorpyrifos, diazinon, (organophosphorus - OPs), carbofuran (carbamate), endosulfan sulfate (organochorine - OC), and the fungicide, vinclozolin (other). Incurred insecticides in the apple sample included carbaryl (carbamate), propargite (other), azinphos-methyl (OP), the anti-scalding agent, diphenylamine, and the fungicide, captan (carboximide) and its degradation product, tetrahydrophthalimide. In the string bean sample, fortified insecticides consisted of chlorpyrifos (OP) and p,p'-DDE (OC), the herbicides, trifluralin (nitroaniline) and dacthal (OC), and the fungicide, quintozene (OC). The OC fungicide, chlorothalonil, was the only incurred pesticide in the string beans. In the case of carrot, no incurred pesticides were found, and the fortified insecticides included bifenthrin (pyrethroid), chlorpyrifos, parathion-methyl (OPs), the fungicide, metalaxyl (other), and the herbicide, atrazine (triazine).

To set spiking levels, the AR reviewed the U.S. tolerance levels for the different pesticides in the chosen commodities (if applicable) and ensured that the concentrations were at or below these values. Then, the limits of quantitation (LOQ) were conservatively estimated for the least sensitive GC/MS instrument used in the study. The lowest spiking level was approximately twice the estimated LOQ and the highest concentration was ≈10xLOQ. The study entailed blind duplicates at 3 concentrations (low, mid, and high) for each fortified pesticide plus a blank. This required 7 test samples per commodity and the spikes and blanks were dispersed among the test samples so that the collaborators would not easily discern the format of the study. The concentration multiplication factors between low, mid, and high concentrations were also varied somewhat for this reason. The fortified pesticide concentrations extended from a low of 30 ng/g for trifluralin to as high as 1000 ng/g for endosulfan sulfate.

2.3 Quality Control

Each step in the analytical method included a QC spike to help ensure the proper functioning of the procedure and isolate the source of problems if they should occur. Prior to the comminution step of the bulk commodities by the AR, 200 ng/g piperonyl butoxide was added to the matrix. This had the dual purpose of determining the homogeneity of the sample preparation step and ability of the method to determine the analyte. This also saved the AR from having to conduct many replicate analyses to determine sample homogeneity because this would be determined by compiling the collaborator results (incurred residues also served this purpose). The next step in the process required the analyst to mix the sample with a drying agent, and the protocol called for the addition of 200 ng/g pendimethalin prior to this step. The variability of replicate analyses of subsamples within a single test sample (matrix blank used for calibration standards) and among the test samples determined if the analyst adequately performed the mixing step. Prior to SFE, 200 ng/g terbufos was added to the subsamples in the SFE vessels to ensure that extraction occurred for each sample and to isolate the variability of the extraction process from other factors. Finally, 200 ng/g of the internal standards, anthracene-d_{10} and chrysene-d_{12}, were added to all extracts to isolate the GC/MS analytical step from the others. It was possible to use one of the other QC standards as a surrogate to account for consistently biased results in the analyses, but this was not done in the final collaborative study. In each case, the QC analytes were selected because they were stable and gave only a single major ion in MS which reduced the chance that they would interfere with the analysis of the other pesticides.

Five-point calibration curves (including the blank) were used in the method covering the expected concentration range of the fortified pesticides. Five subsamples of the matrix blanks were extracted to serve as matrix-matched standards. To compensate for differences caused by the dilution of the extracts by addition of the standard solution, all extracts were brought to the same final volume. Also a reagent blank and mid-level matrix spike were required in the analytical protocol.

2.4 Collaborators and Instrumentation

Of the approximately 40 invited labs, 22 originally agreed to participate in the study. Among these labs were 6 different models of SFE instruments, 5 different GC types, and 7 MS models. The evaluation and comparison of the results from the different instruments became an additional goal of the study. One lab with unique instrumentation dropped out due to the lack of sensitivity of their 1980's generation GC/MS instrument, and 3 other labs dropped out due to a lack of time. One other lab participated in the familiarization round, but had to withdraw from the collaborative study due to a lab relocation. Ultimately, 17 collaborators from 7 countries [USA (10 labs), Germany (2 labs), Spain, Italy, Sweden, Australia, and Japan] returned final results, but one lab was voided by not adequately following the method protocol. A few data sets with poor QC results (from instrument or analyst problems) were also voided. Among the collaborators, 10 were affiliated with federal government labs, 4 were from state government labs, 2 were from industry, and 1 was from academia.

The AR had access to 2 types of SFE and GC/MS instruments each, but the applicability of the method and operation of other types of instruments were not known precisely. The original method development studies tailored the conditions to the SFE and GC/MS instruments in the AR's lab. The AR had to write the protocol to accommodate each of these instruments and any other SFE and GC/MS instruments that may be used with the method. With traditional methods, this is generally not difficult, but in this instrumental-based approach, the task was more complicated due to the many differences among instruments, especially in SFE. Some SFE instruments were automated and some were manually operated; some used solid-phase trapping, others used liquid-based trapping; some had fixed flow restrictors, others had variable flow restrictors; some were capable of higher pressure ratings than others; some required CO_2 cylinders pressurized by He headspace, others cooled the CO_2; some controlled pressure with the restrictor and flow with the pump, while others did the exact opposite; some measured the flow of pressurized fluid in the system, while others measured flow of the depressurized gas. The most complicating feature in the SFE method was that the vessel sizes were different among the instruments, thus subsample sizes were different.

In the case of GC/MS, some labs used quadrupole mass selective detectors (MSD) while others used ion trap instruments. The MSD instruments were operated in selected ion monitoring (SIM) mode to achieve the required sensitivity, particularly for the older models, while the ion traps used full scan operation. A previous study had demonstrated that different GC conditions and detectors had not led to significant differences in results in pesticide analysis among different laboratories.[11] In this study, all collaborators were required to use the same type of GC column, He carrier gas, 1 µL injection, and electron impact ionization. Otherwise, only suggested GC conditions and choice of MS quantitation ions were provided, and the collaborators were free to change the conditions to improve separations and/or increase signal/noise ratios, especially if an interferant appeared. The injection is a critical aspect in GC analysis, but there was no way to unify the method of injection due to the different instrument configurations. The protocol relied upon the skill and experience of the analyst to perform proper maintenance and MS tuning according to manufacturer specifications. As an additional QC measure, the analysts produced a chromatogram and MS spectra of a low level pesticide mixture and blank to help verify GC/MS cleanliness and proper maintenance.

The AOAC Int. collaborative study guidebook[1] did not include instructions for such diverse instrumentation, and the AR resorted to adding a new section to the protocol, "Notes to Collaborators," which gave step-by-step details of the method required for each instrument. The GR and Committee approved the protocol after some revisions and the study proceeded to the familiarization round.

2.5 Familiarization Round

The familiarization round of the study involved triplicate analysis of 13 pesticides fortified in apple at concentrations known to the analysts. Pesticides (and their concentrations in ng/g) included: atrazine (75), bifenthrin (90), carbofuran (75), chlorpyrifos (50), dacthal (60), p,p'-DDE (45), diazinon (60), endosulfan sulfate (150), metalaxyl (75), parathion-methyl (75), quintozene (60), trifluralin (30), and vinclozolin (75). Unbeknownst to the collaborators, these concentrations were the lowest that were to be used in the study. Six of the pesticides were spiked and homogenized in the sample prior to shipment, and the others were spiked by the collaborators. Without the use of a surrogate standard, recoveries were 73-91% with relative standard deviations (RSD) of 16-26% (n = 16-17). With the use of surrogate standards to correct for systematic sources of error, recoveries were 88-104% with RSD of 7-14% (n = 15-17). Two pesticides, diazinon and bifenthrin, gave ≈60% recoveries and ≈25% RSD due to apparent degradation over the course of time prior to their extraction.

The familiarization round served many purposes to prepare for the actual collaborative study. Foremost, it served to provide the chance for the collaborators to practice using the method. Although detailed written descriptions were provided in the protocol, the AR made a training video and sent it to collaborators to help demonstrate the mixing of the samples and packing of the vessels. Even in the case of manual instruments, these were the only steps in the method that depended on the manual skill of the analyst, and practice was essential.

The familiarization round was also designed to answer questions related to the sample comminution procedure used by the AR and mixing techniques of the collaborators. The aged spikes were added by the AR prior to the comminution step and the fresh spikes were added by the collaborator prior to the mixing step with drying agent. The significant improvement in the results for aged and fresh spikes by the use of a surrogate standard showed that systematic sources of error (possibly homogeneity) was a problem. However, the typical repeatability in the analysis of triplicate subsamples was <15% RSD which indicated that the collaborators mixing techniques were not the problem. However, all labs showed a systematic bias in the results for the aged spikes which indicated a problem with the comminution step. The AR used a chopper designed for smaller bulk commodities than the collaborative study entailed, and homogenization was not satisfactory. The protocol was altered with approval of the GR and Methods Committee to permit the direct fortification of the samples by the AR prior to shipment.

Another aspect of the familiarization round was to determine the condition of the samples after shipping and stability of the pesticides. No dry ice was used in the shipments in order to avoid additional shipping charges. Nearly all samples arrived within 1-3 days within the U.S. and the use of cold packs kept most samples frozen. However, no international shipment arrived within 3 days despite the use of "priority" service and high shipping costs. All of these samples arrived thawed, and in at least one case, the samples were turned upside down in the cooler by customs. Despite these problems, the results from the international collaborators were not significantly different from the results of the domestic collaborators, and thawing of the samples did not seem to make a difference. Two of the aged spike pesticides, diazinon and bifenthrin, gave lower recoveries than the others. The AR believed that the losses of diazinon occurred during the comminution step, but the source of the problem with bifenthrin was unknown. The losses of diazinon were not expected to occur in the revised spiking method, and the AR decided not to change analytes and risked unacceptable results in the case of bifenthrin.

Other goals for the familiarization round were to ensure acceptable LOQ by all collaborators, check optimal SFE and GC/MS parameters, verify quality control measures and capabilities of the collaborators, and to provide a preview of the results. If poor results were obtained in the familiarization round, the collaborative study would have been canceled. Fortunately, this was not the case, and only one lab had to drop out due to inadequate LOQ. A few of the labs using the most recently introduced GC/MS instruments could have analyzed much lower concentrations, but the older instruments were not so sensitive. A few labs also obtained lower recoveries of the most nonpolar pesticide, p,p'-DDE. This indicated too much moisture, and the amount of drying agent was increased from a 1:1 sample:drying agent ratio to a 1:1.1 ratio. The QC procedures were deemed satisfactory, and no collaborators were excluded from the study after the familiarization round (the lab that did not follow procedures correctly in the collaborative study had a change in personnel after the familiarization round).

2.6 Results for Fortified Pesticides

At the time of this writing, the final collaborative study report has not been completed, but the AR has evaluated the results. Nearly all results for the fortified pesticides were acceptable within Horwitz equation parameters. At this time, surrogate standards were not used to correct for recoveries, and no outliers were excluded from the results except for QC reasons. Interestingly, no notable losses of diazinon were observed in the apple test samples. However, only slightly better results for bifenthrin were obtained in the carrot test samples than in the familiarization round for apple. This indicated that the method did not work as well for bifenthrin as it did for the other pesticides tested.

The QC approach used in the study was a valuable way to track the performance of each step in the method within and among the labs. For the 16 labs and 3 commodities (48 sets of analyses), the GC/MS QC standard (anthracene-d_{10}) gave 7% RSD on average. In all, 4 sets were dropped with RSD >15% because the GC/MS instruments on those days were not providing acceptable precision. After taking the internal standard into account, the SFE QC standard (terbufos) gave 11% RSD overall. In this case, 2 other sets were dropped with RSD >25% due to unacceptable SFE instrument performance on those days. The mixing step QC standard (pendimethalin) yielded 15% RSD, and no sets of analyses were dropped due to concerns with this step by the collaborators. In essence, the variability grew from 7% to 11% to 15% RSD by working backwards through the method. The limiting source of error in the method appeared to be the analyst mixing and subsampling step. Ideally, all of the collaborators would have used a chopper that the AR's lab had acquired for this mixing step, but the collaborators did not have access to this equipment and a mortar & pestle procedure was used instead.

In the 45 data sets for the fortified pesticides (15 pesticide/commodity pairs at 3 levels), average recoveries among the labs were 73-107% (excluding the most nonpolar pesticides, trifluralin, bifenthrin, and p,p'DDE, which gave 57-68% recoveries). Within laboratory repeatability was 5-18% RSD (except for the low level endosulfan sulfate spike), and among laboratory reproducibility was 11-25% RSD (except for bifenthrin and the low level trifluralin and endosulfan sulfate spikes). The typical reproducibility of results for the fortified pesticides was ≈15% RSD (apparently limited by the mixing step).

2.7 Results for Incurred Pesticides

Incurred pesticides found in the apple test samples included carbaryl, azinphos-methyl, propargite, diphenylamine, and captan and its degradation product, tetrahydrophthalimide. Furthermore, chlorothalonil was found in the green bean test samples. Most of these pesticides have been known to be problematic by any multiclass, multiresidue method.[11] For a variety of reasons, the results for the incurred pesticides in the collaborative study were more difficult to evaluate than the results for the fortified pesticides. First of all, the true concentrations of the incurred pesticides were unknown. Ideally, standard reference materials that contain incurred analytes at known concentrations would be used in collaborative studies, but at this time, no standard reference materials exist for pesticide residues in nonfatty foods. Secondly, the variability of the results was limited by the comminution procedure used by the AR. As previously mentioned, the AR did not have access to a large-volume cryogenic chopper, and the collaborative study results show the inadequacy of the comminution step. Of the 48 sets of analyses, the comminution QC standard (piperonyl butoxide) gave >30% RSD in 12 sets. The remaining 36 sets gave 11 %RSD on average which were used in the determination of the incurred pesticides (if other QC measures were acceptable).

Another problem that developed in the analysis of the incurred pesticides was the use of a method of standard additions rather than the type of calibration curve used for the fortified study. The policies of U.S. regulatory agencies permit the use of standard additions for the analysis of pesticide residues for enforcement actions. The AR had previously used this type of quantitation approach successfully and decided to use this approach for incurred residues in the collaborative study. Unfortunately, a test of this approach was not included in the familiarization round. Whereas the linear correlation coefficients (R^2 values) of the matrix-matched calibration standard plots were nearly always >0.995, no R^2 values were >0.991 in the case of the standard addition calibration plots for the incurred pesticides. The main reason for this was the inadequate comminution step, but also the lack of a true matrix blank was a drawback. There was no sure way of knowing if the calibration plot contained a systematic bias without a matrix blank to check for interferences.

Of the 7 incurred pesticides, only carbaryl, and captan/tetrahydrophthalimide gave reproducibilities <25% RSD. Carbaryl was determined to be 430 ng/g in the apple with 18% RSD, and captan (corrected for degradation) was 298 ng/g with 10% RSD (n = 8 in each case). Diphenylamine was carried over in the SFE method, and GC/MS sensitivity was a problem with azinphos-methyl. Interestingly, despite the poor precision of the incurred pesticide results among the labs, average results for the incurred pesticides compared reasonably well with the results determined using a liquid-based extraction method conducted by the AR.[13]

3 CONCLUSIONS

The practical aspects of carrying out the multiclass, multiresidue SFE and GC/MS collaborative study were formidable, and a significant amount of time, effort, and expense by the AR and collaborators were spent planning and conducting the study. Despite the many complications encountered and risks of failure taken, the results for nearly all of the fortified pesticides were acceptable. Furthermore, the collaborative study was an excellent means to disseminate the SFE and GC/MS approach due to the participation of the collaborators and widespread recognition and publication of AOAC Official Methods. Also, the collaborative study provided experience and satisfaction among the AR and participants. In these respects, the study was successful.

As with any method, analyst skill was just as important in achieving high quality results as instrument performance in SFE and GC/MS. Both well-trained analysts and well-maintained instruments were critical to acceptable performance of the method. Ultimately, sound quality assurance and QC procedures were very helpful to ensure that each step in the method was working properly. The use of multiple QC standards was a convenient and effective way to isolate each step and pinpoint where to focus corrective actions if necessary. The study also demonstrated the importance of proper comminution procedures and that the method of standard additions used was not as precise as expected. The author joins several other scientists in recommending the use of matrix-matched calibration standards to quantify pesticide residue analytes.[10]

In truth, however, the collaborative study was not altogether necessary to demonstrate the performance of the method. The inter-laboratory results overall were not much different from the results obtained during method development and evaluation. Alternative validation procedures involving more rigorous in-house validation or fewer outside labs would have yielded perhaps even better results by including only the most experienced analysts. Furthermore, laboratories wishing to use the inter-laboratory validated method should train the analysts and conduct in-house validation anyway.

Meanwhile, the pace of the technological advancements continues to increase. Even before the completion of the collaborative study, the author was investigating a newer and more advantageous approach to pesticide residue analysis known as direct sample injection (DSI)/GC/MS-MS.[13] The cycle of method development, evaluation, validation, and implementation continues, and the experiences gained from this collaborative study provide a helpful lesson for future validation studies.

References

1. 'Quick & Easy! AOAC Official Methods Program Associate Referee's Manual on Development, Study, Review, and Approval Process', AOAC International, Gaithersburg, MD, USA, 1997.
2. 'Guidelines for Collaborative Study Procedure to Validate Characteristics of a Method of Analysis, Fourth (Final) Draft', *J. Assoc. Off. Anal. Chem.* 1989, **72**, 694.
3. W. Horwitz and R. Albert, *Spec. Publ. - R. Soc. Chem.*, 1984, **49**, 1.
4. S.J. Lehotay, *Inside Lab. Management*, 1997, **1**(3), 22.
5. S.J. Lehotay, *Inside Lab. Management*, 1998, **2**(6), 35.
6. 'Pesticide Analytical Manual, Vol. 1, 3rd Edition', U.S. Dept. of Health and Human Services, Food and Drug Administration, Washington, DC, 1994.
7. S.J. Lehotay and K.I. Eller, *J. AOAC Int.*, 1995, **78**, 821.
8. S.J. Lehotay, N. Aharonson, E. Pfeil and M.A. Ibrahim, *J. AOAC Int.*, 1995, **78**, 831.
9. D.R.Erney, A.M. Gillespie, D.M. Gilvydis and C.F. Poole, *J. Chromatogr.*, 1993, **638**, 57.
10. A. Hill, 'Quality control procedures for pesticide residues analysis - guidelines for residues monitoring in the European Union', Document 7826/VI/97, European Commission, Brussels, 1997.
11. S.L. Reynolds, R. Fussel, M. Caldow, R. James, S. Nawaz, C. Ebden, D. Pendlington, T. Stijve and H. Diserens, 'Intercomparison Study of Two Multi-residue Methods for the Enforcement of EU MRLs for Pesticides in Fruits, Vegetables and Grain' Phase I - Report EUR 17870 EN, 1997, and Phase II - Report EUR 18639 EN, 1998.
12. L.D. Sawyer, *J. Off. Anal. Chem.*, 1985, **68**, 64.
13. S.J. Lehotay, *J AOAC Int.*, in press.

Validation of Analytical Methods – Proving Your Method Is 'Fit for Purpose'

James D. MacNeil, John Patterson and Valerie Martz

CENTRE FOR VETERINARY DRUG RESIDUES, HEALTH OF ANIMALS LABORATORY, CANADIAN FOOD INSPECTION AGENCY, 116 VETERINARY ROAD, SASKATOON, SASKATCHEWAN, CANADA S7N 2R3

ABSTRACT

As part of a national food inspection system and as a laboratory which seeks accreditation of test capabilities under ISO/IEC Guide 25, the validation of analytical methods is a critical issue for us. While in the 1980's, the majority of our analytical methods, with the exception of pesticides, were based on the determination of single analytes, the trend in the 1990's has been to develop multi-residue methods which may be designed to detect, quantitatively or qualitatively, from three to 30 or more compounds. While we have Standard Operating Procedures for the validation of quantitative analytical methods and also for screening tests (test kits), these protocols were designed with single analyte methods in mind. Rigourous application of these protocols to a multi-residue method could generate data requirements which would exceed the available resources, or at least require commitment of staff and resources which would seriously compromise our ability to provide other services. We presently are validating new multi-residue methods for both pesticides and hormones, both involving mass spectrometry. Some of the problems of the application of our existing SOP for such methods and some of the possible solutions, which might be included in a SOP for validation of multi-residue methods, will be discussed. Issues which require consideration include the intended use of the method, the number of matrices to which the method will be applied and the potential that residues of a particular compound will actually be found in samples to be tested.

INTRODUCTION

It might be said that validation, like beauty, is in the eye of the beholder. In other words, method validation may mean different things to different people, depending on the context and the application of analytical science. For those of us who work in food control laboratories, however, there are some relatively clear expectations. The International Organization for Standardization (ISO) defines validation as "confirmation by examination and provision of objective evidence that the particular requirements for a specified intended use are fulfilled".[1] For analytical methods, this includes the establishment of performance characteristics, determining what influences may cause them to change and demonstrating that the method is "fit for purpose".[2]

2 METHOD VALIDATION REQUIREMENTS WITHIN CODEX ALIMENTARIUS

2.1 General Codex Alimentarius Requirements

The Codex Alimentarius Commission has adopted guidelines for testing laboratories involved in the Import and Export Control of Foods which require that such laboratories:[3,4]

a) use internal quality control procedures which comply with the "Harmonized Guidelines for Internal Quality Control in Analytical Chemistry"; [5]
b) participate in proficiency testing schemes designed and conducted in accordance with the "International Harmonized Protocol for Proficiency Testing of (Chemical) Analytical Laboratories";[6]
c) become accredited according to ISO/IEC Guide 25 "General requirements for the competence of calibration and testing laboratories";[7] and
d) whenever available, use methods which have been validated according to the principles laid down by the Codex Alimentarius Commission.

While it is method validation which is the topic of this Workshop, the Codex guidelines, taken together, comprise a framework for a quality system within which a testing laboratory should operate. It may therefore be argued that method validation requirements should be considered within this framework. Therefore, while there should be a minimum uniform set of performance criteria established for purposes of method validation, the requirements for data points (number of concentrations, matrices, analysts and laboratories) may vary according to the laboratory environment within which the method is in use. Put another way, the greater the assurance of the quality practices of the laboratory and the more demonstrable the laboratory performance, based on results in proficiency testing programs, the less onerous may be the requirements for multi-laboratory validation of the method used in that laboratory. Ideally, all methods in use in residue laboratories would be validated via a full collaborative study,[8,9] but the reality is that there simply are not enough laboratories which can devote the necessary resources to achieve such a goal, nor enough scientists willing to undertake the organization of the very large number of method trials which would be required. Other barriers include shipping regulations, import restrictions and analyte stability The objective is therefore to define the minimum acceptable requirements for within-laboratory method validation, assuming the laboratory operates in an environment of test accreditation, internal quality assurance and participation in proficiency testing schemes.

2.2 The Codex Committee on Methods of Analysis and Sampling (CCMAS)

CCMAS have previously recommended the full collaborative study as the desired level of validation for a Codex reference method,[10] which is also reflected in the "Codex Alimentarius Manual of Procedures".[11] The participation requirements for a collaborative study are as follows:[8,9]

♦   a study design involving a minimum of 5 materials (may be reduced to 3 for a single level specification for a single matrix), the participation of 8 laboratories reporting valid data, and should usually include blind replicates or split levels to assess within-laboratory repeatability parameters;
♦   the above requirement may be reduced to 5 laboratories from 8 when very expensive

equipment or specialized laboratories are required;
♦   for qualitative methods, requirements are for 15 laboratories reporting on 2 analyte levels per matrix, 5 samples per level, with 5 negative controls per matrix.

2.3   Codex Committees on Pesticide and Veterinary Drug Residues

The guidelines recommended by CCMAS are more stringent than those which have been applied by the Codex Committee on Residues of Veterinary Drugs in Foods (CCRVDF), which has required testing of the method by a minimum of three analysts, preferably in three different laboratories, in recommending methods as acceptable for Codex purposes.[12] The Codex Committee on Pesticide Residues (CCPR) has "applied the following criteria when selecting methods for inclusion in the List:[13]

a.   published in books, manuals or open literature;
b.   collaboratively studied or known to have been validated in a large number of laboratories;
c.   capable of determining more than one residue, i.e. multi-residue methods;
d.   suitable for as many commodities as possible at or below the specified MRLs;
e.   applicable in a regulatory laboratory equipped with routine analytical instruments."

3   METHOD VALIDATION AND ACCREDITATION

In consideration of the expectation that laboratories operating to the Codex guidelines should meet the ISO/IEC Guide 25 guidelines and the method validation requirements this entails, a laboratory may be recognized as having a suitably validated method for a test included within the scope of the accreditation if that method is considered "appropriate" in that it is suitable for the purpose in terms of accuracy or other required specifications. The Guide states on the issue of method selection that the "laboratory shall use appropriate methods and procedures for all calibrations and tests and related activities", and that "where methods are not specified", it "shall, whenever possible, select methods that have been published in international or national standards, those published by reputable technical organizations or in relevant scientific texts or journals." It further requires that in situations where methods are to be used which have not been established as standard, "these shall be subject to agreement with the client, be fully documented and validated".

4 SOME POSSIBLE OPTIONS FOR DEMONSTRATION OF METHOD VALIDATION

From the above considerations, it may be suggested that a suitably validated method should meet one of the following criteria:
-the method has been published as a standard by a national or international organization, which would usually mean that the method has been collaboratively studied for the application; or
-the method performance has been compared with a recognized reference method and suitably documented; or
-the method has been validated using certified reference materials; or
-the method performance has been suitably documented through participation in interlaboratory proficiency testing; or
-in the absence of the above, the laboratory has performed what is judged at the time by the

auditors to be appropriate validation with supporting documentation.

The above would appear to cover the various criteria which have been referenced respectively by CCMAS, CCRVDF and CCPR in recognizing methods in the past, or recommended by a recent FAO/IAEA Consultation undertaken for the Codex Alimentarius Coimmission.. For Codex purposes, when there is a method that is listed as a "standard" or which has been collaboratively studied, with the results published and available for expert review, there should be no problem in method acceptance. In recognition of the difficulty of conducting full collaborative studies for all required methods, however, committees such as the CCRVDF and CCPR have established the other requirements previously outlined to demonstrate method reliability. These criteria are intended to provide analytical experts with reasonable confidence that a method has been sufficiently validated so that it should perform within the required parameters when implemented in the developing laboratory and that it should be transportable to another suitably equipped and competent laboratory. Any laboratory adopting such a method should, of course, conduct appropriate validation tests with the method to ensure that, in their hands, it is meeting the expected performance standards.

An argument could therefore be made that there are two separate, but highly related, issues to be considered. The first issue is precisely which performance factors should be addressed in method validation to demonstrate expected within-laboratory performance, and what additional measures are required to provide an expectation of method transportability and comparability of results. There are several excellent examples which detail what most analytical chemists would accept constitutes "method validation". These identify performance criteria which must be experimentally demonstrated for a method before it is considered suitable for use. In the case of veterinary drug residues, performance requirements for accuracy, precision and recovery are listed by Codex.[11] More detailed validation requirements have been identified by the European Union,[14] Eurachem[2] and by AOAC International.[9,15] These documents identify the performance requirements which should be demonstrated as being met by the sponsor laboratory (within laboratory validation) and also criteria for multi-laboratory studies (between laboratory variability assessment, or transportability).

The AOAC Method Validation Programs, which include Official Methods, Peer-Verified Methods and Performance Tested Methods, have identified core data requirements which may be considered as integral components of method validation. These data requirements include specificity, sensitivity, false positive and false negative rates, limit of detection, limit of quantitation, precision, linearity, applicable matrix or matrices, ruggedness, recovery, within laboratory repeatability, plus interlaboratory comparison. There are additional requirements with respect to test kits which include review of package inserts and certification of the quality policy of the producer of the kit. Similar requirements are found in other sources, such as Eurachem's "The Fitness for Purpose of Analytical Methods. A Laboratory Guide to Method Validation and Related Topics".[2] As a minimum for within-laboratory method validation for an analyte in a matrix, we would therefore normally expect that the issues identified in these sources would be addressed. For analyses directed at monitoring for compliance with a maximum residue limit (MRL), emphasis is usually placed on demonstrating the method performance at the MRL and bracketing concentrations. For methods used in other applications, such as monitoring dietary exposure, emphasis may instead be on validation at the LOD and LOQ. Such final validation experiments, once parameters such as LOD, LOQ and freedom from interferences have been established during method development and testing, usually require completion of a series of analytical runs involving analytical standards, analyst-spiked and "blind" fortified samples plus analysis of "blind" incurred materials. Completion of such a series of experiments will typically require about 1 month of analyst time to complete validation for 1 analyte in 1 matrix. A similiar validation might also be completed for up to 3 analytes in a simple group-specific method within a similar time.

5   METHOD VALIDATION IN A REGULATORY LABORATORY – A CURRENT APPROACH

Our laboratory has a Standard Operating Procedure for within-laboratory method validation which is designed to determine specificity, freedom from interferences, analytical range, LOD, LOQ, recovery and such other requirements. This final exercise is to show, through a series of experiments, that the method can perform on samples which are received as "unknowns" by the analyst, in addition to "knowns" prepared by the analyst, and to confirm operational standards for the method in routine use. The validation steps include:

- Phase I: The analyst uses chemical standards to prepare standard curves for the analyte using a minimum of 5 concentrations, plus a blank. The experiment is repeated on a separate day and the results must be comparable and within acceptance criteria (suitable analytical range, linearity, etc., for intended use). If the analyst doing the validation is not the developer of the method, then a minimum of 2 practice runs, each with a minimum of 2 matrix blanks and 2 replicates containing the analyte spiked in matrix at the MRL, will also be conducted. Again, these results must be acceptable (freedom from interferences, trueness, repeatability, etc.).
- Phase II: The analyst does 4 complete analytical runs, on different days, each run consisting of 6 blank samples (from 6 different pools, if available), plus 6 replicate samples at each of 0.5x, 1.0x and 2.0x MRL spiked by the analyst into matrix. Again, the results must meet acceptability criteria before the analyst proceeds to the next phase. It should be noted here that obtaining a suitable variety of materials from different sources to prepare 6 different pools can be a problem, but our experience has been that extracts obtained from tissues of animals which are, for example, on different feeding regimens, may differ considerably and that analytical performance can be affected.
- Phase III: The analyst must complete at least one analytical run containing blind replicates at 3 levels, between 0.5x and 2.0x MRL , plus a minimum of 1 run containing 3 replicates of incurred or fortified pool samples within the range of interest. The results must meet acceptabilty criteria.
- Phase IV: A sample storage study must be completed using a minimum of 3 pools of incurred or fortified samples within the analytical range of interest. Twelve sub-samples are prepared and stored at -20^0C, with analysis at $t = 0, 2, 4$ and 6 weeks, reflecting the typical range of storage times for samples awaiting analysis, and also at extended time periods, if this is judged necessary. If the analyte is found to be unstable in the matrix, the study is repeated at -70^0C. The study design includes analysis of samples which have been subjected to two separate freeze-thaw cycles. Other aspects of Phase IV include demonstration of mass spectral confirmatory capability and ruggedness testing of the method, if these have not been done during the initial methods development.
- Once the validation work has been reviewed and approved by the Quality Manager, the method protocol is signed by the developer, approved by the manager of the

Centre and the method is ready for routine implementation. If the method is to be used by another staff member, they must first complete a method familiarization which mirrors Phases I – III of validation before they are approved by the Quality Manager to conduct routine analyses with the method.

6 VALIDATION OF MULTIRESIDUE METHODS – PROBLEMS AND POSSIBLE SOLUTIONS

6.1   A Typical Problem In Method Validation

Increasingly, laboratories are faced with the application of multi-residue methods to multiple matrices. For a veterinary residue control laboratory, for example, a method typically will be validated for at least two tissue types, such as muscle and kidney, from 3-5 species (eg., cattle, swine, sheep, chicken, turkey), for methods which include from 3 to 30 or more analytes. Clearly, it is not feasible for most laboratories to devote approximately 1 month to validate a method for each analyte-matrix combination to which it may be applied, following a validation procedure such as that outlined above. If we take as an example a method which is used to monitor for 20 analytes in 2 tissues from 5 species, validation to the specifications for a single analyte-matrix combination would require approximately 200 person-months (1x20x2x5), with the result that the method would be technologically obsolete before validation has been completed. Our laboratory is currently faced with several such situations, one involving validation of a method for approximately 30 pesticides and metabolites in animal fat from at least 6 different species, while another method is intended to detect a similar number of compounds with growth promoting properties (estrogenics, anabolics, stilbenes, thyreostatics, ß-agonists) in urine, muscle, fat, liver and kidney, for at least 5 species. We must be able to demonstrate suitable validation to support the claimed scope and applicability of such methods, but we also have to be able to generate the data within a reasonable timeframe that enables us to apply the methods to meet the needs of our Agency. In addition, since both methods are based on mass spectral detection, we must deal with the issue of validation requirements for mass spectral methods, a topic to which a separate workshop could be devoted.

6.2   A Practical Approach To Validation Of Multiresidue Methods

6.2.1 A "risk-based approach" to validation. As a practical approach, once method validation has been demonstrated for the target matrices for one species, a simplified approach may be considered to extend the use to other similar matrices, such as tissues from other species, where only the "blind" testing of fortified and/or incurred samples by the analyst may be applied, provided that acceptable results are obtained. Similarly, 3-5 analytes should be included in each set of validation experiments to reduce the number of experiments that are conducted. This still may leave the laboratory and analyst with a significant time investment prior to method implementation.

In our current environment, where analytical methods are often required urgently to implement a new testing program, time is often a very precious commodity. Laboratory managers are therefore faced with the dual problem of implementing a method in a timely fashion, while also ensuring that adequate validation has been carried out. There are many circumstances where both demands cannot be met if a compete validation is to be conducted prior to implementation. However, there can be a reasonable solution to this dilemma, if we borrow some ideas from our colleagues in risk management. In most cases, residue monitoring programs are directed at classes of compounds which are considered likely to produce residues in normal use and at the commodities in which such residues are most likely to appear. When we use a multi-residue method, it is usually the case that all compounds in the class monitored with the method are not all equally persistent nor equally commonly used. In addition, multi-residue methods are also usually applied to serve first as a screening method which detects all target analytes at or above a level of interest. When analytes are detected, they may be also analyzed quantitatively with that method or with an appropriate single residue method. Given such a situation, it may be reasonable to first sufficiently demonstrate the applicability of the method for all the target compounds to which it can be applied qualitatively (specificity, interferences, limit of detection) to detect residues at established levels, working in groups of 3-5 analytes as previously suggested. For those compounds to which it is considered probable that the method may be applied quantitatively because, based on known pattern of use and residue profile, they are judged most likely to be found, a package of validation data to support such use of the method would also be required, again working in groups of 3-5 analytes at a time. Validation for the additional compounds which are considered to be "lower risk" can then be conducted over time, after method implementation, until full validation is complete. In this case, "full validation" may not involve a complete quantitative validation for all compounds, if the method is being used primarily as a screening test at a target concentration for each analyte. Quantitative validation would be required only for those compounds which were found at or above the target level, for which a quantitative result would also be reported with the method.

6.2.2 Constraints which must be observed. If this path of validation in stages is taken, however, there are some clear responsibilities which fall on the laboratory. First, it should be clearly stated in the scope of the method to which compounds the method may be applied only qualitatively and to which it may be applied both quantitatively and qualitatively. Secondly, there should be a schedule and a commitment to complete the full validation for quantitative application to all target compounds, with regular updates to the method scope as additional validation work is completed. Finally, no quantitative results should be reported until full validation has been completed for any analyte on which such information will be reported. In addition, this approach should be fully supported by an internal quality assurance system and participation in appropriate proficiency schemes which may be available. The laboratory must have a very clear record of the validation status of each method that is routinely applied.

CONCLUSIONS

The objective of these proposals is not to suggest that collaborative studies should be considered obsolete or unnecessary, but rather to suggest some alternatives which laboratories may consider when a method which has been collaboratively studied is not available. This frequently is the situation when dealing with residues such as pesticides or veterinary drugs. It is not a viable alternative to suggest that either such products should not be permitted in use, or at least that no testing programs should be instituted until a collaboratively studied method is available. Data packages are developed for regulatory submissions under GLP regulations without the use of collaboratively studied methods and laboratories may be accredited under

ISO/IEC Guide 25 to conduct tests using methods which have not been collaboratively studied. In each of these situations, the test laboratory must demonstrate to auditors that appropriate method validation has been conducted so that there is some ability to ensure the quality of the analytical results. Our challenge is to agree on what, in such cases, constitutes a minimum acceptable set of experiments and data to demonstrate that our within-laboratory validation of a multi-residue method demonstrates that the method, when applied, is indeed fit for purpose. Having done that, we must not forget that the result of the analysis is also dependent on other factors which include the representativeness of the sample, analyst training and quality control measures, such as instrument calibration, in the laboratory. Method validation, of itself, is not a guarantee of a reliable analytical result.

References

1. ISO 8402, "Quality – Vocabulary" International Organization for Standardization, Geneva, 1994.
2. "The Fitness for Purpose of Analytical Methods. A Laboratory Guide to Method Validation and Related Topics", 1999. Access via Eurachem website: http://www2.vtt.fi:82/ket/eurachem.html
3. "Proposed Draft Guidelines for the Assessment of the Competence of Testing Laboratories involved in the Import and Export Control of Foods ALINORM 97/23A", Report of the 21st Session of the Codex Committee on Methods of Analysis and Sampling, Budapest, Hungary, March 10-14, 1997. Food and Agriculture Organization of the United Nations, Rome, Italy, 1997.
4. Report of the 22nd Session of the Codex Alimentarius Commission, Geneva, June 23-28, 1997. Food and Agriculture Organization of the United Nations, Rome, Italy, 1997.
5. "ISO/IEC Guide 25: General requirements for the competence of calibration and testing laboratories", International Organization for Standardization, Geneva, 1990.
6. M. Thompson and R. Wood , *Pure & Appl. Chem.*, 1993, **65,** 2132.
7. M. Thompson and R. Wood, *Pure & Appl. Chem.*, 1995, **67,** 649.
8. W. Horwitz, *Pure & Appl. Chem.*, 1993, **67,** 331.
9. "AOAC International Guidelines for Collaborative Study Procedures to Validate Characteristics of a Method of Analysis", *J. AOAC Int.*, 1995, **78,** 143A.
10. "Codex Alimentarius, Volume 13; Methods of Analysis and Sampling, 2nd ed.", Food and Agriculture Organization of the United Nations, Rome, Italy, 1994.
11. "Codex Alimentarius Commission Procedural Manual", Joint FAO/WHO Food Standards Program, Food and Agriculture Organization of the United Nations, Rome, 1997.
12. "Codex Alimentarius, Volume 3; Residues of Veterinary Drugs in Foods, 2nd ed.", Food and Agriculture Organization of the United Nations, Rome, Italy, 1993.
13. "Codex Alimentarius, Volume 2; Pesticide Residues in Food, 2nd. ed.", Food and Agriculture Organization of the United Nations, Rome, Italy, 1993.
14. Heitzman, R.J. (ed.), "Veterinary Drug Residues. Residues in Food Producing Animals and Their Products: Reference Materials and Methods. 2nd ed.", Blackwell Scientific Publications, Oxford, 1994 .
15. "AOAC® Peer-Verified Methods Program, Manual on Policies and Procedures", AOAC International, Gaithersburg, MD, 1993.

Validation of a Multi-residue Method for Analysis of Pesticides in Fruit, Vegetables and Cereals by a GC/MS Iontrap System

M.E. Poulsen and K. Granby

INSTITUTE OF FOOD RESEARCH AND NUTRITION, DANISH VETERINARY AND FOOD
ADMINISTRATION, MØRKHØJ BYGADE 19, DK-2860 SØBORG, DENMARK
e-mail: MEP@fdir.dk

1 INTRODUCTION

Contamination of the food with pesticide residues has become a growing source of public concern. Some pesticides seem to have endocrine-disrupting effects[1] and additive effects may be expected for pesticides with similar biochemical reaction mechanism[2]. In order to follow the exposure of the population to pesticides it is relevant to have multimethods, which are able to measure many pesticides within a reasonable time. Furthermore it is relevant to have low limits of detections (LODs) so residues of pesticides at low concentrations will be reported. The low maximum residue limit (MRL) for infant food of 0.01 mg/kg[3] also requires LODs at or below this level.

Iontrap GC-MS is a technique that may be applied for identification and quantification of multiple pesticide residues in food [4,5,6]. The iontrap GC-MS has been proven to be a sensitive and selective detector for pesticide residue determinations[4,6]. In the present paper a GC multimethod using iontrap MS with Electron Ionisation (EI) for detection is validated.

The Danish Veterinary and Food Administration has been responsible for monitoring programmes on pesticide residues in food for many years and they have performed the development and validation of new methods before these have been implemented for routine analyses at regional laboratories. The validation of the GC-MS multimethod presented here has been performed as an in-house method validation.

Before validating the multiresidue method, which is applied for many pesticides applicable for many different fruit, vegetable and cereal matrices it was necessary to consider the extent of the experimental work[7]. It is impossible to validate all combinations of sample matrix, pesticide and concentration of the residues.

Concerning the choice of sample matrices for validation, they should be characteristic for fruit, vegetable and cereals. In the present work it was chosen to include apple as an acid fruit, avocado as a fatty commodity, carrot as a root vegetable with many interferences and potato as a root vegetable with a high consumption and which is easy to analyse. Salad is chosen as a leaf vegetable and wheat as a cereal with a high consumption. This comprises validation on 6 matrices out of approximately 105 matrices included in the Danish Monitoring programme[8, 9].

The present investigation includes validation of 82 pesticides analysed within the Danish Monitoring Programme. The GC multimethod with nitrogen phosphorous (NP)- and electron capture (EC) detection include approximately 130 pesticides. The 82 pesticides have been chosen from the 130 pesticides after a test run of calibration standards. Some of the pesticides excluded are difficult to chromatograph on a GC and should probably be analysed by a LC-MS/MS multimethod. Table 1 shows the pesticides that were excluded before the validation because they had unacceptable calibration curves.

Table 1 *Pesticides found 'not suited' for iontrap GC/MS.*

acephate	deltamethrin	diphenylamine	methamidophos	tetradifon
azinphos-ethyl	demeton-S-methyl	fenvalerate	mevinphos	tolclofos-methyl
azinphos-methyl	demeton-S-methyl	flucytrinate	monocrotophos	triadimenol
captafol	sulfone	formothion	myclobutanil	triazophos
captan	demeton-S-methyl	heptachlor	nuarimol	trichlorfon
carbaryl	sulfoxide	epoxide	omethoate	trichloronat
cyfluthrin	dichloran	imazalil	phosphamidon	vamidothion
cypermethrin	dimethoate	Mecarbam	phoxim	

In the study it was tested whether the pesticides need to be separated in test mixtures (A, B and C) as it is common when running GC-EC or GC-NP methods or if it was possible to add all 82 pesticides in one spike solution (D). The latter will save laboratory time.

The validation presented here has been performed using three spiking levels at 0.02, 0.1 and 0.4 mg/kg. For each matrix, the three spiking levels were made as double determinations and repeated three times. This makes it possible to determine repeatability and reproducibility and LODs can be calculated based on the standard deviations of the spiked concentrations (n=6) at the lowest accepted spiking level.

The time consumed on validation of one sample matrix by the method described and by addition of one spike solution with all 82 pesticides was approximately 10 working days of 6.5 hours/day.

2 EXPERIMENTAL

2.1 Chemicals

2.1.1 Pesticide standards. Pesticide reference standards were all purchased from Ehrensdorfer. Most of the standards had a purity of 98-100%. Pesticides investigated are listed in Table 3. Official common names were adopted from The Pesticide Manual [10].

2.1.2 Pesticide solution. Pesticide stock solutions (1mg/g) were prepared by dissolving pesticide standards in toluene and storing in a freezer at −18 (C in ampoules (~1 ml) with argon atmosphere. Four pesticides spike solutions A, B, C and D (Figure 1) with concentrations of 0.02, 0.09 and 0.4 µg/ml were prepared by dilution of stock solution with ethyl acetate:cyclohexane (1:1). The spike solutions were added to the matrix after weighing the sample for extraction.

2.1.3 Matrix matched calibration standards. Calibration standards were made by diluting pesticide solutions. To prevent matrix effect causing over- or underestimation of recoveries matrix matched standards were prepared by mixing standards with blank sample of the respective matrix. The concentrations of the matrix matched calibration standards were approximately 0.01, 0.05, 0.1, 0.5 and 1.0 μg/ml.

2.1.4 Organic solvents and reagents. Acetone (Rathburn, Glass distilled grade), ethyl acetate (Rathburn, HPLC Grade), cyclohexane (Rathburn, HPLC) and anhydrous sodium sulphate (granular, BHD 10398, heated to 500 C for 5 hours and conditioned to room temperature covered with alu-foil).

2.1.5 Sample matrices. Samples of apple, avocado, carrot, potato, salad, and wheat were organically grown and purchased a local stores. The matrices were homogenised (the grain were milled) and stored frozen at −18°C in small portions until chemical analysis.

2.2 Equipment

2.2.1 Homogenization equipment. Chopper (Weisner), Mill (Retsch Ultra Centrifugal Mill ZM1000, grind size 0.5 mm), Ultra-Turrax mixer (Janke & Kunkel).

2.2.2 Stirring equipment. Vibrofix mixer, Janke & Kunkel, VF1.

2.2.3 Centrifuge. Hereaus Sepatech Megafuge 3.0 R

2.2.4 Clean up system. Sample clean-up was performed by an automatic Gel Permeation Chromatography (GPC) system consisting of a Gilson 233 XL fraction collector, a Gilson 402 Syringe Pump, a Waters 510 HPLC pump and a column (Omnifit) at 45 cm x 15 mm (i.d.) of styrene-divinylbenzene (Bio-bead SX-3, 200-400 mesh, 3% cross-linked). Sample extracts of 2.3 ml were injected and the column was eluted at a flow of 1.4 ml/min with ethyl acetate:cyclohexane (1:1).

2.2.5 MS system. An iontrap system consisting of a Finnigan GCQTM, a Finnigan GCQTM Mass Detector Gas chromatograph and a Finnigan MAT AS200 autosampler. The splitless injector was temperature programmed from initial 180°C held for 0.2 min. increasing to 280°C at a rate of 180°C/min. Purge off time was 1.00 min and 2μl of the sample was injected. The oven was temperature programmed from initial 60°C held for 1 min. increasing to 180°C at a rate of 30°C/min and increasing til 280 °C at a rate of 5°C. This temperature was held for 20 min. The column used was a DB5-MS from J&W Science (30m x 0.25 i.d., film thickness 0.025μm). The velocity of the helium carrier gas was constantly at 42 cm/sec. The temperature of the transfer line was 282 °C and the temperature of the ion source was 180°C. The iontrap was scanning in EI-mode from 50-450 amu. In order to maintain the system the liner in the injector and the ion volume was cleaned after every 200 injections.

2.3 Extraction and clean up.

2.3.1 Fruit and vegetables. The sample was homogenised with a chopper for 1 minute and a 25.0g portion was added 50 ml of acetone and extracted for 2 min. by an Ultra-Turrax mixer. 50.0 ml ethyl acetate:cyclohexane (1:1) and 50g anhydrous sodium sulphate were added and the extraction went on for further 2 min. The sample was centrifuged; 20.0ml extract was transferred to a sample tube, added dodecane as a keeper and evaporated using a gentle stream of nitrogen to nearly dryness. 1g of anhydrous sodium sulphate and 4ml of ethyl acetate:cyclohexane (1:1) were added and the extract was stirred by a mixer for 1 min.. The extract was filtered through a 0.4 μm filter (Minisart, SPR 15, Sartorius) and cleaned up using gel permeation chromatography (see 2.2.4). The eluent fraction from 30-52 min. was collected (the fraction collected was defined as the eluate where the markers isophenphos, fenvalerate and pentachlorobenzene were included). The eluate was added dodecane as a keeper and concentrated to nearly dryness using a gentle stream of nitrogen. The sample was diluted with ethyl acetate:cyclohexane (1:1) to1.25ml.

2.3.2 Cereals. The sample was milled (grind size 0.5 mm) and a 5.0g portion was added 2 g anhydrous sodium sulphate and extracted by an Ultra-Turrax mixer for 3 min. with 50.0ml ethyl acetate. The sample was centrifuged; 30.0ml extract was transferred to a sample tube and evaporated using a gentle stream of nitrogen to a volume of 0.5-1.0ml. This volume was quantitatively transferred to a 5.00ml volumetric flask using ethyl acetate:cyclohexane (1:1) for rinsing and diluting to 5.00ml. The extract was filtered through a 0.4 μm filter (Minisart, SPR 15, Sartorius) and cleaned up using gel permeation chromatography (see 2.2.4). The eluent fraction from 21-50 min. was collected (the fraction collected was defined as the eluate where the markers lambda-cyhalothrin and isophenphos was included). The eluate was added dodecane as a keeper and concentrated to approximately 0.5 ml using a gentle stream of nitrogen. This volume was transferred to a volumetric flask with ethyl acetate:cyclohexane (1:1) and diluted to1.0ml.

2.4 MS analysis

Aliquots of 2μl were analysed by the iontrap GC/MS. The software screens the chromatogram for the selected mass spectrum within the appropriate retention time windows. A pesticide is detected, when the mass spectrum in the sample is identical (with a fit factor above 700) to the mass spectrum in the library. The quantification is then performed on one ion. The seven sample extracts from each analytical series were analysed together and quantified using bracketing calibration curves. Calibration standards at five concentration levels at approximately 0.01, 0.05, 0.1, 0.5 and 1.0 μg/ml were prepared in matrix. According to Guidelines for Residues Monitoring in the European Union, the response from the bracketing calibration curves must not differ by more than 20 %[11]. In general, this value was used as an acceptance criterion for the GC/MS-analysis.

3. RESULTS AND DISCUSSION

The validation included five sample matrices that were spiked with 82 pesticides. The recovery analyses of each standard solution were performed in three analytical series (three repetitions) and covered double determination of three concentration levels 0.02, 0.1 and 0.4 mg/kg and one blank, a total of seven samples. Four matrices (apple, carrot, potatoes and salad) were spiked separately with solution A, B and C each with approximately 30 pesticides. Three matrices (avocado, potatoes and wheat) were spiked with solution D containing all 82 pesticides. In total 45 analytical series of seven samples were performed by five different skilled technicians.

When validating and reporting an analytical method it is important to define the accuracy and precision. In the present work calculations of accuracy and precision have been made for each matrix separately. No outlier test was performed due to the relatively small number of comparable data. Recovery for every sample and the mean recovery for the individual matrix at the three spiking levels were determined. The repeatability and in-house reproducibility were calculated as described by ISO5725-2[12] using the double determinations for calculations of repeatability and the repetitions for calculation of reproducibility.

The results were accepted if the following conditions were fulfilled. [1] The mean recoveries were between 70-110%. [2] Both the repeatability and reproducibility were lower than $RSD_{Horwitz}$ - the relative standard deviation proposed by Horwitz[13]. $RSD_{Horwitz}$ is given by a function of the analyte concentration, which allows larger variation for low concentration levels than for higher ones. Although $RSD_{Horwitz}$ is based on proficiency tests and describes 'between laboratory variation' it appears to be reasonable to use for 'within laboratory variation' of pesticide multimethods in foodstuff. In a proficiency test, arranged by the European Commission[14], it was shown that the $RSD_{Horwitz}$ was a stricter criterion for selection compared to RSD_{Robust} - robust standard deviation – which was used as the target value. $RSD_{Horwitz}$ were for most of the pesticides 50-75 % of RSD_{Robust}

Table 2. *Recovery, repeatability (RSD_r), reproducibility (RSD_R) and limits of detection (LOD) for the apple/chlorpropylate combination. $RSD_{Horwitz}$ is used as the upper limits for acceptance of RSD_r and RSD_R. Recovery has to be between 70-110%. LOD is calculated as 3 times SD at lowest accepted concentration level.*

Matrix Pesticide	Apple Chloropropylate		
Spike level, mg/kg	0.02	0.1	0.4
No. Results	6	6	6
Repetitions	3	3	3
Recovery, %	79	85	82
RSD_r, %	4.6	5.5	6.9
RSD_R, %	10.1	8.0	7.7
$RSD_{Horwitz}$, %	29.4	23.1	18.8
LOD, mg/kg	0.004		

When the results were examined and the lowest concentration level, which fulfilled the acceptance criteria, was found - LOD were calculated as 3 times SD (standard deviation) of the six results at the lowest acceptable spiking level (n=6). The calculations for apple matrix and the pesticide chloropropylate are illustrated in Table 2.

3.1 Results from validation of the 6 matrices.

The results from the validation of 6 matrices including mean recoveries and LODs are listed in Table 3. All results, for which the validation data were accepted, have repeatability and reproducibility lower than the RSD$_{Horwitz}$ as described above. Spiking levels not accepted are marked with dashes. However, results for some of those pesticide/crop combinations were qualitative, as it was possible to identify the pesticide, but not to quantify within the limits. For those pesticides the methods may be used as semi quantitative. Pesticides which have high vapour pressures (e.g. pentachlorbenzene) were accepted although the recoveries were below 70%. These recoveries are marked with italic.

In total 41 pesticides were validated for all six matrices. Five pesticides, dicofol, thiometon and the three endosulfans, were not validated at all. Dicofol is known to be very rapidly degraded to p'p'-dichlorobenzophenone in the GC and when matrix is present. An HPLC-method would diminish the problem to some degree. Thiometon showed good calibration curves with standards in pure solvent but was degraded to the sulfoxide and sulfone, when matrix was present. The endosulfans have a very fragmented mass spectrum and small amounts of interfering compounds from the matrix demolished the identification.

Some pesticides with high vapour pressures, such as dichlorvos and pentachlorbenzene gave low recoveries, properly due to losses during concentration steps. A more gently concentration would properly give higher recoveries. As mentioned above, these pesticides are accepted as validated if the repeatability and reproducibility were lower the RSD$_{Horwitz}$. Some of the organohalogens, such as the DDTs and endrin, were not detectable at the low spike level and showed high RSD at the high level. As for the endosulfan it may be due to the weak capability of the iontraps to analyse those compounds. Because the compounds become very fragmented in the iontrap, only small amounts of interfering compounds can 'hide' the mass spectrum, both in the matrix-matched calibration standard and in the samples. These pesticides are therefore not suited for analyses in EI-mode. Negative chemical ionisation, NCI has been shown to improve the detection limit and standard deviation[6].

Table 3 *Relative recoveries (Rec., %) and LODs (mg/kg) for apple, avocado, carrot , potato, salad and wheat at three spiking levels (0.02, 0.4 and 1.0 mg/kg). All relative standard deviation are lower the $RSD_{Horwitz}$. Recoveries below 70% are marked with italic.*

Recoveries,% and LODs, mg/kg	Mix		Apple 0.02	Apple 0.1	Apple 0.4	Avocado 0.02	Avocado 0.1	Avocado 0.4	Carrot 0.02	Carrot 0.1	Carrot 0.4	Potato 0.02	Potato 0.1	Potato 0.4	Salad 0.02	Salad 0.1	Salad 0.4	Wheat 0.02	Wheat 0.1	Wheat 0.4
Organohalogens:	*Mix:*																			
Aldrin	B	Rec.	-	-	92	-	-	84	-	91	77	-	83	85	86	77	82	-	96	94
		LOD	0.177			0.160			0.039			0.044			0.012			0.054		
Bromopropylate	C	Rec.	87	86	83	-	-	-	83	86	88	70	87	88	98	79	82	-	88	87
		LOD	0.006						0.012			0.009			0.005			0.018		
Chlorfenson	C	Rec.	-	87	84	81	73	74	-	88	87	-	85	82	-	-		-	-	
		LOD	0.025			0.005			0.029			0.013								
Chlorobenzilate	C	Rec.	93	92	88	-	-	70	82	78	82	72	83	88	91	81	78	78	90	88
		LOD	0.006			0.059			0.005			0.002			0.006			0.009		
Chloropropylate	A	Rec.	79	85	82	-	79	73	75	84	90	80	86	88	90	102	103	88	95	92
		LOD	0.004			0.044			0.010			0.004			0.010			0.012		
Chlorthalonil	C	Rec.	70	80	73	-	-	84	80	76	80	73	79	84	82	73	72	78	103	101
		LOD	0.003			0.140			0.004			0.006			0.008			0.006		
DDD-*p,p'*	C	Rec.	83	86	85	-	74	73	81	79	84	76	87	89	96	79	79	-	98	93
		LOD	0.004			0.032			0.006			0.006			0.007			0.034		
DDE-*p,p'*	C	Rec.	-	95	86	-	-	-	78	76	82	71	85	87	79	74	76	-	91	89
		LOD	0.017						0.007			0.004			0.006			0.052		
DDT-*o.p'*	B	Rec.	-	-		-	74	73	-	-	81	84	87	89	-	70	82	-	100	93
		LOD				0.030			0.139			0.005			0.030			0.036		
DDT-*p,p'*	C	Rec.	-	98	86	-	-	-	-	-	108	-	-	-	-	-		-	-	
		LOD	0.033						0.205											
Dichlofluanid	A	Rec.	-	-	-	-	-	80	-			-			-			-		
		LOD	-			0.112														
Dicofol	B	Rec.	-						-			-			-			-		
		LOD	-																	
Dieldrin	A	Rec.	-	70	70	-	-		-	94	87	-	83	89	-	95	86	-	-	
		LOD	0.026						0.043			0.021			0.056					
Endosulfan, alfa	A	Rec.	-			-			-			-			-			-		
		LOD	-																	
Endosulfan, beta	B	Rec.	-			-			-			-			-	-	95	-		
		LOD	-												0.130					
Endosulfan, sulfate	B	Rec.	-			-			-			-			-			-		
		LOD	-																	
Endrin	B	Rec.	-	-	-	-	-	74	-	-	79	-	87	87	-	76	84	-	-	87
		LOD				0.151			0.133			0.026			0.028			0.142		
Fenson	A	Rec.	72	78	79	-	-	78	-	78	75	-	-	79	-	93	87	-	95	90
		LOD	0.008			0.109			0.038			0.062			0.037			0.037		
HCH – alfa	C	Rec.	80	87	82	-	90	84	77	74	80	61	75	76	57	76	74	66	92	90
		LOD	0.009			0.043			0.006			0.011			0.009			0.010		
HCH – beta	B	Rec.	-	-	96	79	85	81	93	76	84	-	77	81	-	69	83	78	96	93
		LOD	0.206			0.009			0.008			0.018			0.009			0.010		
HCH-gamma	A	Rec.	81	84	82	-	88	83	78	74	83	-	75	79	74	96	87	70	94	92
		LOD	0.009			0.040			0.007			0.016			0.011			0.006		
Heptachlor	A	Rec.	78	77	88	-	93	88	71	76	76	75	79	86	111	101	85	95	105	98
		LOD	0.010			0.050			0.008			0.009			0.016			0.013		
Hexachlorbenzene	A	Rec.	53	55	55	-	86	79	72	73	79	61	63	69	59	70	49	68	90	88
		LOD	0.004			0.051			0.004			0.010			0.009			0.008		
Methoxychlor	C	Rec.	-	93	77	-	-	76	-	-	102	82	81	86	-	-		72	88	90
		LOD	0.049			0.090			0.151			0.004						0.007		
Pentachloranilin	C	Rec.	80	86	82	-	93	87	78	76	79	55	78	84	79	78	75	-	106	99
		LOD	0.009			0.058			0.006			0.026			0.010			0.044		
Pentachloranisol	C	Rec.	82	79	74	-	96	85	86	68	68	59	74	76	-	67	71	-	97	96
		LOD	0.007			0.053			0.005			0.011			0.031			0.040		
Pentachlorbenzene	A	Rec.	63	71	68	-	74	74	69	62	59	47	56	62	63	70	62	-	83	84
		LOD	0.003			0.047			0.011			0.007			0.003			0.047		
Quentozene	C	Rec.	-	80	65	-	94	85	-	68	70	-	68	79	-	70	69	-	98	97
		LOD	0.040			0.050			0.023			0.030			0.035			0.043		
Tecnazene	C	Rec.	-	73	71	-	94	79	-	63	64	-	59	58	-	72	72	-	80	84
		LOD	0.033			0.046			0.028			0.030			0.033			0.024		
Tetradifon	B	Rec.	-	85	77	-	-		-	76	74	-	84	82	-	86	88	-	-	82
		LOD	0.052						0.153			0.013			0.035			0.112		
Tetrasul	C	Rec.	79	82	78	-	82	77	80	81	85	78	85	91	88	76	78	92	98	89
		LOD	0.006			0.044			0.004			0.003			0.010			0.006		
Trichloronate	C	Rec.	84	95	88	-	91	85	70	85	83	80	87	88	78	81	78	99	101	95
		LOD	0.008			0.049			0.005			0.006			0.005			0.016		
Organonitrogens:																				
Carbofuran	A	Rec.	85	85	92	79	83	81	89	85	90	75	79	84	93	92	91	70	95	93
		LOD	0.006			0.000			0.009			0.010			0.002			0.006		
Fenarimol	A	Rec.	-	-		-	-		86	86	84	76	80	83	-	106	87	-	91	86
		LOD	-			-			0.009			0.007			0.042			0.024		

Table 3 continuation

Recoveries,% and LODs, mg/kg	spike, mg/kg		Apple 0.02	0.1	0.4	Avocado 0.02	0.1	0.4	Carrot 0.02	0.1	0.4	Potato 0.02	0.1	0.4	Salad 0.02	0.1	0.4	Wheat 0.02	0.1	0.4
Fenpropimorph	A	Rec.	97	80	70	-	-	81	-	-	72	72	76	82	92	91	96	-	-	-
		LOD	0.015			0.121			0.248			0.004			0.005			-		
Folpet	B	Rec.	-	-	-	-	-	79	-	-	-	-	78	85	-	-	-	-	97	105
		LOD	-			0.121			-			0.025			-			0.023		
Iprodione	A	Rec.	84	78	85	-	-	-	99	84	82	-	93	96	117	109	96	91	97	90
		LOD	0.008			-			0.011			0.018			0.009			0.007		
Metalaxyl	B	Rec.	-	82	88	72	81	82	-	-	-	76	80	87	-	79	86	90	97	97
		LOD	0.050			0.007			-			0.007			0.018			0.009		
Pirimicarb	B	Rec.	-	-	-	-	83	79	-	-	80	77	81	86	-	73	83	81	98	94
		LOD	-			0.032			0.114			0.006			0.012			0.007		
Procymidone	A	Rec.	79	78	80	-	-	-	78	81	82	-	86	89	107	96	93	73	93	89
		LOD	0.008			-			0.006			0.010			0.008			0.011		
Propham	C	Rec.	-	89	75	-	97	77	-	73	70	-	58	68	-	73	70	-	83	82
		LOD	0.022			0.032			0.035			0.019			0.029			0.023		
Tolyfluanid	C	Rec.	102	102	92	-	70	74	111	96	96	79	80	88	108	90	77	78	96	93
		LOD	0.015			0.021			0.016			0.003			0.009			0.005		
Triadimefon	A	Rec.	-	110	94	-	83	82	-	94	92	-	83	87	-	103	101	-	106	91
		LOD	0.059			0.025			0.032			0.021			0.035			0.050		
Vinclozolin	B	Rec.	91	85	100	-	88	84	-	86	77	87	80	86	-	80	85	78	96	93
		LOD	0.009			0.033			0.041			0.007			0.030			0.006		
Organophosphorous:																				
Bromophos	B	Rec.	-	-	-	-	90	82	-	-	79	-	79	90	-	77	88	-	96	92
		LOD	-			0.044			0.120			0.018			0.012			0.037		
Bromophos-ethyl	C	Rec.	83	92	85	-	89	80	78	80	83	81	84	89	84	83	78	85	98	91
		LOD	0.009			0.038			0.004			0.007			0.006			0.013		
Carbophenothion	A	Rec.	75	76	74	-	-	-	97	82	80	-	-	91	120	101	94	-	85	81
		LOD	0.008			-			0.013			0.133			0.019			0.020		
Chlorfenvinphos	B	Rec.	87	85	90	-	83	81	-	93	82	72	90	91	-	85	88	96	104	100
		LOD	0.007			0.040			0.042			0.005			0.019			0.005		
Chlormephos	B	Rec.	-	-	69	-	-	-	-	-	-	-	-	59	-	58	63	-	86	74
		LOD	0.157			-			-			0.107			0.033			0.042		
Chlorpyrifos	C	Rec.	-	99	86	-	99	86	91	81	84	-	79	90	83	76	75	-	-	89
		LOD	0.032			0.052			0.004			0.021			0.005			0.161		
Chlorpyriphos-methyl	A	Rec.	76	78	79	-	96	86	72	79	81	70	82	89	80	88	88	82	103	100
		LOD	0.005			0.060			0.009			0.008			0.006			0.011		
Dialifos	C	Rec.	-	83	81	-	-	-	80	89	89	86	84	86	-	83	87	77	93	87
		LOD	0.017			-			0.011			0.005			0.035			0.008		
Diazinon	B	Rec.	-	-	-	-	90	87	-	85	78	70	81	85	-	-	86	80	97	95
		LOD	-			0.039			0.042			0.009			0.100			0.008		
Dichlorvos	A	Rec.	-	69	66	-	65	66	-	-	-	-	-	-	-	-	-	-	-	72
		LOD	0.047			0.027			-			-			-			0.161		
Ditalimfos	C	Rec.	-	88	81	-	72	72	-	83	89	-	79	78	91	85	80	-	-	92
		LOD	0.036			0.030			0.053			0.039			0.009			0.184		
Ethion	A	Rec.	79	79	76	-	72	74	83	86	88	81	84	89	95	99	101	86	90	90
		LOD	0.008			0.026			0.012			0.003			0.014			0.010		
Etrimfos	A	Rec.	73	80	81	-	101	89	75	84	83	71	82	90	79	89	87	-	106	104
		LOD	0.005			0.049			0.010			0.009			0.006			0.056		
Fenchlorphos	C	Rec.	74	94	84	-	94	88	84	83	82	70	80	88	79	76	75	86	101	97
		LOD	0.002			0.045			0.005			0.009			0.009			0.009		
Fenitrothion	C	Rec.	85	90	85	-	-	83	91	85	86	92	83	90	86	77	79	94	108	104
		LOD	0.008			0.143			0.003			0.007			0.010			0.008		
Heptenophos	B	Rec.	-	89	95	-	83	81	106	87	75	-	80	79	-	-	79	70	89	88
		LOD	0.034			0.024			0.011			0.042			0.082			0.008		
Iodofenphos	B	Rec.	84	81	89	-	78	76	-	-	-	70	81	85	-	78	84	-	-	-
		LOD	0.012			0.043			-			0.005			0.012			-		
Isofenphos	A	Rec.	87	86	87	-	82	80	84	89	92	-	101	92	93	94	99	99	101	92
		LOD	0.006			0.034			0.009			0.037			0.010			0.012		
Malathion	B	Rec.	-	88	99	-	84	79	-	108	95	-	-	-	-	84	91	-	-	-
		LOD	0.051			0.040			0.000			-			0.013			-		
Methidathion	A	Rec.	74	78	79	-	-	-	92	91	88	75	84	84	-	-	80	80	92	88
		LOD	0.009			-			0.010			0.009			0.153			0.005		
Parathion	C	Rec.	100	93	88	-	93	84	-	-	84	-	86	91	102	79	81	-	101	98
		LOD	0.006			0.053			0.083			0.039			0.010			0.020		
Parathion – methyl	C	Rec.	90	89	83	96	90	88	87	82	85	80	82	89	84	82	78	-	114	103
		LOD	0.006			0.008			0.007			0.008			0.007			0.030		
Phenthoat	B	Rec.	-	-	-	-	-	-	-	-	-	81	86	89	-	80	90	-	91	96
		LOD	-			-			-			0.006			0.018			0.021		
Phorat	B	Rec.	-	76	76	-	77	75	-	81	74	-	72	71	106	57	64	-	-	-
		LOD	0.039			0.036			0.038			0.039			0.010			-		
Phosalone	C	Rec.	-	89	87	-	-	-	104	88	94	81	85	85	-	88	86	82	88	88
		LOD	0.020			-			0.016			0.005			0.053			0.010		
Phosmet	B	Rec.	88	71	78	-	-	-	-	78	81	75	80	85	-	96	96	75	94	89
		LOD	0.002			-			0.043			0.003			0.038			0.004		
Pirimiphos – ethyl	C	Rec.	-	70	84	95	84	74	83	76	85	76	87	86	83	76	73	-	109	103
		LOD	0.043			0.011			0.003			0.004			0.009			0.033		
Pirimiphos – methyl	B	Rec.	-	-	81	-	99	89	-	83	77	80	82	93	-	76	86	78	110	102
		LOD	0.174			0.051			0.023			0.007			0.010			0.012		

Table 3 continuation

Recoveries,% and LODs, mg/kg			Apple 0.02 0.1 0.4			Avocado 0.02 0.1 0.4			Carrot 0.02 0.1 0.4			Potato 0.02 0.1 0.4			Salad 0.02 0.1 0.4			Wheat 0.02 0.1 0.4		
	spike, mg/kg		0.02	0.1	0.4	0.02	0.1	0.4	0.02	0.1	0.4	0.02	0.1	0.4	0.02	0.1	0.4	0.02	0.1	0.4
Profenofos	B	Rec.	91	89	88	78	73	75	-	-	-	-	87	92	-	90	94	-	95	89
		LOD	0.007			0.013			-			0.017			0.020			0.060		
Prothiofos	A	Rec.	77	94	90	-	86	83	83	84	89	-	-	-	86	90	94	-	112	96
		LOD	0.007			0.050			0.013			-			0.007			0.078		
Pyrazophos	B	Rec.	-	-	80	-	-	-	114	87	88	83	87	86	-	94	76	76	88	88
		LOD	0.229			-			0.012			0.004			0.038			0.011		
Quinalphos	C	Rec.	-	86	80	-	89	73	81	80	84	-	80	86	92	78	78	-	100	90
		LOD	0.022			0.045			0.008			0.022			0.008			0.033		
Sulfotep	A	Rec.	70	71	71	-	94	89	-	78	77	-	72	82	75	86	82	75	95	93
		LOD	0.004			0.059			0.040			0.040			0.006			0.009		
Tetrachlorvinphos	B	Rec.	94	87	87	-	82	76	-	-	-	74	88	95	-	-	-	91	95	89
		LOD	0.007			0.048			-			0.002			-			0.005		
Thiometon	B	Rec.	-	-	-	-	-	-	-	-	-	-	-	-	-	-	-	-	-	-
		LOD	-			-			-			-			-			-		
Pyrethroids:																				
Bifenthrin	A	Rec.	79	80	80	-	80	70	80	86	90	81	81	85	95	103	101	73	93	90
		LOD	0.006			0.038			0.010			0.004			0.012			0.008		
Fenpropathrin	A	Rec.	92	82	79	-	-	-	83	88	91	73	84	86	112	113	97	76	90	85
		LOD	0.005			-			0.013			0.008			0.016			0.004		
Permethrin – trans	A	Rec.	74	76	74	-	-	-	86	88	90	79	84	84	85	99	94	-	-	81
		LOD	0.009			-			0.011			0.012			0.013			0.100		

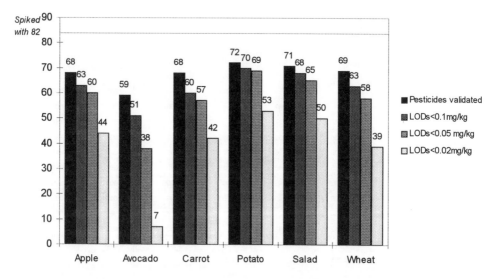

Figure 1. *For each matrix is shown the number of pesticides validated (out of 82 pesticides investigated) and the number of pesticides with LOD lower than 0.1; 0.05 and 0.02 mg/kg, respectively.*

Figure 1 shows the number of validated pesticides for each matrix and the number of pesticides with LOD below 0.1 mg/kg, 0.05 mg/kg and 0.02 mg/kg. Potato and salad were the matrices with most validated pesticides (71-72 out of 82) and lowest LODs, 70-74% had LOD<0.02 mg/kg. Apple, carrot and wheat were more difficult with 2-4 fewer pesticides validated and only 56-65% of the LOD below 0.02 mg/kg. The fat containing matrix, avocado was the most difficult to analyse as only 59 pesticides were validated and only 12% of those had LODs below 0.02 mg/kg.

The differences among the matrices were not only the number of validated pesticides and the level of LODs. The chromatograms of avocado matrix showed many co-eluting substances indicating that the clean up was not efficient enough for this matrix. For the other 5 matrices it was a general pattern that the pesticides not accepted varied from matrix to matrix, except for some of the organohalogens mentioned above.

3.2 Comparison between results of potatoes spiked separately with standard solution A, B; C and potatoes spiked with standard solution D containing all 82 pesticides.

In the beginning of the validation analyses, the matrices apple, carrot, potato and salad were spiked separately with the standard solutions A, B and C, each containing up to 30 pesticides. In total this gave 9 analytical series per matrix. In order to minimise the time consuming laboratory work, it was tested if it was possible to combine all 82 pesticides in one spike solution. The experiments were made on potato matrix.

When comparing the experiments no significant differences are displayed (Figure 2). The same pesticides were accepted as validated and the detection limits were at the same levels. The only differences are 1) DDT-*o,p'* which is accepted in the analytical series with standard solution D, and 2) prothiophos, which is not accepted.

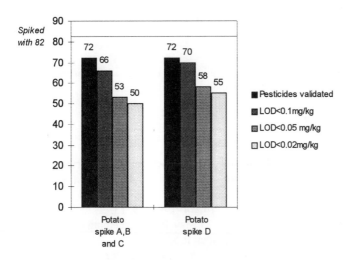

Figure 2. *For potato matrix spiked with standard solution A,B,C or standard solution D is shown the number of pesticides validated (out of 82 pesticides investigated) and the number of pesticides with LOD lower than 0.1; 0.05 and 0.02 mg/kg, respectively.*

4. CONCLUSIONS AND IDEAS ABOUT FUTURE WORK.

An iontrap GC/MS multimethod for pesticide analyses of fruits, vegetables and cereals has been validated using the matrices apple, avocado, carrot, potato, salad and wheat. Roughly 70 out of 82 pesticides were accepted as quantitative and the mean recoveries and LODs were calculated. Among the remaining 12 pesticides some were found to be semi quantitative. Different results for the different matrices necessitate the validation on several matrices.

Organohalogens and pyrethroids were not easy to detect on iontrap GC/MS with EI. However using GC/MS with NCI the detection of these substances is expected to improve. This will be validated in the near future. Also validation of pesticides without characteristic atoms as N, P or halogens will be included in the validation.

Comparison between addition of three spiking solutions A, B, C and a spike solution D with all 82 pesticides showed no significant differences with respect to repeatability, reproducibility and LODs. Validation of multimethods and performance of standard calibrations in monitoring analyses are much easier when all pesticides are added or calibrated with one standard solution. This will spare a lot of time compared to validation or calibration using GC-NP or GC-EC multimethods. The time consumption on 10 working days per matrix validated including double determinations and three repetitions at tree spiking levels seems reasonable.

Acknowledgements

Laboratory technician G. Andersen, M. B. Ludwigsen, I. Schröder, S. Shanmugam and L. J. Hansen are gratefully acknowledged for their technical assistance.

References
1. A.M. Vingaard, V. Breinholt and J.C. Larsen, Screening of selected pesticides for oestrogen receptor activation *in vitro*. *Food Additives and Contaminants,* 1999, **12**, 533.
2. J.P. Croten *et al., Fundam. and Appl. Toxicol.* 1997, **36**, 15.
3. EU Commission Directive 1999/50/EF, 25 May 1999 concerning change of Directive 91/321/EØF about infant formula and infant food.
4. R.S. Sheridan and J.R. Meola, *J. of AOAC Int.,* 1999, **82**, 982.
5. A Gelsomino, B. Petrovicová, S. Tiburtini, E. Magnani and M. Felici, *J. of Chromatography A.,* 1997, **782**, 105.
6. H. Obana, K. Akutsu, M. Okihashi, S. Kakimoto and S. Hori, *Analyst,* 1999, **124**, 1159.
7. A.R.C. Hill and S.L. Reynolds, *Analyst,* 1999, **124**, 953.
8. R.K. Juhler, M.G. Lauridsen, M.R. Christensen and G. Hilbert, *J. of AOAC Int.,* 1999, **82**, 337.
9. J.A. Andersen, K. Granby and M.E. Poulsen, 1999. Pesticide Residues in Fruits, Vegetables and Cereals in Denmark-1998. Report concerning directives 90/642/EEC,

86/362/EEC and Commission Recommendation 97/822/EC. The Danish Veterinary and Food Administration, November 1999.

10. The Pesticide Manual, 11th Ed., Ed. C.D.S. Tomlin, British Crop Protection Council Publication Sales, Berks RG42 5QE, UK, 1998.

11. Quality Control Procedures for Pesticide Residue Analysis- Guidelines for Residues Monitoring in the European Union, Document 7826/VI/97, European Commission, Brussels, 1997.

12. ISO 5725-2:1994. Accuracy (trueness and precision) of measurement methods and results – Part 2. First edition. December 1994.

13. W. Horwitz, *Anal. Chem.,* 1982; **54**, 67A.

14. European Commission's Proficiency Test on Pesticide Residues in Fruit and Vegetables, Test 3, 1999. National Food Administration, Uppsala Sweden Sept. 1999.

Development and Validation of a Generic Gas Chromatographic Method for the Determination of Organophosphorus Pesticide Residues in Various Sample Extracts

H. Botitsi, P. Kormali, S. Kontou, A. Mourkojanni and D. Tsipi*

GENERAL CHEMICAL STATE LABORATORY, C DIVISION, PESTICIDE RESIDUES
LABORATORY, 16, AN. TSOHA, 11521 ATHENS, GREECE
e-mail: gxk-pest@ath.forthnet.gr

1 INTRODUCTION

Method validation is an important part of a laboratory Quality Assurance program and establishes that the performance characteristics of the method meet the specifications related to the intended use of the analytical results. Quality Control techniques demonstrate on a regular basis the compliance of the analytical method with the documented performance characteristics of the validation study [1,2].

Multi-residue methods of pesticide residues analysis generally comprise the following steps:

- Sample extraction
- Clean-up
- Chromatographic separation / measurement

The pretreatment to which the sample should be subjected prior to the instrumental determination step depends on the nature of the substrate (water and fat content).

In cases where no matrix-effect is apparent, the gas chromatographic detection and quantitation process of a certain class of analytes in the final extracts, obtained by various procedures, is identical. In order to avoid repetitive studies, the generic approach of validating separately the GC-determination step is proposed [3].

The aim of the present study is to develop and validate a generic gas chromatographic method (GC-FPD-NPD) for the determination of organophosphorous pesticides in food and water extracts. System precision, linearity, detection limit and specificity were established and the internal Quality Control scheme which pertains to the routine application of the method was defined. The method is routinely applied in conjunction with four different extraction and cleanup procedures [4,5,6,7].

2 MATERIALS AND METHODS

2.1 Apparatus and analytical conditions

2.1.1 Gas chromatographs: (1) Carlo Erba Model Mega 2 equipped with FPD, split/splitless injection port and autosampler Model A200S, with program for the evaluation of GC runs (Chrom-Card). (2) Hewlett-Packard Model 5890 II, equipped with NPD, split/splitless injection port and autosampler Model 7673, with program for the evaluation of GC runs (HPCHEM).

2.1.2 GC columns: (1) 30m x 0.53mm id DB-1701 fused silica capillary column (J&W Scientific Inc.), film thickness 1μm. (2) 50m x 0.32mm id CP-Sil 13 CB fused silica capillary column (Chrompack) fused silica capillary column (Chrompack), film thickness 0.4μm. The chromatographic conditions for the two columns are listed in Table 1.

Table 1 *Chromatographic conditions for two different columns.*

Conditions	DB1701 FPD	CP-SIL 13CB NPD
TEMPERATURE PROGRAM		
Initial temp. (^{o}C)\ Time (min)	100\1	100\1
Programming rate (^{o}C/min)	35	30
1st step temp (^{o}C)\ Time (min)	140\0	150\2
Programming rate (^{o}C/min)	5	3
2nd step temp (^{o}C)\ Time (min)	220\0	205\0
Programming rate (^{o}C/min)	10	2
3rd step temp (^{o}C)\ Time (min)	255\25	260\1
Inlet temperature (^{o}C)	250	250
Detector temperature (^{o}C)	250	280
FLOW RATES Carrier gas ((He)ml/min)	6.85	1.19
Make up gas ((N2)ml/min)	33	30
SPLITLESS INJECTION Valve time on (sec)	90"	60"

2.2 Materials

2.2.1 Reference standards in neat (undiluted) form, preferably certified. If neat standards are not available, certified solutions of standards may be used.

2.2.2 Certified solution of standard ULTRA Scientific, Organophosphorous Pesticides Mixture, concentration C=200ppm, approximately. (Code : Part Number SPM-614), in 1ml ampoules, J.T.Baker.

2.2.3 Pesticide Residue Quality Solvents : Ethyl Acetate ($CH_3COOC_2H_5$), Acetone (CH_3COCH_3), Methanol (CH_3OH), Acetonitrile(CH_3CN), of Pesticide grade.

2.3 Extraction and clean-up

The extraction methods employed in this study were based on four different extraction procedures reported previously.

2.3.1 Fruits and vegetables- Ethyl acetate extraction [4] *(OP3A)*. Weigh 50 g of sample into a high speed blender jar and add 60 g $NaSO_4$ and 100 ml ethyl acetate. Blend at high speed for 2-3 min. Transfer the homogenate through a Buchner funnel with glasswool and 20g sodium sulfate into a 200 ml measuring cylinder. The blender and the filter cake are washed with 2*25ml ethyl acetate. Determine the volume. Take an aliquot of one fourth and reduce the volume to 2.5 ml. The concentrate is transferred quantitatively to a 5ml volumetric flask.

2.3.2 Cereals- Acetone partition [5] *(OP4)*. Mix the sample very well. Transfer 25g sample to a conical flask with stopper. Add 100 ml of acetone:dichloromethane (50:50)v/v and allow to stand at least overnight for static extraction. Tranfer the homogenate through a funnel with glasswool and 20g sodium sulfate into a 200 ml measuring cylinder. The flask and the filter cake are washed with 2*25ml acetone:dichloromethane (50:50)v/v. Concentrate to a few ml in a rotary evaporator at a bath temperature of 45°C The concentrate is transferred quantitatively using acetone to a 5ml volumetric flask.

2.3.3 Olive oil- Hexane-acetonitrile partition [6]*(OP2)*. The sample of oil is allowed to take the room temperature and mixed well before the test portion is removed for analysis.

Weigh 10g (±0.2g) of the sample into a 100ml beaker. The sample is mixed with 50ml of hexane saturated with acetonitrile. The mixture is transferred to a separatory funnel. 100ml of acetonitrile saturated with n-hexane and 1ml water reagent are added. The funnel is shaken for 1min and left for 15min to allow the phases to separate.The lower acetonitrile phase is collected in a second separatory funnel containing 25ml of n-hexane saturated with acetonitrile and 1ml water reagent and the above procedure is repeated. A second 50ml volume of acetonitrile saturated with n-hexane is added to the first separatory funnel, followed by 0.5ml of reagent water and the extraction procedure is repeated. The acetonitrile phases from the two separatory funnels are combined in a round-bottomed flask which is connected to the vacuum evaporator. The combined extracts are rotary evaporated to almost dryness. 5ml of acetone is added. The produced solution is evaporated to ≈ 1 ml. The residue is dissolved in acetone and is transferred quantitatively into a 2ml volumetric flask.

2.3.4 Water- Dichloromethane partition [7] *(OPT1)*. A 500ml of water sample is extracted with three separate 150ml portions of dichloromethane using 1L separatory funnel and vigorous shaking. The organic phases are combined and concentrated in a rotary evaporator to a few ml. The extract is evaporated to dryness with a gentle stream of nitrogen and residue is taken up in 0.2 ml ethyl acetate.

The condensed residue extracts *(2.3.1-2.3.4)* can be directly injected to the GC/FPD and GC/NPD.

3 RESULTS AND DISCUSSION

3.1 Gas chromatographic analysis

3.1.1 GC analysis sequence. All food extracts must be analysed with a valid analysis sequence as given in Figure 1. The sequence of Figure 1 is repeated after the analysis of 12 samples.

Figure 1 *GC analysis sequence for standard and sample extracts*

1. Solvent (Instrument blank)
2. Quality Control Sample (QCS)
3. Low Calibrated Level (LCL)
4. Medium Calibrated Level (MCL)
5. High Calibrated Level (HCL)
6.Spike Shooting Solution
7. Recovery
8. Blank
9. Sample 1
10. Sample 2
....
20. Sample12
21. Quality Control Sample (QCS)

Depending on the basis of the nature of the extract (*2.3.1-2.3.4*) analyzed the appropriate mixtures of standard working solutions are chosen at three concentration levels LCL($\approx$ 0.05ppm), MCL($\approx$ 0.3ppm), and HCL($\approx$1.25ppm)   The Lowest Calibrated Level corresponds to the Reporting Limit (RL) which is the lowest concentration level at which residues will be reported as absolute numbers. RL can be specified as a fraction of the MRL.

3.1.2 Performance evaluation. Performance evaluation is carried out on the QCS check sample before starting the analysis of samples.

a) Calculation of peak resolution

The quality of the separation achieved is based on the evaluation of resolution of the worst resolved peak pair. If the peaks of interest are separated by less than 1 min the worst resolved peak pair is identified and the peak resolution factor RF is calculated using the following equation:

$$RF= 2(t1-t2)/(w1+w2)$$

t^1 = Peak 1 retention time (min)
t^2 = Peak 2 retention time (min)
w^1 = Peak 1, peak width at the baseline (min)
w^2 = Peak 2, peak width at the baseline (min)

Peak resolution acceptance criteria:

Resolution factor> 0.7

b) Calculation of peak symmetry

The Peak Gaussian Factor (PGF) of the most asymmetrical peak is calculated using the equation:

PGF=1.83xW(1/2)/W(1/10)

Where W(1/2) is the peak width at half height and W(1/10) is the peak width at tenth height.

Peak symmetry acceptance criteria:

0.80<PGF<1.20

3.1.3 Analytical calibration. Before any samples are analyzed, the calibration factors are calculated   (Correlation Coefficient R^2,slope, intercept), the Retention Times RTs and the windows of RTw).

Analytical calibration acceptance criteria

- Correlation Coefficient R^2 ≥0.98
- The values of the intercept should not exceed the lower and the upper limits listed in Table 3.
- The calibration line should not be forced through the origin.
- The QCS must bracket the sequence during which sample data are collected in order to calculate the drift and to ensure that the response remains constant.

Acceptance criterion drift< 20%

Where the response differs by more than 20%, the determinations should be repeated in smaller batches. Repetition of the determinations is not necessary for samples which contain residues <LCL, if the LCL response remains measurable throughout the batch. Extracts containing high-level residues may be diluted to bring them within the calibrated range.

3.1.4 Identification of the peaks The identification of the pesticide residues in Gas Chromatography is based on the Retention Times (RTs) data. Analytes are identified when peaks are observed in the RT window for the compound on the GC columns.

In case of negative results the chromatograms from the analyses of the sample extract must use the same scaling factor as was used for the LCL standard solution, of the calibration associated with those analyses.

3.2 Confirmation of the results

3.2.1 Results which are below the LCL, should not be reported as absolute numbers and are considered confirmed if the recovery and LCL data for the batch are acceptable.

3.2.2 Results at or above the LCL require additional support in order to be considered confirmed (qualitative confirmation).

3.2.3 Where the batch of analyses was performed without calibration for the particular pesticide(s), the confirmation is essential. In this case, the re-analysis of the extracts, with appropriate calibration for the pesticide(s) detected is demand. As the pesticide(s) involved should be detected infrequently and may therefore be unusual, re-analysis of the sample with concurrent recovery determination is to be preferred.

When the quantity of the pesticide is greater than the MRL the confirmation of the analyte detected should be quantitative and qualitative.

Qualitative confirmation is required, when the presence of a new, non-typical pesticide is reported.

The confirmation methods, which are used, are the following:
a) Different chromatographic columns of significantly different polarity.
b) Use of special detectors in gas chromatography (FPD,NPD).
c) Analysis by one or more separate, replicate subsamples
d) Mass Spectrometry (GC-MS): In this case, sample extracts are subjected to concentration of higher degree.

The results on the same extract (cases a,b) should not differ more than 25% . In these cases the minor result is referred as final result.

3.3 Validation

The laboratory selected 20 representative pesticides (Table 1) for validating the generic GC-FPD-NPD method. On the basis of the extraction procedure (2.3.1-2.3.4) applied in each commodity the group of pesticides determined for the validation are as follow:

OP3A: chlorpyrifos, chlorpyrifos-me, omethoate, pirimiphos-me, triazophos
OP4: chlorpyrifos, chlorpyrifos-me, diazinon, dichlorvos, etrimphos fenitrothion, malathion, methacrifos, pirimiphos-me
OP2: chlorpyrifos, diazinon, dimethoate, fenthion, methidathion, parathion
OPT1: chlorpyrifos, demeton, diazinon, dichlorvos, disulfoton, ethion, fenitrothion, fenthion, malathion, parathion, parathion-me

3.3.1 Precision. The precision measures include repeatability standard deviation s_r and reproducibility standard deviation s_R. Six replicate injections were carried out within a day for 15 different days and relative standard deviations (Table 2) were calculated for retention times and peak area. Results showed good repeatability and reproducibility of the chromatographic system.

Table 2 Data on the specificity (expressed as peak RT) precision (expressed $\%RSD_r$ and $\%RSD_R$ of peak RT and peak Area) and Limits of Detection (LOD) of the Chromatographic system GC-FPD (column : DB1701).

A/ A	COMPOUND	RT (min)	RT %RSD$_r$	RT %RSD$_R$ *	Area %RSD$_r$	Area %RSD$_R$ *	LOD (ppb) **
1	chlorpyrifos-et	18.58	0.11	0.79	1.10	3.60	1.0
2	chlorpyrifos-me	17.06	0.10	0.18	1.62	8.45	2.1
3	demeton -s	12.86	0.12	0.47	3.36	4.86	8.2
4	demeton -o	11.28	0.14	0.50	4.26	11.01	4.5
5	diazinon	14.50	0.14	0.22	0.64	2.19	0.7
6	dichlorvos	5.38	0.14	0.29	2.75	13.27	1.5
7	dimethoate	17.08	0.16	0.17	2.80	8.72	6.1
8	disulfoton	15.43	0.11	0.42	1.64	4.78	0.6
9	ethion	23.89	0.07	0.29	1.65	6.70	1.1
10	etrimfos	15.45	0.04	0.13	1.53	3.24	0.5
11	fenitrothion	19.71	0.03	0.11	1.58	5.32	1.7
12	fenthion	19.50	0.10	0.15	3.73	16.70	3.5
13	malathion	19.41	0.03	0.11	1.06	3.45	1.1
14	methacrifos	9.41	0.05	0.19	1.29	3.33	1.0
15	methidathion	22.14	0.08	0.13	2.52	9.04	5.2
16	omethoate	14.66	0.53	0.36	11.43	23.86	85
17	parathion-et	20.38	0.08	0.14	1.32	3.53	7.2
18	parathion-me	18.99	0.09	0.33	2.27	10.65	1.5
19	pirimiphos-me	17.95	0.04	0.17	1.11	4.93	1.1
20	triazophos	26.13	0.05	0.16	3.35	13.24	5.6

* Under reproducibility conditions (15 different days, 3 measurements per day)
** LOD= k_{LOD} x h_n x c_s/h_s , J.E. Knoll, Estimation of the limit of detection in chromatography, J. Chrom. Sci., Vol. 23, 422-425, 1985.

3.3.2 Linearity. Calibration graphs of the peak areas vs the concentrations were constructed for each pesticide (5 points/3 measurements per point). The calibration factors, the equation y=ax+b, the correlation coefficient R^2, the standard errors of the slope and the intercept were calculated for each detector. The results for FPD are shown in Table 3

Table 3 Linearity data of OP pesticides in FPD detector.

A/A	y=ax+b	Range (x) (ppm)	R^2	a±SEa	b±SEb
1	$y=716.3*10^4 x-6.1*10^4$	0.0370-1.480	0.9997	$(716.3\pm 4.5)*10^4$	$(-6.1 \pm 2.3)*10^4$
2	$y=716.7*10^4 x-15.5*10^4$	0.0490-2.525	0.9997	$(716.7 \pm 4.1)*10^4$	$(-15.5 \pm 4.6)*10^4$
3	$y=571.5*10^4 x-2.2*10^4$	0.0330-3.210	0.9987	$(571.5 \pm 6.1) *10^4$	$(-2.2 \pm 8.5)*10^4$
4	$y=138.8*10^4 x-6.7*10^4$	0.0330-3.210	0.9984	$(13.9 \pm 1.7) *10^4$	$(-6.7 \pm2.3)*10^4$
5	$y=1551*10^4 x-44.6*10^4$	0.0330-3.210	0.9994	$(1551 \pm 11.8)*10^4$	$(-44.6 \pm 16.4)*10^4$
6	$y=932.2*10^4 x-11.1*10^4$	0.0710-3.740	0.9996	$(932.2 \pm 6.8)*10^4$	$(-11.1 \pm 11.8)*10^4$
7	$y=733.2*10^4 x-20.1*10^4$	0.032-1.280	0.9968	$(733.2 \pm 15.7)*10^4$	$(-20.1 \pm 7.0)*10^4$
8	$y=974.8*10^4 x-34.6*10^4$	0.033-3.210	0.9989	$(974.8\pm 9.7)*10^4$	$(-34.6 \pm 13.4)*10^4$
9	$y=1095*10^4 x-29.8*10^4$	0.033-3.210	0.9988	$(1095 \pm 11.6)*10^4$	$(-29.8\pm 16.2)*10^4$
10	$y=1339*10^4 x -4.8*10^4$	0.059-3.070	0.9997	$(1339 \pm 7.6)*10^4$	$(-4.8 \pm 11.6)*10^4$
11	$y=812.5*10^4 x + 0.7*10^4$	0.0580-3.055	0.9990	$(812.5\pm 9.1)*10^4$	$(0.7 \pm 12.8)*10^4$
12	$y=872.1*10^4 x-8.9*10^4$	0.037-1.480	0.9999	$(872.1\pm 4.0)*10^4$	$(-8.9 \pm 2.1))*10^4$
13	$y=583.4*10^4 x-14.4*10^4$	0.330-3.2080	0.9990	$(583.4 \pm 5.6)*10^4$	$(-14.4 \pm 7.8)*10^4$
14	$y=1333*10^4 x-18.6*10^4$	0.045-2.345	0.9998	$(1333\pm 7.1)*10^4$	$(-18.6 \pm 7.7)*10^4$
15	$y=579.6*10^4 x-12.2*10^4$	0.029-1.170	0.9983	$(579.6\pm 9.0)*10^4$	$(-12.2\pm 3.7)*10^4$
16	$y=377.0*10^4 x-57.2*10^4$	0.076 – 3.9720	0.9931	$(377.0\pm 10.0)*10^4$	$(-57.2\pm 17.9)*10^4$
17	$y=723.4*10^4 x-19.0*10^4$	0.033-3.210	0.9989	$(723.4 \pm 7.4)*10^4$	$(-19.0\pm10.2)*10^4$
18	$y=672.0*10^4 x-22.3*10^4$	0.033-3.210	0.9985	$(672.0 \pm 7.9)*10^4$	$(-22.3\pm10.9)*10^4$
19	$y=928.2*10^4 x -9.3*10^4$	0.0690-3.610	0.9998	$(928.2 \pm 4.7)*10^4$	$(-9.3 \pm 7.9)*10^4$
20	$y=484.0*10^4 x-30.5*10^4$	0.054 – 2.835	0.9984	$(484.0 \pm 6.0) *10^4$	$(-30.5 \pm 7.8)*10^4$

3.3.3 Limits of Detection. The Limits of Detection and Quantification were determined (Table 2). The Limit of Detection (LOD) is expressed by the concentration of the analyte that produces a chromatographic peak having a height equal to three times the standard deviation (SD) of the baseline noise of the blank sample. For the calculation the J.E.Knoll equation was used (LOD= $K_{LOD} . h_n.C_S/h_s$).

3.4 Internal Quality Control

The techniques which are used for the internal quality control are the followings:

3.4.1 For the control of the interferences the blank analysis is used.

3.4.2 For the control of the reliability and good performance of the chromatographic system for routine analyses, quality control samples are used which are prepared weekly from standards of high accuracy. The construction of the control charts was based upon the data obtained over 15 days for peak area and concentration values. Each day of the study (n=15) the same solution (QCS) was injected three times.

3.4.3 For the routine analyses, the sequence of standards and samples is injected in the Gas Chromatograph. The calibrations factors (linearity,intercept, R^2), Retention Time RT, and the window of RT are calculated before the analysis of the samples. The criteria for accepted calibration should be as those setting in section 3.1.3.

The analysis of the samples doesn't proceed if the criteria of the linearity, quality control and blank test are not accomplished. In this case, the system is inspected for problems and corrective actions are taken to achieve the acceptance criteria.

3. 5 External Quality Control

The Laboratory has successfully participated in the appropriate proficiency testing schemes such as, AQUACHECK, CHECK, FAPAS, SMT, International Olive Oil Council.

3.6 Conclusions

The GC-FPD-NPD generic system was found suitable for routine analysis of organophosphorus pesticide residues in various sample extracts while method performance characteristics are in compliance with the acceptability criteria.

This method is UKAS accredited.

References

1. ISO 5725-Parts 1,2,3,4,6(1994) "Accuracy, (trueness and precision) of measurement methods and results.
2. UNITED KINGDOM ACCREDITATION SERVICE (UKAS), "Analytical Quality Assurance for a Trace Organics Laboratory" November 1997.
3. EPA Contract Laboratory Program, Statement of Work for Organic Analysis. Multi-Media, Multi- Concentration Section III. GC/EC Analysis of Pesticides.
4. Method "R": Extraction with ethyl acetate of organophoshorus compounds... CEN prEN 12393-2:1996 p.41-47.
5. Analytical Methods for Pesticide Residues in Foodstuffs 6th Edition 1996, Multi-Residue Method 1, Ministry of Public Health and Welfare and Sport, The Netherlands Part1 p.30
6. Hiskia A.E., Atmajidou M.E. and Tsipi D.F, J.Agric. Food Chem., 46, 570-574 (1998).
7. EPA Manual of Analytical Methods for the Analysis of Pesticides in Human and Environmental Samples 1980 p.431-456.

Validation of Gas Chromatographic Databases for Qualitative Identification of Active Ingredients of Pesticide Residues

János Lantos[1], Lajos Kadencki[2], Ferenc Zakar[3] and Árpád Ambrus[4]

[1] PLANT HEALTH AND SOIL CONSERVATION STATION OF SZABOLCS-SZATMÁR-BEREG COUNTY, NYÍREGYHÁZA, PF. 124, H-4401. HUNGARY
[2] PLANT HEALTH AND SOIL CONSERVATION STATION OF BORSOD-ABAUJ-ZEMPLÉN COUNTY, MISKOLC, PF. 197, H-3501, HUNGARY
[3] PLANT HEALTH AND SOIL CONSERVATION STATION OF VAS COUNTY, TANAKAJD, AMBRÓZI SÉTÁNY 2, H-9762, HUNGARY
[4] FAO/IAEA TRAINING AND REFERENCE CENTRE FOR FOOD AND PESTICIDE CONTROL, FAO/IAEA AGRICULTURE AND BIOTECHNOLOGY LABORATORY, A-2444 SEIBERSDORF, AUSTRIA

1 INTRODUCTION

Pesticide residue analysis of samples of unknown origin requires the identification of a number of pesticides potentially present in the sample extract. Multi residue methods frequently apply gas chromatographic databases for that purpose. The database of Pesticide Analytical Manual comprises relative retention data of more than 700 substances on four packed chromatographic columns of different polarities (1). Unfortunately, the tabulated relative retention data are valid only for the specified packing materials working under isotherm conditions. Any set of analytical conditions requires the establishment of updated databases for the correct identification of active ingredients. The elution order of analytes does not change very much when similar conditions are used. In this respect the relative retention data is very advantageous for analytical chemists.

The application of capillary columns has become widespread in pesticide residue analysis. Analytical laboratories frequently apply narrow bore and/or wide bore columns of the same type of coating material. Published databases, like that of Stan and Linkerhägner (2) comprising more than 400 active ingredients, or that of Papadopoulou-Mourkidou et al. (3) help the identification of potential interfering materials on a specific gas chromatographic capillary column. Similar databases were established by other authors too (4). The absolute retention times depend on the analytical conditions used e.g., manufacturer and size of column, type of injector, temperature program, flow rate of carrier gas etc., and they are not directly applicable in another laboratory for identification purposes. Even small changes in analytical conditions may substantially alter the retention time.

As many laboratories do not have the possibility of mass spectrometric detection, information on the correct retention time and the knowledge of potentially interfering active ingredients are of utmost importance. Preliminary data of Zakar and Lantos measured on dimethyl-siloxane phases (5) showed that relative retention data, determined on different capillary columns of different manufacturer, may be correlated. If such a correlation could be established, the transformation of an existing retention database to another analytical circumstance (e.g. different columns and column temperature programmes) or another laboratory would be possible. It would significantly reduce the time (number of analytical standards injected) required to establish the appropriate database of the laboratory based on published results. Moreover, it would enable the laboratory to adjust its own retention data to the slightly, but continuously, changing column performance.

2 OBJECTIVES

The objectives of this study were to establish a technique for the determination of the mathematical function between two sets of gas chromatographic retention data of active ingredients of pesticides; and to determine the retention time window necessary for the tentative identification of analytes using this equation. The procedure should be applicable for validation of gas chromatographic databases. The study was restricted to apolar coating materials of chemically bonded capillary columns.

3 METHOD

Retention data of pesticides on narrow and wide bore capillary columns were compared for 100 % dimethyl-polysiloxane (CP-Sil 5 CB and OV-1) as well as for (95 % dimethyl + 5 % biphenyl) polysiloxane coating materials (HP-5, DB-5 or CP-Sil 8 CB). Both liquid phases are widely used for multi residue analysis. Data were taken from databases of 3 laboratories of the Hungarian Pesticide Analytical Network, and from the literature (2, 3). The number of individual data points was between 29 and 400. Original data were given in absolute retention time, or in relative retention (Chlorpyrifos=1.00). Temperature programmes for the injector and oven depended on the instrumental conditions, like type of injector, length of columns etc. References and important chromatographic conditions used are given in *Table 1*. The last two lines in this table refer to the 100 % dimethyl-polysiloxane phases, while the others to the 5% biphenyl phases.

The actual retention data in two independent databases (t_1 and t_2) were tabulated, the linear, the power and the polynomial functions providing the best fit between the two data sets and its correlation coefficient were determined using Microsoft Excel®. The function with the best correlation coefficient was selected for additional evaluation. Applying the function describing the relationship, the expected retention data of compounds in the second data base (Rt_{2i}) were calculated from the corresponding retention value (t_{1i}) of the first retention data set, and the calculated retention data were compared with the actually measured ones (t_{2i}). The relative deviations of the calculated and actually measured retention values were calculated. Basic equations used for these calculations are given in *Table 2*. When the correlation coefficient was less than 0.99, data points with large relative deviation were removed and the regression analysis repeated.

An example of calculation procedure is shown in *Table 3*. The original databases contained 84 corresponding data pairs. The cells of x and y columns were filled with the numeric values taken from Stan and Kadenczki (databases 1 and 5). Polynomial parameters and r^2 were determined by the Excel programme. Other numeric data (columns of Rt_2, D%, $Rt_{2,min}$ and $Rt_{2,max}$ in Table 3) were calculated using the equations given in *Table 2*. The last 3 data pairs were removed from the data set in order to improve the correlation coefficient. Numeric data underlined are outside the w_2% retention time window (w_2%= 9.27%)

Table 1 *Analytical conditions of the retention data bases evaluated*

Reference	Capillary column	Injection system	Oven temperature	Number of ai measured	Data set in Table 4
Stan and Linkerhägner[1]	HP-5 25m*0.32mm* 0.17 μm	PTV	50 °C 2 min, 30°C/min to150 °C, 3 °C/min to 205 °C, 10 °C/min to 250 °C	more than 400	1
Papadopoulou[2]	DB-5 MS 29m * 0.2mm *0.25 μm + guard column	split, 230 °C	80 °C 1 min, 6°C/min to200 °C, 3 min, 6 °C/min to 260 °C, 8min	95	2
Zakar 1	DB-5.625 30m*0.245 mm*0.25 μm	PTV	120 °C 1min, 5 °C/min to 250 °C 5min	33	3
Kadencki	CP-Sil 8 CB 30m*0.32mm* 0.25 μm	splitless, 220°C	70 °C 2min, 20 °C/min to 160 °C, 4 °C/min to 250 °C	72-81	5
Lantos 1	CP-Sil 8 CB 30m*0.32mm*0.25 μm	cold on-column, 70 °C	70 °C 1 min, 15 °C/m to 250 °C 21min	21-89	6
Lantos 2	CP-Sil 8 CB 24.9m*0.53mm*1 μm	wide bore, 220 °C	140 °C 1 min, 15 °C/min to 250 °C 40min	21	7
Lantos 3	CP-Sil 8 CB 25m*0.53mm*1 μm	wide bore, 220 °C	140 °C 1 min, 15 °C/min to 250 °C 40min	21	8
Lantos 4	CP-Sil 5 CB 10 m* 0.25mm * 0.25 μm	split, 200 °C	120 °C 1 min, 10 °C/min to 250 °C 8min	120	9
Zakar 2	OV-1[a] 25m * 0.25mm * 0.25 μm	Split250 °C	60 °C 1min, 25 °C/min to 250 °C 26min	150	4

Note: (a) Ohio Valley

The function was calculated based on the original form i.e. in absolute retention times or in relative retention. When it was possible, estimated relative retention values were also calculated and the corresponding correlation coefficients determined. The results of the comparison are also given in *Table 4*. Examples of graphical presentation of corresponding data points obtained are given in *Figures 1-4*.

The number and percentage of data pairs of pesticides within 3%, w_1% and w_2% retention time window was determined.

The required number of data points to establish the mathematical function was tested by selecting a specified number of compounds, covering the retention range, from one of the two comparable databases. The number of data pairs was gradually decreased and the correlation coefficient of the new mathematical function determined.

Table 2   *Definitions and equations used*

Definitions
n: number of compounds matching in data sets
t_1 : measured retention time in first data set   (absolute or relative)
t_2: measured retention time in second data set (absolute or relative)
Rt_2: calculated retention time in second data set
Linear equation
$y = ax + b$ or $Rt_2 = a*t_1 + b$
Polynomial equation (the actual quotient of power depends on the databases)
$y = ax^4 + bx^3 + cx^2 + dx + e$ or $Rt_2 = a*(t_1)^4 + b*(t_1)^3 + c*(t_1)^2 + d*(t_1) + e$
Correlation coefficient between two sets of data: r^2
Relative deviation of measured and calculated retention data
$D = Abs((t_2-Rt_2)/t_2)$ $D\% = Abs(100 * (t_2-Rt_2)/t_2)$
Mean of relative deviations
$D_{mean} =: (\Sigma(D))/n$ $D\%_{mean} =: (\Sigma(D\%))/n$
SD_D: standard deviation of D
$SD_D\%$: standard deviation of D%
Retention time window %:
$w_1 = D_{mean} + SD_D$ $w_2 = D_{mean} + 2 * SD_D$
$w_1\% = D\%_{mean} + SD_{D\%}$ $w_2\% = D\%_{mean} + 2 * SD_{D\%}$
Range of retention time window required based on function determined:
$Rt_2(1-w_2) < Rt_2 < Rt_2(1+w_2)$

4   RESULTS

Data of analysis show that analytical capillary columns containing similar composition of functional groups result in similar separation sequence of pesticide active ingredients. The mathematical function providing the best correlation coefficient between two sets of retention data depends on the analytical conditions used. The correlation coefficient of mathematical function and the mean relative deviation of data points depend on the "quality" of databases. The validity of the function is limited to the range evaluated. In order to establish the mathematical function, either the original absolute retention time or the calculated relative retention times may be used, the value of correlation coefficient is the same (e.g., *Table 4*, databases 1 and 6).

In some instances the corresponding data points did not fit in the trend of other compounds. This irregularity was observed in case of multi-peak compounds such as pyrethroids, dinocap, and thermally labile compounds like linuron, metobromuron. In some cases no obvious explanation could be found for the deviations. In such cases the identity of the analyte and the retention values of outlier data points should be checked carefully under the actual conditions of the laboratory and the confirmed value used in further measurements. If the source of outlier value is the adapted or reference data base the outlier value should be eliminated, in order to improve the accuracy of data transformation indicated by the correlation coefficient. The elimination of outliers greatly improves the correlation coefficient as can be seen in *Table 4*, databases 1 and 3; where by eliminating 3 data points the r^2 changed from 0.9533 to 0.9957.

Examples of correlation of databases including equations and correlation coefficients are demonstrated in *Figures 1, 2, 3 and 4*. Data in *Table 4* show additional details on the comparison of correlation coefficient calculated for absolute and relative retention times.

Linear correlation may be expected in case of isotherm conditions or similar analytical conditions. Even in the latter cases the polynomial regression usually results, better correlation than the linear one (0.9980 vs. 0.995 (*Table 4.*), or 0.9957 vs. 0.9896 (*Table 4.* and *Figure 1-2.*). In case of significantly different analytical conditions (Stan HP column and PTV injector and Lantos CP-Sil 8CB column and on-column injector), only the polynomial function provides an acceptable correlation coefficient (*Figure 3*). The extrapolation of data outside the established region is not recommended.

The findings in *Table 4* indicate that data of carefully established databases can be transformed with 3-5% retention time window into another closely similar conditions, like columns of the same manufacturer and similar temperature programme. Examples of this situation are given in *Figures 1-2*. Normally the correlation coefficients were better than 0.998 when within laboratory databases were used for the calculation. Two examples are given in *Figure 4* for narrow and wide bore columns using the same temperature programme. The lower dotted line with $r^2=0.9999$ was the result of measurements on two independent wide bore columns of the same size. The upper straight line with $r^2=0.9961$ comes from the data pairs of one wide bore and one narrow bore column working at similar oven temperature programme.

Beside the correlation coefficient, the mean deviation of all data points (D_{mean}) as well as the standard deviation (SD_D) of the relative deviations influence the required retention time window. The calculation of retention time window is given in Table 2. The smaller the values of D_{mean} and SD_D the smaller is the retention time window. For example in databases 1 and 2, a window of 3.5 % covers 96 % of the data points, while in databases 4 and 9, a window 10.9 % covers only 94 % of the data points.

The true retention time of a compound (Rt_{2i}) may be expected with 95% confidence to be within the retention time window:

$$Rt_{2i}(1-w_2) < Rt_{2i} < Rt_{2i}(1+w_2)$$

Data in column w_2% of *Table 4.* clearly prove this assumption, and practically all data sets fulfil this condition.

The correlation equations were repeatedly calculated based on gradually decreased number of data pairs from databases of Stan and Kadenczki. The parameters of the equations and the correlation coefficients remained nearly the same until 69 of 89 data points were removed. The lowest and largest retention values were always kept in this process. The results indicate that he retention data of approximately 20 compounds, uniformly covering the retention interval of interest, should be available under the actual chromatographic conditions intended to be used for the establishment of the relationship between the reference data base and the new retention data. (*Figure 4.*)

Table 3 *Example of calculation*

Data of x column were taken from Stan, those of y column were taken from Kadencki; but only 23 of 81 data points were selected in this example.

Polynomial parameters determined by Microsoft Excel®:
$Rt_2 = a*x^4 + b*x^3 + c*x^2 + d*x + e$
a=0.000286   b=-0.018416   c=0.409479   d=-2.705125   e=12.41518 r^2=0.9919

Data pairs:

Active ingredient	x	y	Rt_2 calc'd	D%	$Rt_{2,min}$= $Rt_2*(1-w_2)$	$Rt_{2,max}$= $Rt_2*(1+w_2)$
Methamidophos	7.18	7.998	8.0455	0.6	7.300	8.791
Dichlorvos	7.22	7.668	8.0757	5.3	7.327	8.824
EPTC	8.25	9.893	8.9521	9.5	8.123	9.782
Propachlor	12.15	14.189	13.198	7.0	11.975	14.420
Cymoxanil	12.29	14.012	13.357	4.7	12.120	14.595
Dimethoate	14.78	16.436	16.072	2.2	14.583	17.561
Simazine	15.04	16.548	16.336	1.3	14.823	17.850
Carbofuran	15.10	16.527	16.396	0.8	14.877	17.916
Monolinuron	15.19	16.428	16.486	0.4	14.959	18.014
Atrazine	15.26	16.592	16.556	0.2	15.022	18.090
Fenithrothion	20.41	21.115	20.833	1.3	18.903	22.763
Pirimiphos-methyl	20.59	21.113	20.963	0.7	19.021	22.905
Dichlofluanid	20.68	21.624	21.028	2.8	19.080	22.976
Malathion	21.09	21.486	21.324	0.8	19.348	23.300
Penconazole	23.51	24.016	23.212	3.3	21.061	25.363
Dinocap-III	32.15	39.021	42.266	8.3	38.350	46.182
Furathiocarb	32.94	44.073	46.112	4.6	41.840	50.384
Dinocap-V	33.09	45.019	46.904	4.2	42.558	51.250
Azinphos-methyl	33.11	44.834	47.011	4.9	42.656	51.367
Phosalone	33.12	44.735	47.065	5.2	42.504	51.426
Dialifor	34.93	52.787	58.430	**10.7**	53.017	63.844
Bitertanol-I	35.75	56.291	64.771	**15.1**	58.770	70.772
Bitertanol-II	36.00	58.782	66.869	**13.8**	60.673	73.064

$D\%_{mean}$	3.1
$SD_{D\%}$	3.1
$w_2\% = D\% + 2* SD_{D\%}$	9.3
percentage of data within 3%	59 %
percentage of data within w_1%	88 %
percentage of data within w_2%	95 %

3 additional data points were deleted from the original 84 data pairs due to their large D%:

Hexaconazole	26.19	21.814	26.166	**20.0**
Fenarimol	34.26	42.585	53.825	**26.4**
Benalaxyl	29.64	24.426	33.168	**35.8**

Note: underlined values are outside the w_2% retention time window

Table 4 *Characteristics of regression analyses*

Databases		Regression analysis		Time base	Correlation coefficient	No. of data pairs	Deviations of data points from calculated regression line		window required	Percentage of data point within the range specified		
t1	t2	linear	polynomial	t1-t2	r^2	n	$D\%_{mean}$	$SD_{D\%}$	w_2 %	3%	w_1 %	w_2 %
1	2		+	Rt-Rt	0.9980	83	1.1	1.2	3.5	96		
1	2	+		Rt-Rt	0.9955	83	1.3	1	3.3	99	88	98
1	3		+	RRT-Rt	0.9957	29	1.7	1.66	5.0	90	90	97
2	1	+		RRt-RRt	0.9976	83	1.9	1.4	4.7		94	99
1	6		+	Rt-Rt RRt-RRt	0.9968	40	2.3	1.6	5.5	67	85	97
1	5		+	Rt-Rt	0.9846	84	2.8	4	10.8	67	94	95
1	5		+	Rt-Rt	0.9919	81	3.1	3.1	9.3	59	89	95
4	9		+	RRt-RRt	0.9931	109	3.18	3.85	10.9	64	89	94
1	3	+		Rt-Rt RRt-RRt	0.9896	29	5.88	2.92	11.7	17	79	96
1	3		+	Rt-Rrt	0.9533	32	17.8	23.4	64.6	19		

Details of analytical conditions used for establishing the databases of 1-9 are given in Table 1. Definitions for $D\%_{mean}$, $SD_{D\%}$, $w_1\%$ and w_2 % are given in *Table 2*

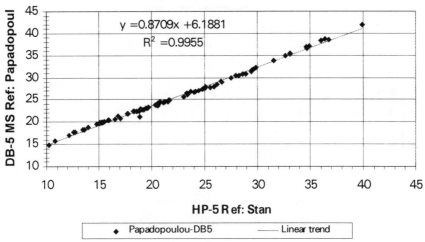

Figure 1

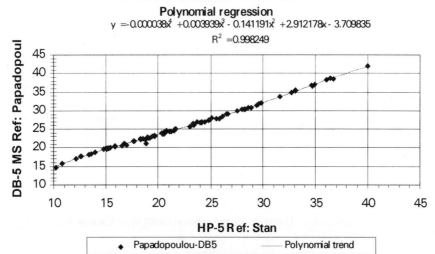

Figure 2

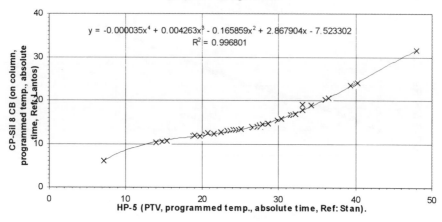

Figure 3

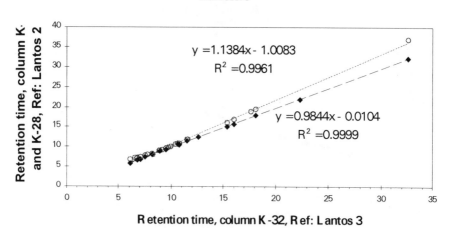

Figure 4

5 CONCLUSIONS

Regression analysis of retention data provides a valuable tool for the application of an external or internal retention database for different analytical conditions. In order to establish the correlation between two retention databases reflecting different analytical conditions, a minimum of 20-25 active ingredients, approximately evenly covering the retention time range of interest, should be selected from the reference database. The selected compounds should be injected (separately or in mixtures) using the new analytical conditions. For establishing the new database, the reference retention data should be taken as x values and the correlation equation determined. The expected retention data for the new condition can then be calculated for the other compounds with the parameters of the correlation equation.

Linear correlation may be expected in the case of isotherm conditions or closely similar column temperature programmes. The polynomial regression usually results in better correlation than the linear one. The function providing the best correlation coefficient depends on the analytical conditions used.

The transformed data are suitable for tentative identification of pesticide residues taking into account the retention time window. It is emphasised that the identification and qualitative confirmation of suspected analytes should be carried out with analytical standards. The combination of reference database (e.g., published by Stan and Linkerhagner) and the transformation technique presented here may be used to establish a tentative local database, without injecting several hundred pesticides, instead using the transformed data of reference database.

The principle may also be applied for the identification of active ingredient of an unknown pesticide formulation.

References

1. US FDA: Pesticide Analytical Manual, Vol. 1 Appendix 1.
2. H. J. Stan and M. Linkerhägner, Multimethod applying GC-AED to water samples, in: H. J. Stan (Ed.) Analysis of Pesticides in Ground and Surface Water II, Springer, (Series of Plant Protection Vol. 12), pp 71.
3. E. Papadopoulou-Mourkidou, J.Patsias, A. Kotopoulou, Determination of Pesticides in Soils by Gas Chromatography-Ion Trap Spectrometry, J. A.O.A.C International, Vol. 80, No 2, 1997 pp. 447-454
4. L. Kadencki , A. Ambrus, I. Korsos, Application of the combination of relative response factors and relative retention times for identification of pesticide residues, Poster No. 513, 8.th IUPAC Congress of Pesticide Chemistry (IUPAC), Washington, 1994.
5. J. Lantos and F. Zakar, unpublished research data 1993-1995.

Acknowledgement

The technical assistance of Iren Gyarmati, Ibolya Korsós and Shahram Mashayekhi, in the establishment of local databases is highly appreciated.

Estimation of Significance of 'Matrix-induced' Chromatographic Effects

E. Soboleva*, N. Rathor, A. Mageto and Á. Ambrus

FAO/IAEA TRAINING AND REFERENCE CENTRE FOR FOOD AND PESTICIDE CONTROL, FAO/IAEA AGRICULTURAL AND BIOTECHNOLOGY LABORATORY, A-2444 SEIBERSDORF, AUSTRIA
e-mail: E.Soboleva@iaea.org

1 INTRODUCTION

"Matrix-induced" chromatographic response effects have been widely reported for both fat-containing and plant product extracts analysed with gas chromatography[1-5,8,9]. Analyte peak enhancement can be observed when matrix components, or other pesticides in a complex mixture, compete with analytes for active sites on the injector parts (mostly glass liner) and protect susceptible pesticides from adsorption or decomposition. Response enhancement up to 1000 % was reported for some pesticides[1]. Consequently, inaccurate false positive results can be obtained if calibration solutions in pure solvent are used.

Another problem, opposite to that described above, is the "matrix-induced" response diminishing effect. Polar active sites, originated from non-volatile matrix components that have accumulated in the inlet or in the front section of a capillary column from matrix deposit, were found to be responsible for adsorption and decomposition of analytes in the injector port. Though the matrix enhancement effect was investigated and reported more often than the diminishing effect, the latter might cause even more serious problems during analysis since it leads to wrongly reported low values[2].

It was shown that the pesticides with higher polarity are more liable to matrix effect. Stronger effect was observed for compounds with multiple P=O bounds, rather than single P=O or P=S bonds, so as for pesticides containing amine and amide groups, and thermolabile compounds[2,3].

Contradictory results were reported concerning the parameters that influence matrix response enhancement, such as nature and amount of matrix[4], structure of pesticide, matrix and analyte concentration, geometry of the GC system (injector port), surface activity of liners, and types of detector used. For instance, an increase of matrix enhancement effect for lower concentrations of analytes, as well as a lack of differences or even lower matrix enhancement effect for lower concentration for certain pesticides, were described[2,3]. Some researchers reported the matrix effect to be independent of commodity[6]. As a result, extract

---

* Author correspondence to be addressed.

of one vegetable found to be free of target analytes can be used for preparation of matrix matched solution for analyses of different commodities. Others observed that the extent of matrix effect is related to both the type of matrix and its concentration. In this case, the use of the matrix matched standard solutions, prepared from one vegetable to quantify residues in the extracts of other vegetables, is most likely a wrong practice.

Sensitivity of different GC systems was reported to be different for matrix enhancement effect for the same extract, because of the different types of injector port, detectors, and column diameter used[3], that leads to the incomparability of the results even within a laboratory if at least two different GC systems are used.

The most common approach to eliminate systematic errors arising from matrix induced chromatographic response enhancement is to use matrix matched solution for calibration[4,5]. It was shown that their use provides not only more accurate results but also leads to better linear fit for the calibration plot with the intercept closer to the origin than that in neat solvent. Unfortunately, preparation of standard matched solutions is time-consuming and it depends on the availability of an analyte-free matrix.

Another possibility to eliminate matrix effect is to "contaminate" the GC system before use by several injections of uncleaned samples. This method is based on the approach that the number of active sites on liner surface is limited, therefore once covered they should not affect the degree of analyte transfer to the column. On the other hand the number of active sites can increase as the liner ages, causing changes in the analyte response. Moreover, continuous system contamination, as mentioned above, leads to a iminishing matrix effect, peak shape distortion, and pure separation.

The standard addition method, suggested as a possible way of eliminating the matrix effect, might not work if the matrix effect strongly depends on analyte concentration.

The most straightforward way of matrix effect reduction is the application of an extensive clean-up procedure. Unfortunately, even very thorough clean-up involving several SPE cartridges was shown to be insufficient for eliminating the matrix enhancement effect completely, but it increases the analysis time and costs significantly[3]. Besides that, use of cartridges is often not practical for multiresidue procedure because of high selectivity causing possible loss of analytes. Limited availability of certain cartridges in many laboratories and the necessity to validate additional clean-up procedures make this approach impractical.

Different types of GC injection techniques were shown to have a strong influence on the degree of matrix effect. It was reported to be significantly less with on-column injection compared to temperature programmed vaporisation, but the latter was better than hot-splitless injection. Despite being the best technique for pesticide analysis cool on-column injection was reported to be limited only for relatively clean samples since the column was contaminated in a short time[1]. Pulse splitless injection, allowing introduction of a large sample amount with high column flow, was reported to eliminate matrix induced effect[1]. The disadvantage of using this method is a rapid decrease of chromatographic resolution caused by carrying non-volatile matrix materials far into the column, and the limitation for GC/MS system that can not handle very high column flows. Moreover it is not available in many laboratories equipped with old GCs.

Contradictions reported concerning the matrix effect might be a source of serious problems in achieving comparability of laboratory data. Since it strongly depends on purity of extracts, controversial results are generated by laboratories employing different either extraction or clean-up methods. Even if the same clean-up procedure is used, e.g. GPC, the matrix effect will depend on the calibration and conditions of a particular GPC column and may also result in a different extent of matrix effect even within one laboratory where two or several GPC systems are used. Performing GC analysis with different systems (company, type of column, injection technique, detector, etc.) and under different GC conditions might also cause uncertain results.

Therefore, it is necessary to establish statistically-based procedure to justify or avoid the use of matrix-matched calibration solutions.

The aims of this study were:
- to test the applicability of various statistical methods for the evaluation of matrix effect;
- to investigate the influence of matrix effect on the response of pesticides as a function of both the concentration of pesticide residues and the matrix content of the calibration solution;
- to estimate the significance of matrix effect for different types of matrices.

2 EXPERIMENT

2.1 Materials

Solvents of organic trace analysis grade obtained from Merck were used. Several pesticides, that in previous investigations were shown to be the most sensitive to matrix-induced chromatographic enhancement or diminishing were selected for this study. Individual stock solutions, as well as mixed standard solutions, were prepared in isooctane.

Representative species from three different commodity groups were chosen: lettuce, orange, and wheat grains. Selection of these commodity groups were based on a literature survey, and was aimed to represent troublesome matrices giving significant matrix effect (wheat, orange) and relatively "simple" ones (lettuce). Crops were purchased in the local supermarket. Blank extracts were tested for possible interfering compounds under the same condition as the calibration solutions were analysed.

2.2 Extraction and clean-up

Extraction was performed with ethylacetate followed by clean-up with GPC columns according to the EU Standard procedure[7]. Final fractions were combined and evaporated under a gentle nitrogen stream to obtain 10 g of sample equivalent per 1 ml of solvent.

2.3 Preparation of the calibration solutions

Calibration solutions covering a wide range of concentration in neat isooctane, and solvent containing different matrix equivalent, were prepared and analysed with two different GCs equipped with nitrogen/phosphorous detector (NPD) and electron capture detector (ECD) to estimate the matrix effect for different analyte concentrations covering the "working" range of these detectors.

In previous investigations, matrix effect was reported to be significant from 0.1 g matrix equivalent per ml and higher. Therefore, calibration solutions of pesticides were prepared in iso-octane and matrix extract with sample equivalent of 2.5, 1.25, and 0.83

g/ml for NPD and 0.25, 0.12, and 0.08 g/ml for ECD, that represent X, X/2, and X/3, where X is a sample equivalent in purified final extract.

Previously the matrix effect was studied mostly at two concentration levels with relatively narrow range (20-200 pg/μl or 100-500 pg/μl). Calibration plots based on 7 concentration levels in the range 10 to 1000 ng/ml (up to 3000 for compounds with lower response) and 5-200 (up to 400 for compounds with lower response) ng/ml for NPD and ECD, respectively, were used in our study. Ranges of the analyte concentrations in the calibration solutions are given in Table 1. Therefore, working solutions represented analyte per matrix concentration in a range approximately 0.02-1.2 (up to 3.6) mg/kg for ECD and 0.003-1 mg/kg for NPD.

Table 1 *Pesticides concentration in the calibration solutions*

Calibration levels		c1	c2	c3	c4	c5	c	c7
Concentration	ECD	4-18	16-73	40-182	80-364	100-455	140-636	200-909
range, ng/ml	NPD	9-24	35-96	173-480	345-960	518-1440	777-2159	1035-2879

2.4 Equipment

Two GC systems, HP 5890 Series II with ECD, and Varian 3800 with NPD, were equipped with capillary columns CP-SIL 8 CB, 25 m / 0.32 mm ID, 0.25 DF. Deactivated glass line with 4 mm ID was installed on the HP GC. High performance deactivated liner with 0.5 mm ID was used with the Varian GC for lettuce experiments and 2 mm ID liner was installed for other commodities. Hot splitless injectors with constant temperature of 270°C were used in both cases, with split valve opened 2 min after injection. Constant 2 ml/min flow and constant head pressure 12 psi (2.6 ml/min flow rate at 70°C) were used with the Varian GC and HP GC, respectively.

The oven temperature program was 70°C held for 1 min, 20°C/min until 160°C, and 4°C/min until 270°C held for 10 min. NPD detector was optimised for highest phosphorus response. Bead current was 3.025 A, temperature 320°C, make up gas flow 28 ml/min, H_2 flow 5.2 ml/min, Air flow 175 ml/min.

Peak areas were used for quantification of ECD results and peak height were used for NPD detected compounds because some of them exhibited tailing.
Sequence of injections is presented in Table 2.

Table 2 *Injection sequence.*

Number of injection	Sample description
1	isooctane
2	matrix blank 0.8 g/ml matrix equivalent
3	matrix blank 1.2 g/ml matrix equivalent
4	matrix blank 2.5 g/ml matrix equivalent
5	1st calibration point in isooctane, 1st replicate
6	1st calibration point in 2.5* g/ml matrix equivalent solution, 1st replicate
7	1st calibration point in 1.25* g/ml matrix equivalent solution, 1st replicate
8	1st calibration point in 0.8* g/ml matrix equivalent solution, 1st replicate
9	1st calibration point in iso-octane, 2nd replicate
10	1st calibration point in 2.5* g/ml matrix equivalent solution, 2nd replicate
11	1st calibration point in 1.25* g/ml matrix equivalent solution, 2nd replicate

Number of injection	Sample description
12	1st calibration point in 0.8* g/ml matrix equivalent solution, 2nd replicate
etc. All 7 calibration levels were injected sequentially, 3 replicates of each solution were made.	

* - 10-folds less matrix equivalent was used with ECD.

2.5 Estimation of matrix effect

The paired t-test and linear regression[10] were applied to estimate if the responses of analyte differed significantly in neat isooctane solutions and matrix matched extracts, with different amount of matrix equivalent. Differences of analyte responses in matrix matched solutions, with different amount of matrix equivalent, were also compared to each other to estimate if 2 or 3-fold diluted matrix solutions could be used instead.

2.5.1 Pared T-test. The paired t-test was performed with the average values of three replicate injections. A disadvantage of using t-test in this case is its application over a wide concentration range; calibration solutions with 60-fold and 100-fold differences in concentrations between the first and the last calibration points for ECD and NPD were used, respectively (Table 1). There is a great limitation of the t-test application to our results because validity of the t-test is based on the hypothesis that the random or systematic errors are independent of concentration[10], while experimental results indicate that matrix-induced effect might depend on concentration of analyte. In order to make application of t-test more justifiable, it was also applied for a narrower concentration range including the first and last four calibration points.

2.5.2 Linear regression. As an alternative to the t-test, linear regression was used to compare responses of compound in isooctane and matrix matched solution. X-axis of a regression graph was used for the responses of analyte in solvent, and the y-axis for the responses in matrix matched solutions. When responses in matrix matched solution containing different amount of matrix equivalent were compared with each other, responses in more diluted solutions were assigned to x-axis.

Slope, intercept, and their standard deviations were calculated with Excel programme to establish the confidence limit for slope and intercept at 95 % significant level. Significance of intercept and slope differences from zero and one, respectively, was estimated. If 1 was included in the calculated confidence interval of the slope and 0 was included in the calculated confidence interval of the intercept, then it was concluded that the slope and intercept did not differ significantly from the ideal regression line values of 1 and 0 and that there were no systematic differences between responses of two sets of calibration solutions.

Although the linear regression method can be applied for a wide range of concentrations, there are still several limitations why it might not be totally justifiable to use it for the estimation of matrix effect. First, regression line of y on x is calculated on the assumption that errors in the x values are negligible. As it was mentioned before responses of analyte in pure solvent were assigned to x-axis as more precise values, though random errors are inevitable in this case. Second, the linear regression works well when ≥ 10 data points equidistantly covering the analytical range of interest are used for the comparison. Therefore the analyte responses obtained from 3-5 calibration levels may not provide suitable information for testing the matrix effect with linear regression, even if a few replicate injections are made.

2.5.3 Relative differences. Relative difference (RD) between responses in pure solvent and matrix matched solutions were calculated as:

$$RD = \frac{y - x}{A}, \quad (1)$$

where x is a response in pure solvent, y is a response in matrix matched solution, and A is average of two responses.

Relative differences were plotted against concentrations to observe any trend of matrix effect along the concentration range.

2.5.4 Percentage of matrix effect. Percentage of matrix effect was calculated by dividing the analyte response in matrix matched solution by the response in neat solvent and multiplying on 100%.

Neither relative difference nor percentage of matrix effect can be used to estimate statistical significance of matrix effect but they provide clear visual information.

3 RESULTS AND DISCUSSION

3.1 T-test and linear regression

Table 3 and 4 summarise the results calculated by t-test and linear regressions for calibration solutions prepared in orange and wheat matrix extracts and measured with NPD and ECD, respectively. Results of t-test and regression calculated for other matrices are not presented here since they were similar to those for orange and wheat extracts. Responses of analytes in isooctane were compared with responses of analytes in matrix matched solutions with different concentration of matrix equivalent: 2.5 g/ml, 1.25 g/ml , and 0.8 g/ml. The differences between the responses in matrix matched solutions with 0.8 and 1.25 g/ml were compared to those with 2.5 g/ml matrix equivalent. "Plus" indicate that differences were shown to be significant. Results calculated for 7 calibration points (7clp) are listed in the first line for each compound. Results for the first and the last four calibration points of calibration plot are presented in the second and the third line for each compound.

Comparing results of linear regression and t-test, one can notice that the two tests do not give the same estimation of matrix effect for the same set of results, and that regression seems more "sensitive" to the differences. Limitations for the applications of these tests described in section 2.5 can be the reason of this inconsistency.

Calibration plots for several analytes based on 7 and 4 calibration points (Fig. 1-8) illustrate the applicability of t-test and linear regression.

A diminishing matrix effect can be very well observed from Dimethoate calibration plots (Fig. 1-3). Linear regression applied to these results also shows significant matrix effect (slope confidence limits do not include 1 and they are less than 1) for matrix matched solution with different concentration of orange matrix equivalent (Table 3). T-test applied for 7 calibration points and last four calibration points also indicate significant matrix effect (calculated t value is larger than the critical one). However, the t-test does not indicate significance of matrix effect for the first four calibration levels. Referring to the Table 5 where percents of matrix effect in orange are presented, one can notice slight enhancement effect on two first calibration levels for Dimethoate in solutions with 1.2 and 0.8 g/ml matrix equivalent. In this case t-test might not indicate the differences since differences

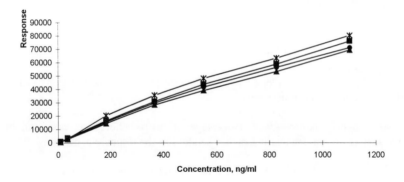

Figure 1 *Dimethoate calibration plot based on 7 calibration levels in iso-octane and orange matrix-matched solutions analysed with NPD*

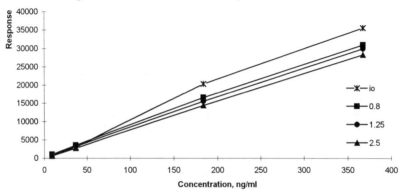

Figure 2 *Dimethoate calibration plot based on the first 4 calibration levels in iso-octane and orange matrix-matched solutions analysed with NPD*

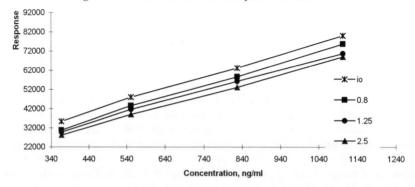

Figure 3 *Dimethoate calibration plot based on the last 4 calibration levels in iso-octane and orange matrix-matched solutions analysed with NPD*

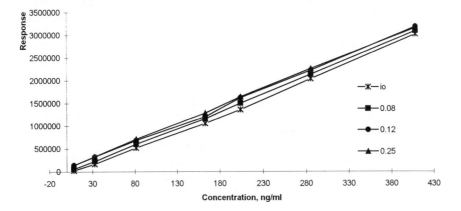

Figure 4 *Chlorothalonil calibration plot based on 7 calibration points in iso-octane and wheat matrix matched solutions analysed with ECD*

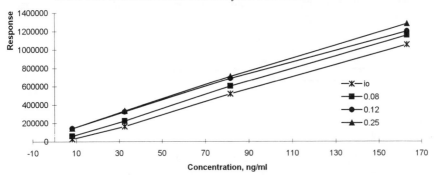

Figure 5 *Chlorothalonil calibration plot based on the first 4 calibration points in iso-octane and wheat matrix matched solutions analysed with ECD*

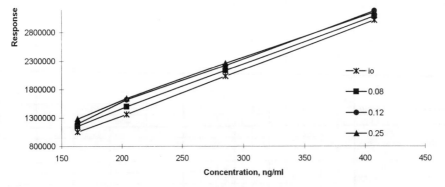

Figure 6 *Chlorothalonil calibration plot based on the last 4 calibration points in iso-octane and wheat matrix matched solutions analysed with ECD*

between enhanced and diminished responses eliminate each other giving very low average and consequently low t-value. On the other hand there is no indication of significant differences between responses in isooctane and matrix matched solution with 2.5 g/ml matrix equivalent, although the diminishing effect was observed on all calibration levels. In this case, the calculated t-value is low because SD of response differences is very high. It can be noticed that in many cases when differences in slopes can be visually observed, t-test does not indicate significant difference for the same reason. However, when matrix effect is more or less uniform along the calibration plot as in the case of Dimethoate plot based on the last 4 calibration points (Fig. 3), Chlorothalonil (Fig. 4-6, Table 4), or Fenvalerate (Fig. 8, Table 4) it is indicated by the t-test.

Therefore, t-test might fail to estimate significance of matrix effect since it is not sensitive to changes in the slope and can not indicate differences if type of matrix effect changes with concentration from enhancement to diminishing.

Calibration plots of Cypermethrin analysed with NPD in orange matched extracts are shown on Fig. 7. There is a notable diminishing effect in both solutions containing 2.5 and 1.2 g/ml matrix equivalent as it can be seen from the figure 7 and results presented in Table 5. However, neither t-test nor linear regression reflected significant differences between plots based on the first and the last four calibration points where about 70% matrix diminishing effect was observed.

Table 3 *Results calculated using regression line and t-test for orange matrix matched calibration solutions measured with NPD.*

Compound		Regression										T-test				
		io/2.5 a*	b	io/1.2 a*	b	io/0.8 a*	b	0.8/2.5 a*	b	1.2/2.5 a*	b	io/2.5	io/1.25	io/0.8	0.8/2.5	1.2/2.5
Acephate	7 clp **		+						+		+			+	+	+
	First 4 clp								+		+					
	Last 4 clp														+	+
Dimethoate	7 clp		+		+		+		+		+	+	+	+	+	+
	First 4 clp		+		+		+		+		+					
	Last 4 clp		+		+		+		+			+	+	+	+	+
Chloro-thalonil	7 clp	+		+		+	+		+			+	+	+		
	First 4 clp															
	Last 4 clp	+		+								+	+	+		
Vinclosolin	7 clp		+		+				+		+					
	First 4 clp															
	Last 4 clp															
Phosphamido	7 clp	+		+		+						+	+	+	+	
	First 4 clp								+		+					
	Last 4 clp	+	+				+									
Parathion Ethyl	7 clp											+	+	+		
	First 4 clp		+				+									
	Last 4 clp											+	+	+		
Procymidon	7 clp						+		+		+	+	+			
	First 4 clp				+		+							+		
	Last 4 clp													+	+	
Imazalil	7 clp	+	+	+	+	+						+	+			
	First 4 clp															
	Last 4 clp													+	+	
Iprodione	7 clp		+				+		+		+					
	First 4 clp		+		+				+							
	Last 4 clp		+						+		+					
Lcyhalothrin	7 clp		+						+						+	
	First 4 clp		+				+		+		+					
	Last 4 clp										+					

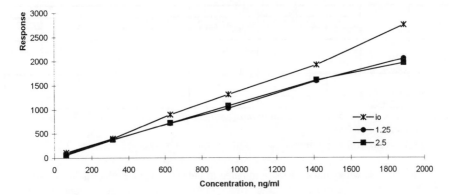

Figure 7 *Cypermethrinl calibration plot based on 7 calibration points in iso-octane and orange matrix matched solutions analysed with NPD*

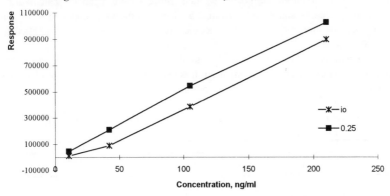

Figure 8 *Fenvalerate (Sum of two isomers) calibration plot based on the first 4 calibration points in iso-octane and wheat matrix matched solutions analysed with ECD*

Compound		Regression										T-test				
		io/2.5		io/1.2		io/0.8		0.8/2.5		1.2/2.5		io/2.5	io/1.25	io/0.8	0.8/2.5	1.2/2.5
		a*	b	a*	b	a*	b	a*	b	a*	b					
Coumaphos	7 clp	+		+		+		+			+	+	+	+	+	+
	First 4 clp	+		+		+		+								
	Last 4 clp	+									+	+		+		+
Cypermethri		+		+						+			+			
	First 4 clp															
	Last 4 clp			+										+		

Note: io - isooctane; 2.5, 1.2, 0.8 g/ml matrix equivalent;
 *a-intercept, b-slope;
 **clp - calibration points

Another case where linear regression failed to indicate a matrix effect is presented on Figure 8 for Fenvalerate in wheat extract where more than 350 and 200 % enhancement were observed on the first two calibration levels (Table 6).

As mentioned above, the application of linear regression to the results with a limited number of degrees of freedom should be avoided. Since t-value used for calculation of confidence limits increases significantly from 2.57 for 7 calibration points to 4.30 for 4 calibration points, the liner regression test becomes much less sensitive when applied to a part of calibration plot.

Therefore neither of the two statistical methods tested can be reliably used to indicate significance of matrix effect on their own, and if it is needed to be estimated both tests have to be applied to avoid false results as much as possible.

Table 4 *Results calculated using regression line and t-test for wheat matrix matched calibration solutions measured with ECD.*

Compound		Regression										T-test				
		io/0.25		io/0.12		io/0.08		0.08/0.25		0.12/0.25		io/0.25	io/0.12	io/0.08	0.08/0.25	0.12/0.25
		a*	b*	a	b	a	b	a	b	a	b					
Dichlorvos	7 clp**		+	+		+						+	+	+		
	First 4 clp	+		+								+		+		
	Last 4 clp											+	+	+		
Lindane	7 clp	+	+								+		+	+		
	First 4 clp		+									+				
	Last 4 clp							+			+			+		
Chloro-thalonil	7 clp	+	+			+		+				+	+	+	+	
	First 4 clp	+	+			+		+			+	+	+	+	+	
	Last 4 clp	+				+		+				+	+	+	+	
Vinclozolin	7 clp		+							+						
	First 4 clp							+	+		+			+		
	Last 4 clp															
Aldrin	7 clp		+·					+		+					+	
	First 4 clp								+	+						
	Last 4 clp				+											
Chlor-fenvinphos	7 clp			+		+						+	+	+		
	First 4 clp											+	+	+		
	Last 4 clp											+	+	+		
Alfa-endosulfan	7 clp	+		+				+		+						+
	First 4 clp	+				+		+		+			+			
	Last 4 clp	+	+	+												+
p,p-DDT	7 clp	+		+		+						+	+	+		+
	First 4 clp	+		+		+										
	Last 4 clp											+	+	+		+

Compound		Regression										T-test				
		io/0.25		io/0.12		io/0.08		0.08/0.25		0.12/0.25		io/0.25	io/0.12	io/0.08	0.08/0.25	0.12/0.25
		a*	b*	a	b	a	b	a	b	a	b					
Endrin	7 clp	+		+	+		+	+				+	+	+	+	
	First 4 clp	+		+			+					+	+	+	+	
	Last 4 clp											+	+	+		
Permethrin (Sum)	7 clp					+										
	First 4 clp							+								
	Last 4 clp															
Fenvalerate (Sum)	7 clp		+									+				
	First 4 clp											+				

Note: io - isooctane; 0.25, 0.12, 0.08 g/ml matrix equivalent
*a-intercept, b-slope;
**clp - calibration points

3.2 Relative differences and percentage of matrix effect

Application of the Relative Difference (RD) for estimating the matrix effect was found to be very illustrative. When plotted against concentrations in the same scale, RD can adequately indicate a type and degree of matrix effect. Several graphs presented in Figures 9-10 can be compared with percents of matrix effect presented in Tables 5 and 6 for NPD and ECD, respectively. From these figures it is evident that dramatic matrix effect either diminishing or enhancement occur on the first two calibration levels for analytes concentration from 4 to 70 ng/ml and it is much less significant for higher concentrations.

RSDs calculated for the percentages of matrix effect in matrix matched solutions with different matrix content are presented in Tables 5 and 6. RSD values above 12 % are shadowed. Results show that the amount of matrix present in the solution analysed with ECD cause a serious change on the responses of some analytes only between 4-18 ng/ml concentration (Table 6). High RSD values were observed for several compounds analysed with NPD on different calibration level. However, generally higher RSD were calculated for two first calibration levels (10-100 ng/ml).

Figures 11-12 represent RSDs calculated for the averaged percentages of matrix effect in the solutions prepared with three different matrices with 2.5 and 0.25 g/ml matrix equivalent for NPD and ECD, respectively, at 7 calibration levels. As it can be seen, the nature of the matrix is important for three lowest calibration levels in case of NPD and two lowest calibration levels in case of ECD. Therefore special care concerning preparation of matrix matched solutions for quantitative determination of analytes has to be taken if matrix content in analysed solution is high, as in the case of NPD-analysed solutions, and analyte concentration is low.

3.4 Preparation of the working calibration solutions for multi residue/multi matrix analytical method

Based on the previous conclusions, the quantitative determination of an analyte has to be performed using a matrix matched calibration solution covering narrow concentration range especially for low analyte content. We recommended the calibration range between 0.3 and 3 times accepted limit (AL)*. In order to avoid the tedious procedure of matrix matched solution preparation for all analytes of interest, preliminary screening may be done

---

* Accepted limit (AL) - maximum residue limit (MRL) or maximum permitted limit (MPL).

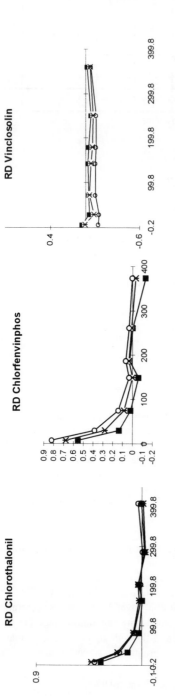

Figure 9 *Relative differences in lettuce matrix matched solutions analysed with ECD.--o-- 0.25 g/ml;--*-- 0.12 g/ml;--■-- 0.08 g/ml matrix equivalent*

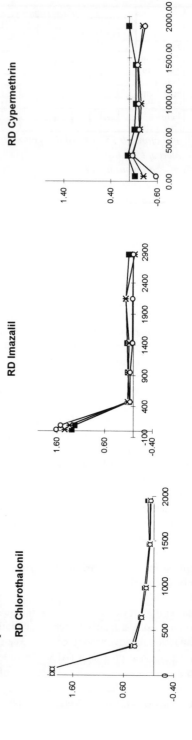

Figure 10 *Relative differences in orange matrix matched solutions analysed with NPD. --o-- 2.5 g/ml; --*-- 1.25 g/ml; --■-- 0.8 g/ml matrix equivalent.*

Table 5 *Percentages of matrix effect (response in matrix/response in iso-octane×100%) for calibration solution prepared in orange, wheat, and lettuce matrix matched extracts measured with NPD.*

Compound	matrix content	Orange c1	c2	c3	c4	c5	c6	c7	Wheat c1	c2	c3	c4	c5	c6	c7	Lettuce c1	c2	c3	c4	c5	c6	c7
Acephate	0.8	292	1464	215	156	107	117	104	532	1074	101	127	87	98	93	*	1224	48	71	71	61	62
	1.2	295	1360	191	150	91	106	91	457	*	95	122	80	91	88	*	1566	55	86	81	75	82
	2.5	204	875	160	129	78	85	85	441	822	83	106	73	83	78	*	1169	60	87	97	84	87
	RSD. %	20	28	▨	10	▨	▨	▨	10		10	9	9	8	9	▨	▨	12	11	▨	▨	▨
Dimethoate	0.8	133	117	82	87	91	93	95	135	119	93	93	99	97	97	233	112	94	92	94	96	92
	1.2	114	107	77	84	87	89	88	100	0	91	96	95	92	95	504	119	103	100	95	100	98
	2.5	77	90	71	80	81	84	86	99	106	91	92	93	92	93	329	116	105	101	103	105	103
	RSD. %	26	2	7	4	6	5	5	10		10	9	9	8	9	39	3	6	5	5	4	6
Chlorothalonil	0.8	*	*	155	129	119	110	113	132	449	116	109	112	109	104	*	776	105	108	105	104	105
	1.2	*	*	148	127	116	108	108	82	*	96	99	101	99	100	*	716	100	101	89	94	93
	2.5	*	*	145	127	115	108	107	93	357	88	109	108	107	104	*	719	101	99	95	97	97
	RSD. %			3	1	2	1	3	▨		▨	5	5	5	2		3	3	4	8	5	6
Vinclozolin	0.8	104	105	102	101	104	99	99	128	114	99	98	107	103	96	*	173	92	102	102	101	101
	1.2	104	102	96	100	101	99	97	92	*	95	100	104	101	96	*	92	90	92	82	85	85
	2.5	117	106	96	99	101	99	95	98	102	95	99	104	102	98	*	92	91	90	90	85	93
	RSD. %	7	2	3	1	1	0	2	▨		2	1	2	1	2	▨	39	1	7	11	10	8
Phosphamidone	0.8	*	2721	133	120	113	106	105	864	1185	119	110	111	107	105	*	504	98	101	100	101	103
	1.2	*	2643	130	119	110	107	100	781	*	116	109	105	103	105	*	476	96	95	86	92	92
	2.5	*	2663	127	115	108	101	99	841	1117	120	108	107	107	105	*	483	96	96	93	94	95
	RSD. %		2	3	2	1	3	4	5		2	1	3	2	0		3	1	3	8	4	6
Parathion Ethyl	0.8	116	114	112	108	113	102	106	114	108	104	101	109	103	102	106	109	106	105	106	104	103
	1.2	109	115	110	105	111	102	103	99	*	104	105	107	102	102	112	101	96	92	85	88	86
	2.5	114	117	106	108	108	102	101	102	100	103	103	107	103	101	150	100	97	93	91	91	91
	RSD. %	3	1	3	1	2	0	2	8		0	2	1	0	1	▨	5	5	8	12	9	9
Procymidon	0.8	*	110	105	105	107	106	105	91	106	104	101	111	106	100	*	49	104	110	106	101	103
	1.2	*	110	106	106	108	106	101	*	*	103	104	107	106	100	*	136	94	94	81	84	85
	2.5	*	117	104	108	107	105	99	82	97	102	103	109	105	100	*	137	91	93	88	86	89
	RSD. %		4	1	1	0	1	3	8		1	1	2	0	0		47	7	10	8	11	10
Imazalil	0.8	459	414	109	112	114	117	110	226	427	88	101	108	104	88	*	206	66	78	76	83	84
	1.2	609	502	113	111	110	118	98	127	*	82	104	91	95	82	*	353	65	66	58	67	68
	2.5	941	736	107	107	102	102	100	87	280	64	103	77	75	68	*	341	67	68	65	74	76
	RSD. %	▨	▨	3	2	6	8	6	▨	▨	0	1	2	0	1	▨	▨	1	9	▨	11	11
	0.8	83	114	102	101	99	102	103	99	108	103	102	110	105	101	*	110	102	101	99	93	93

Compound	matrix content	Orange							Wheat							Lettuce						
		c1	c2	c3	c4	c5	c6	c7	c1	c2	c3	c4	c5	c6	c7	c1	c2	c3	c4	c5	c6	c7
Iprodion	1.2	98	105	97	93	102	99	93	94	*	100	104	103	98	99	*	116	103	91	74	80	78
	2.5	84	102	85	86	89	89	84	88	80	92	87	96	92	91	*	112	105	94	83	83	83
	RSD %	9	6	9	8	7	7	10	6	12	6	9	6	7	5	10	3	1	5	14	8	9
L-cyhalothrin	0.8	*	100	99	99	106	98	109	*	98	105	103	113	107	104	*	289	121	109	99	102	93
	1.2	*	89	99	94	101	100	94	*	*	98	107	105	95	101	*	233	96	88	68	74	73
	2.5	*	77	87	85	88	89	80	*	81	90	94	102	91	94	*	197	94	87	77	78	79
	RSD %	15	13	7	8	9	6	16	10	8	7	5	8	5		19	19	19	19	20	18	19
Coumaphos	0.8	89	89	88	92	89	86	91	113	100	96	93	99	94	89	122	96	89	90	84	81	82
	1.2	89	84	83	81	87	89	79	74	*	89	96	91	85	88	139	96	85	80	63	67	65
	2.5	79	78	68	74	74	75	69	78	75	81	81	83	76	81	115	92	88	83	71	70	70
	RSD %	7	6	13	11	10	9	14	24	18	8	9	9	10	5	10	2	2	6	15	10	12
Cypermethrin	0.8	*	89	104	90	87	87		*	85	88	88	98	92	87	*	137	99	81	80	77	77
	1.2	*	74	95	80	78	83	75	*	*	*	80	81	76	75	*	85	88	73	60	62	58
	2.5	*	56	93	81	82	84	71	*	89	69	68	72	66	69	*	135	98	79	76	68	77
	RSD %	20	14	6	7	6	3		17	17	17	17	16	17	12	25	25	6	6	15	11	16

Note: c1-c7 - calibration levels (see table 1 for concentrations)

*-response was not measured or can not be calculated.

Table 6 Percentages of matrix effect (response in matrix/response in iso-octane×100%) for calibration solution prepared in wheat, orange, and lettuce matched extracts measured with ECD.

Compound	matrix content	Wheat							Orange							Lettuce						
		c1	c2	c3	c4	c5	c6	c7	c1	c2	c3	c4	c5	c6	c7	c1	c2	c3	c4	c5	c6	c7
Dichlorvos	0.08	177	132	118	109	109	115	110	160	127	117	112	113	112	127	143	115	104	96	106	118	104
	0.12	186	134	115	118	113	114	113	171	127	117	116	111	111	122	159	119	104	97	102	112	103
	0.25	196	133	116	107	113	121	110	197	130	121	117	118	115	113	182	126	104	99	102	116	103
	RSD %	5	1	1	5	2	3	2	11	1	2	2	3	2	6	12	5	0	1	2	3	1
Lindane	0.08	116	105	101	103	105	102	101	96	105	105	103	102	102	96	106	101	98	99	101	95	100
	0.12	154	113	103	102	104	100	100	122	103	102	104	100	98	100	107	103	100	99	100	96	97
	0.25	159	112	101	101	103	99	96	92	95	95	95	101	99	97	104	103	99	97	98	94	99
	RSD %	16	4	1	1	1	2	3	16	5	5	2	1	2	2	1	1	1	1	2	3	1
Chloro-thalonil	0.08	226	136	116	109	110	105	102	148	130	111	105	105	104	91	142	113	102	100	103	98	100
	0.12	535	197	132	114	119	109	105	96	125	106	104	100	100	97	155	123	107	102	103	97	99
	0.25	535	201	137	121	121	111	105	170	121	105	98	102	100	98	150	121	107	101	101	99	103

Compound	matrix content	Wheat							Orange							Lettuce						
		c1	c2	c3	c4	c5	c6	c7	c1	c2	c3	c4	c5	c6	c7	c1	c2	c3	c4	c5	c6	c7
Vinclozolin	0.08	104	100	100	105	106	100	102	94	102	103	105	103	100	100	105	96	97	98	97	95	99
	0.12	151	112	102	104	105	98	100	95	101	100	99	100	99	102	101	91	96	95	93	93	96
	0.25	144	111	99	101	102	99	97	98	100	97	99	101	100	99	87	86	90	90	89	89	94
	RSD %	20	6	8	5	5	3	2	2	3	3	4	3	2	4	10	5	4	4	4	3	2
Aldrin	0.08	98	98	99	102	102	102	100	89	97	99	99	100	99	100	97	97	97	98	100	95	99
	0.12	101	97	97	103	102	100	100	89	97	98	100	99	102	103	94	99	99	98	100	96	98
	0.25	98	95	94	97	99	99	95	93	86	93	96	97	95	93	90	97	98	97	98	94	99
	RSD %	2	6	2	3	1	2	3	3	7	3	2	2	3	5	3	1	1	0	1	3	1
Chlor-fenvinphos	0.08	219	158	126	112	125	118	114	211	149	130	116	117	114	98	177	114	102	94	103	100	88
	0.12	249	227	152	122	131	122	113	225	147	127	114	117	104	103	202	133	109	100	104	101	97
	0.25		152	150	125	131	124	112	230	135	116	112	112	106	104	239	148	115	104	107	104	100
	RSD %	3		10	6	3	3	7	5	5	6	2	2	3	5			6	5	2	2	7
A-endosulfan	0.08	106	103	101	101	104	97	100	92	97	98	99	101	99	94	99	98	98	99	100	96	99
	0.12	204	120	104	105	103	100	98	93	85	98	101	101	99	98	99	99	100	98	100	96	97
	0.25	180	108	92	93	95	92	90	83	86	87	87	89	88	89	95	99	98	98	98	95	99
	RSD %		8	7	6	5	3	6	6	8	7	8	7	7	5	3	1	1	0	1	1	1
p,p-DDT	0.08	112	115	117	116	108	113	111	107	109	105	102	104	103	97	116	105	102	100	104	97	101
	0.12	118	133	123	121	125	115	111	105	107	106	105	104	101	101	118	113	106	100	106	101	102
	0.25	112	124	115	113	119	114	108	106	110	109	107	94	95	94	118	115	106	104	106	101	104
	RSD %	3	7	3	4	8	1	2	1	1	2	2	6	4	3	1	5	3	2	1	2	2
Endrin	0.08	138	119	109	105	109	102	106	124	123	112	110	108	109	97	123	102	101	99	102	96	100
	0.12	304	155	119	112	115	107	109	125	122	112	108	109	99	102	122	114	104	105	104	102	102
	0.25	274	154	118	112	114	107	106	139	126	113	109	110	107	103	119	112	102	102	101	98	101
	RSD %	5		5	3	3	2	1	6	2	1	1	1	5	3	2	6	1	3	1	3	1
Permethrin	0.08	102	103	100	105	103	95	100	139	162	120	98	93	96	93	117	119	108	105	109	97	101
	0.12	97	107	95	97	98	94	95	266	186	127	113	100	92	94	101	111	105	101	101	93	97
	0.25	112	106	98	102	108	103	97	147	172	127	107	106	101	100	99	111	105	98	99	92	96
	RSD %	7	2	3	4	5	5	3		7	3	7	6	5	4	10	4	2	4	5	3	3
Fenvalerate	0.08	*	*	*	*	*	*	*	*	*	*	*	*	*	*	76	105	110	97	109	102	102
	0.12	*	*	*	*	*	*	*	*	*	*	*	*	*	*	95	131	127	105	114	106	106
	0.25	365	233	141	114	130	125	0	653	395	284	173	182	161	137	139	149	143	117	127	120	118
	RSD %																		10	8	9	7

Note: c1–c7 – calibration levels (see table 1 for concentrations)
*-response was not measured or can not be calculated.

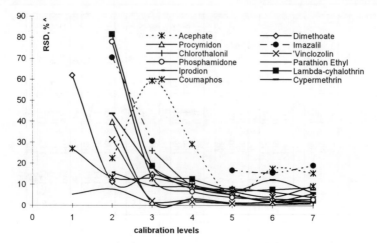

Figure 11 *RSD of matrix effect averaged percentages of lettuce, orange, and wheat calibration solutions containing 2.5 g/ml matrix equivalent analysed with NPD*

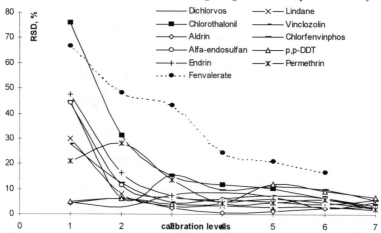

Figure 12 *RSD of matrix effect averaged percentages of lettuce, orange, and wheat calibration solutions containing 0.25 g/ml matrix equivalent analysed with ECD*

using one point calibration solution prepared in pure solvent, and only those analytes which will be detected during screening procedure will be the subject of further analysis. It is crucial to avoid false negative results in this case that can arise from possible matrix diminishing effect, that caused 20-40 % of signal suppression for some of the pesticides (Table 5). In order to detect a compound at 0.3 of AL with possible 40 % diminished response in matrix, the lowest calibrated level (LCL) prepared in pure solvent must be at = 0.15 AL. There is no danger to underestimate concentration of analytes experiencing matrix enhancement effect during screening procedure, but the "bracketing" range for preparation of matrix-matched calibration solutions might be wrongly estimated if percent of matrix effect was not determined previously. Therefore, the type and degree of matrix effect has to be assessed during method validation for each compound to be analysed in the laboratory.

5 CONCLUSION

Two statistical tests, the t-test and linear regression were applied to estimate significance of matrix induced chromatographic response enhancement within the working concentration range of ECD and NPD. It was shown that the two tests do not provide equivalent results.

The t-test was shown to be not sensitive enough for the estimation of significance of matrix effect if its type changes from enhancement to diminishing with concentration and if differences between the responses in matrix matched solution and pure solvent are large for different calibration levels. Linear regression appeared to be more sensitive to matrix effect than the t-test, but it can not be reliably used for a limited number of calibration points. Hence, if significance of matrix effect has to be estimated within the laboratory, both test have to be used.

Relative Difference plotted against concentration can be successfully used to indicate type and degree of matrix effect, though it can not be associated with any statistical significance.

Both types of matrix effect, diminishing and enhancement, were observed during the experiment. More then 1000% enhancement and 60 % diminishing were calculated for some of the pesticides. Dramatic matrix effects occurred mostly at concentration levels from 4 to 100 ng/ml which is the most important range for pesticide residue analysis. The matrix effect becomes less significant at higher concentrations from 100 to 2000 ng/ml. The amount and nature of matrix present in the solutions were shown to be important for analytes concentrations from 4 to 73 ng/ml for ECD and from 9 to 480 ng/ml for NPD analysis where up to 0.25 g/ml and 2.5 g/ml matrix equivalent was used, respectively. Hence, matrix matched calibration solutions have to be prepared with the same matrix type and concentration in the sample if relatively low concentrations have to be detected. For higher concentrations, different types of matrices and three-fold diluted matrix matched solution can be used.

References
1.   P. L. Wylie and K. Uchiyama, J. AOAC Intern., 1996, **79** (2), 571-577.
2.   J. Hajslova, K. Holadova, V. Kocouek, J. Poustka, M.Godula, P. Cuhra, and M. Kempny, J. Chromatogr. A.,1998, 800, 283-295.
3.   S. J. Lehotay and F. J. Schenk, J. Chromatogr. A., (in press).

4.   D. R. Erney and C.F. Poole, J. High Resol. Chromatogr.,1993, **16**, 501-503.
5.   R. Erney, T. M. Pawlowski, and C. F. Poole, J. High Resol. Chromatogr.,1997, **20**, 375-378.
6.   A. Andersson, H. Palsheden, and B. Aren, 1st European Pesticide Residue Workshop, Alkmaar, The Netherlands, June 10-12, 1996.
7.   EN 12393-2 (final draft), Non-Fatty Foodstuffs - Multiresidue methods for the gas chromatographic determination of pesticide residues, Part 2: Methods for extraction and clean-up, Method "P", CEN/TC 275 N **245**, May 1997.
8.   M. Godula, J. Hajslova, and K. Alterova, J. High Resol. Chromatogr.,1999, **22** (7), 395-402.
9.   D. R. Erney, A.M. Gillespie, D. M. Gilvydis, and C. F. Poole, J. Chromatogr., 1993, **638**, 57-63.
10.   J.C. Miller and J. N., "Statistics for Analytical Chemistry" Ellis Horwood, London, 1988 Chapters 3, 5, and 7.

Worked Example for Validation of a Multi-residue Method

Á. Ambrus

FAO/IAEA TRAINING AND REFERENCE CENTRE FOR FOOD AND PESTICIDE CONTROL,
FAO/IAEA AGRICULTURE AND BIIOTECHNOLOGY LABORATORY, A-2444 SEIBERSDORF,
AUSTRIA
e-mail: a.ambrus@iaea.org, http://www.iaea.org/trc

1 INTRODUCTION

The aim of this worked example is to show a possible way to apply the principles described in the Practical Procedures and the corresponding Guidelines for Single-laboratory Validation of Analytical Methods for Trace-level Concentrations of Organic Chemicals revised by the AOAC/FAO/IAEA/IUPAC Expert Consultation. (referred to further on as Guidelines[a].

Naturally, other options are also possible, and the practical application of the principles depends on the particular situation, available information and results of the tests.

The robustness test performed according to the Youden procedure is not included in this worked example as it is considered part of the method development/optimisation procedure, and it is very well explained elsewhere[1].

The tests recommended for determining the homogeneity of analyte and the "well mixed" condition of processed samples are discussed in elsewhere[2,3]. However it is pointed out, that those tests are independent from the analyte. Once they are properly carried out, the results can be used for any other analytical method based on the same sample processing.

The abbreviations used are specified in the Glossary of the Guidelines[4].

---

[a] Since the principles of the Guidelines remained the same, the worked examples are applicable for the revised document as well.

2 ASSUMPTIONS

The laboratory has to determine the residues of some 150 pesticide active ingredients amenable to direct GC analysis, and has to report residues present in samples at ≥ 0.05 mg/kg concentration. The laboratory had used a method for several years based on Ethyl Acetate extraction and GPC cleanup with additional cleanup on silicagel for ECD determination of residues if necessary. The final cleaned extract contains 2.5 g sample equivalent, of which 2.5 mg is injected to the GC.

The method was recently optimised and its robustness was tested in the laboratory with the participation of 2 technicians who will also apply the method in the future. The optimisation work was documented according to GLP and all records (e.g. sample blanks, calibration curves, recovery data, relative retention data of analytes of interest on 3 columns of different polarity) are available.

In addition, information on stability of residues during sample processing and analysis is available from previous analysis of samples, storage stability data, and results of comparative extraction of incurred residues and stability of analytical standards in stock and intermediate solutions are available for a large number of pesticide from the JMPR evaluations, submissions for registration and literature references[5,6,7], respectively.

The laboratory had to validate the method to characterise and record its performance parameters. (Availability of validated methods is also a requirement for formal accreditation according to either ISO 17025 and/or GLP.)

3 VALIDATION OF THE METHOD

The work was conducted by two technicians both *being familiar with the method*. Each technician analysed all representative compounds in each sample matrix. The validation was performed on several days. Thus the precision determined reflects the within laboratory reproducibility.

3.1 Selection of representative analytes and representative compounds

The laboratory selected 12 representative pesticides (Table 1) and 5 representative commodities (Tables 3 - 7) for validating the method and to determine its typical characteristics.

The analysts chose representative compounds with a wide range of physico-chemical properties: water solubility from 0.065 mg/l to 200 g/l; logP$_{ow}$: from - 0.8 to 6.1; vapour pressure: from 5×10^{-4} to 12 mPa; and hydrolytic stability from 1 to 800 days at pH 7.

> Notes: Naturally other or more compounds could also be selected to represent the analytes intended to be determined with the method (represented analytes), taking into account three important criteria:
>
> 1. As many analytes as it is possible to quantify in one mixture should be included in the method validation;

2. The analytes critical for the purpose of the study for which the method will be used should always be included;

3. The physico-chemical properties of selected compounds should represent a sufficiently wide range to include the properties of the 'represented analytes';

Table 1 *Summary of physico-chemical properties*[8,9] *of selected representative compounds*

Active ingredient	Water solubility		LogP$_{ow}$		Vapour pressure		Hydrolysis	
	mg/l	pH; °C		°C	MPa	°C	DT50 [day]	pH; °C
DDE-p,p	0.065	24						
Permethrin	0.2	30	6.1	20	0.045	25	>720	4, 50
Endosulfan a	0.32	22	4.74	PH 5	0.83	25[a]		
Chlorothalonil	0.81		2.89		0.076	25		
Chlorpyrifos	1.4	20	4.7		2.7	25	water, 1.5	8, 25
Lindane	7.3	25			0.051	25	191	7, 22
Iprodione	13	20	3	PH 3, 5	0.0005	25	1 to7	7
Dimethoate	23.3	5, 20	0.704		1.1	25	12	9, -
Azinphos-Methyl	28	20	2.96		0.18	20	87	4, 22
Diazinon	60	20	3.3		12	25	0.49 / 185	3.1, 20 / 7.4, 20
Progargite	632	25	3.73		0.006	25	800	7, -
Methamidophos	200,000	20	-0.8	20	2.3	20	657	4, 22

Notes: (a) 2:1 mixture of a and b isomers

3.2 Analytical range, calibration

The method was validated at 4 residue concentrations within the range of 0.02 mg/kg - 2 mg/kg.

Note: for MRM the concentrations of representative analytes do not necessarily relate to their actual Accepted Limits (ALs).

Calibration mixtures were prepared for ECD and NPD detection at 7 concentration levels both in solvent and in cleaned sample extract containing the equivalent of 2.5 g blank sample. The standards were injected into on-column injector in duplicate (number of injections in one detection system k =2*7=14). The GC was prepared according to the daily maintenance routine and its operating conditions, checked with appropriate test mixtures, were within the acceptable range. No special procedure was applied before the validation. In order to demonstrate the effect of the number of calibration points, a three-level calibration (k=2*3=6) is simulated in this example with the omission of 4 levels from the calculations.

As an example, the injected amount and average response (Y) relationship (rounded to whole number) is given hereunder for dimethoate and chlorothalonil. The shaded values indicate the three-level calibration. The standard solution prepared in iso-octane containing 2.5 mg/ml matrix equivalent is indicated by m2.5 mg.

Injected [pg]	9.2	36.8	184	368	552	827	1103
Dimethoate	385	3337	20006	40669	59065	82929	109285
Dimethoate m2.5mg	898	3739.33	18834	37550	55553	79397	100381
$Y_{matr}-Y_{solv} = d_I$	512.66	402.333	-1171.67	-3118.67	-3512	-3532	-8904.67

Mean of differences ($\bar{d}$): -2760, standard deviation of differences (S_D): 3321.8

Injected [pg]	16.3	65.2	326	652	979	1468	1958
Chlorthalonil	0	61.3	2318	4653	7007	10215	13521
Chlorthalonil m2.5	40.66	476	2434	5010	7336	10665	14164
$Y_{matr}-Y_{solv} = d_I$	40.66	414.7	116	357	328	450	643

Mean of differences ($\bar{d}$): 335.9, standard deviation of differences (S_D): 203.8

The calibration charts of dimethoate and chlorothalonil corresponding to 7 levels and 3 levels are shown in Figures 1, 2 and 3, 4, respectively. The linear regression equations are given in Table 2.

Table 2 *Summary of calibration parameters and calculated decision limits at $\alpha=0.05$*

Analyte	Level of Calibr.	Regression equation	R^2	S_b	Decision limit CCα [pg]
Dimethoate	7	Y= 99.5x + 1318	0.9975	878	15.7
Diethoate in m.	7	Y= 92.2x + 1798	0.9967	906	17.5
Dimethoate	3	Y= 98.6x+ 1459	0.9987	1612	34.8
Diethoate in m.	3	Y= 90.1x + 1815	0.9979	1412	33.4
Chlorothalonil	7	Y= 7.03x - 97	0.9987	150.5	38.1
Chlorothalonil in m.	7	Y= 7.26x + 67.8	0.9993	128.4	31.5
Chlorothalonil	3	Y= 6.95x-17.7	0.9980	179.3	55
Chlorothalonil in m.	3	Y= 7.24x + 69.7	0.9950	310.1	91

The significance of matrix effect was tested for all compounds with paired t-test.

The t value is calculated with equation:

$$t = \left| \frac{\bar{d}\sqrt{n}}{S_D} \right|$$

eq. 1

where $\bar{d}$ is the average and S_D is the standard deviation of differences (d_i):

$$S_D = \sqrt{[\Sigma(d_i - \bar{d})^2/n-1]}$$

eq. 2

with degrees of freedom of $\nu=n-1$, where n is the number of data pairs.

Surprisingly, the test did not indicate significant matrix effect for dimethoate (tabulated value for $t_{\nu=6,\ 0.05}$ is 2.45) (t_{calc}: 2.267), while it did for chlorothalonil (t_{calc}: 4.36). The matrix effect can also be seen from the substantially increased intercept compared with

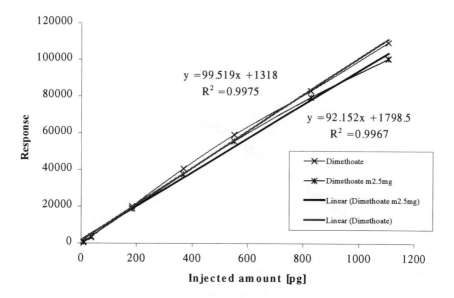

Figure 1 *Calibration chart for dimethoate in isooctane and matrix solution containing 2.5 g sample equivalent/m.*

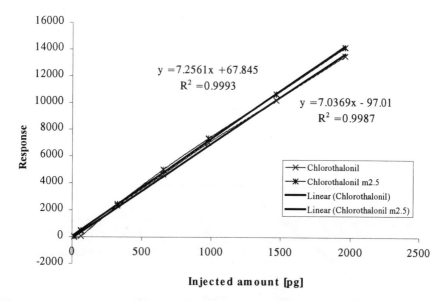

Figure 2 *Calibration chart for chlorothalonil in isooctane and matrix solution containing 2.5 g sample equivalent/ml*

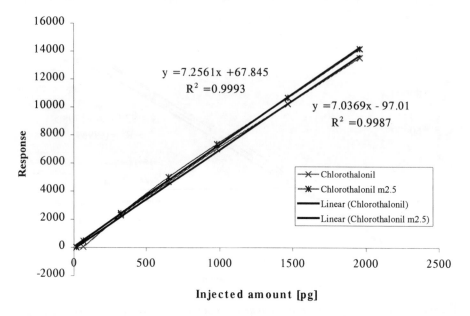

Figure 3 Calibration curve of dimethoate based on three-level calibration

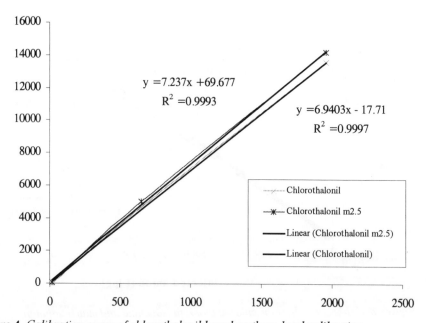

Figure 4 *Calibration curve of chlorothalonil based on three-level calibration*

standards without matrix (Table 2) in case of chlorothalonil, and a lesser extent for dimethoate.

Notes: 1Chlorothalonil is used here to illustrate the matrix effect. It was not included in the method validation programme as it undergoes significant degradation during sample processing[10].

2. Although the difference in slope is larger in case of dimethoate, the paired t-test shows no matrix effect (Figures 1 & 2). Other tests performed under varying conditions indicated that the **paired t-test** is not a sensitive indicator of matrix effect and should not be used and relied on alone. The authors provide alternative possibilities for testing the matrix effect[11].

No significant matrix effect was observed with the paired t-test for the other representative compounds which is in accordance with the practical experience gained with on-column injection, which generally induce smaller matrix effects than split/splitless injectors. Since the average recoveries were within the acceptable range (70-110%) for all compounds, the quantification of the chromatograms was carried out with analytical standard mixtures prepared without matrix according to the guidance given in the Guidelines.

Note: If the recoveries measured with standards prepared in pure solvent were outside the acceptable range, then matrix matched standards would be used for quantification. Practical experience indicates that the matrix effect may vary from day to day during the regular use of the method depending on the condition of the gas chromatographic system. Consequently, the extent of matrix effects found during method validation is only for information. It must be checked regularly as part of internal quality control. As suggested in the Guidelines, when residues measured during the application of the method for analysis of samples are between 0.7 AL (0.5AL) and 2AL the confirmation of the analyte concentration should always be carried out with matrix matched standard solutions. Alternatively, standard addition calibration may be used, which will also dispel any doubts that the blank matrix extracts available produce effects identical to those of the sample extract.

3.3 Calculation of Decision Limit ($CC\alpha_{0.05}$) and Limit of Detection (LD)

The terms limit of detection (**LD**), limit of determination (LOD) and limit of quantification (LOQ) are often used as synonyms, and there has been very little agreement in their calculation in the practice of residue analysts. Therefore the analysts should always report how the various quoted parameters were determined or calculated. Though they may be the subject of justified criticism, and more precise procedures are available and used in other areas of chemometrics[12,13] the procedures described hereunder may be used for estimation of approximate LD and LOD. These values provide an appropriate starting point for the experimental confirmation of the LOQ. It is emphasised that the reported LOQ must always be supported with mean recovery and its relative standard deviation obtained with spiking of samples at the reported LOQ.

As the blank response was often negligible and not easily measurable, the decision limits, $CC\alpha$[14], (IUPAC recommends[13] the term *Detection decision L_C*) were calculated with equations 3-5. This was possible because a reasonably good linear relationship was observed at the low concentration range and the intercepts were >0, especially with NPD.

Linear regression equation:

$$Y = ax + b \qquad \text{eq. 3}$$

The upper 95% confidence level of the intercept:

$$Y_{x=0,0.95} = b + t_{0.05} * S_b \qquad \text{eq. 4}$$

The corresponding analyte concentration:

$$CC\alpha = L_C = (Y_{x=0,\,0.95}-b)/a = t_{0.05}S_b/a \qquad \text{eq. 5}$$

where $Y_{x=0,\,0.95}$ is the response at the upper 95% confidence level, $t_{0.05}$ is the tabulated value of one tail Student distribution with degree of freedom $v=k-2$ ($t_{v:12}= 1.782$; $t_{v:4}=2.132$), S_b is the standard deviation of the intercept (b). For testing the presence of banned substances $\alpha=0.01$ may be chosen, allowing only 1 percent probability for false positive detection.

The regression equations for 7 and 3- level calibrations were reasonably similar, but the calculated $CC\alpha$ is strongly influenced by the number of calibration points and the consequent degrees of freedom, which affects the t value. The calculated values for dimethoate and chlorothalonil are shown in Table 2.

Assuming normally distributed peak areas/heights and similar standard deviations at the Decision Limit and at the LD, the LD can be approximately calculated with the following equation:

$$LD = C_B + t_{v,1-\alpha}S_B + t_{v,1-\beta}S_B \qquad \text{eq. 7}$$

where LD is the limit of detection expressed as concentration (or mass) of the analyte;
C_B is the analyte equivalent of the average blank response
S_B is the standard deviation of the analyte equivalent of blank response.

$t_{v,\,1-\alpha}$ and $t_{v,\,1-\beta}$ are the <u>one tailed</u> Student t values corresponding to a specified degree of freedom (v) and probabilities (α indicating the probability of false positive detection or β indicating the probability of false negative detection).

A number of points should be considered in applying the above equations:

In computing the detection and quantification limits, the effects of analyte concentration, matrix effect and matrix interference on the variance of the estimated quantity, as well as the degrees of freedom, and the acceptable probability of false positive and negative detection must be taken into account.

The response of the analyte is most sensitively effected by the matrix and detection conditions at the low concentration range. The linear calibration function is often distorted at the lower end. Consequently **the best estimate of the CCα and LD may be obtained from the replicate analysis of different blank samples** (e.g. different variety and/or stage of maturity of apples).

The other alternative may be to estimate the LD from the calibration points:

$$Y_{LD} = Y_{0,\,0.95} + t_{v,\,1-\alpha}S_b + t_{v,\,1-\beta}S_b \qquad \text{eq. 8}$$

$$LD = (Y_{LD}-b)/a = \delta_{\alpha,\beta}S_b/a \qquad \text{eq. 9}$$

Where $\delta_{\alpha,\beta} = t_{v,\,1-\alpha} + t_{v,\,1-\beta}$

The estimation of the true standard deviation, and consequently the LD, based on a few measurements is very imprecise. The number of measurements is partly taken into account by the increasing t value, and reflected by the nearly doubled CCα values obtained from the 3-level (6 points) calibration compared with those calculated based on a 7-level calibration. Further on, the calculated LD is uncertain by the ratio σ/s^{13}.

The calculation of LD based on equations 7 - 9 provides the necessary flexibility for the analysts to estimate the limit of detection to fit best for the purpose of the analysis. (e.g. selecting α and β) and experimental conditions applied.

The estimation of Limit of Detection for dimethoate based on the 7 level calibration curve with equation 9 at $\alpha=\beta=0.05$ probability level ($S_b = 878$, $t_{12,\,0.05} = 1.78$, $a=99.51$) resulted in 31.4 pg. Based on 3 level calibration, the estimated LD was about 70 pg. The estimated LD values correspond to 0.014-0.028 mg/kg residue concentration. Based on 7-level calibration, the range of estimated LD-s for the other compounds was similar.

3.4 Accuracy and precision of the method

Taking into account the estimated LD values and prior experience with the method, the accuracy and precision of the method were determined by the analyses of 5 replicate analytical portions of each representative commodity spiked at about 0, 0.02, 0.3, 1 and 2 mg/kg. The spiking solutions contained all of the twelve representative compounds which were determined together.

The summary of the validation results obtained is given in Tables 3-9. The results indicate that the performance of the method with the representative commodities and representative pesticides meets the requirements of a quantitative method (mean recovery 70-100%, reproducibility CV_A 23-32%) specified in the Guidelines.

Table 3 *Summary of recovery (Q%) data in cabbage*

	Fortification level [mg/kg]								
	0.02		0.3		1		2		
	Q	CV_A	Q	CV_A	Q	CV_A	Q	CV_A	Qa
Methamidophos	73	15	72	14	78	12	77	11	75.0
Dimethoate	77	15	75	10	78	9	76	11	76.5
Lindane	78	10	80	6	85	12	90	10	83.3
Chlorpyrifos	86	13	87	12	80	13	85	12	84.5
Propargite	80	14	88	11	86	12	84	9	84.5
Azinphos-methyl	87	11	85	10	84	12	83	14	84.8
Ethion	82	9	90	13	88	14	90	12	87.5
Endosulfan I	88	12	85	8	92	11	89	14	88.5
Iprodione	85	12	95	16	92	9	82	14	88.5
Diazinon	82	10	90	14	93	14	90	13	88.8
p,p' DDE	88	12	95	6	90	11	92	10	91.3
Permethrin	95	12	105	7	99	10	95	13	98.5
Average	83.4	12.1	87.3	10.7	87.1	11.6	86.1	11.9	86.0

	Fortification level [mg/kg]								
	0.02		0.3		1		2		
	Q	CV_A	Q	CV_A	Q	CV_A	Q	CV_A	Qa
Cvtyp		12.6		11		12		12.4	
Lsd	13.5		11.78		13.4		13.7		

Notes: Q is the average of replicate recovery tests at one level.

Qa is the average of recoveries at all fortification levels.

Lsd: Least significant difference (P=0.05)

3.5 Statistical tests for the evaluation of recovery data

The average recoveries, standard deviation, coefficient of variation and variances of recoveries were calculated for each analyte/sample matrix for each fortification level independently. Detailed results for cabbage fortified at 0.3 mg/kg level are given in Tables 8 and 9.

Table 4 *Summary of recovery (Q%) data in apple*

	Fortification level [mg/kg]								
	0.02		0.3		1		2		
	Q	CV_A	Q	CV_A	Q	CV_A	Q	CV_A	Qa
Methamidophos	72	14	74	12	75	11	82	12	75.8
Chlorpyrifos	80	11	78	6	82	9	84	14	81.0
Dimethoate	80	13	81	13	86	13	87	11	83.5
Propargite	84	12	91	11	88	9	91	11	88.5
Ethion	84	12	93	10	93	15	89	12	89.8
Azinphos-methyl	83	9	95	8	95	12	90	9	90.8
Lindane	87	10	95	8	91	13	92	12	91.3
Endosulfan I	92	8	91	6	93	11	90	13	91.5
Diazinon	85	12	97	9	92	12	93	11	91.8
p,p' DDE	88	11	100	12	94	14	92	12	93.5
Permethrin	95	10	103	10	100	14	92	13	97.5
Iprodione	92	14	110	11	89	12	99	15	97.5
Average	85.2		92.3		83.8		90.1		92.4
CV_{typ}		12.7		11		12		13	
Lsd	13		12.32		14.3		14.3		

Notes: see Table 3

Table 5 *Summary of recovery (Q%) data in orange*

	Fortification level [mg/kg]								
	0.02		0.3		1		2		
	Q	CV_A	Q	CV_A	Q	CV_A	Q	CV_A	Qa
Methamidophos	72.5	14.5	75	11	77.5	10.5	85.2	13.2	77.55
Dimethoate	78.5	14	79	12	83	10	84.8	12.7	81.33
Propargite	80	13	85	8	84	9.5	83.5	12	83.12
Chlorpyrifos	83	12	88	9	82	10	86.8	11.7	84.96
Azinphos-methyl	85	10	92	12	90.5	11	82.6	11.2	87.52
Ethion	83	10.5	93	13	91.5	13.5	84.9	12.7	88.11
Diazinon	83.5	11	95	13	93.5	12	82.3	12.2	88.57
Iprodione	88.5	13	96	14	91.5	9.5	82.2	12	89.54
Lindane	82.5	10	102	8	89	11.5	85.9	11.5	89.84
Endosulfan I	90	10	98	10	93.5	10	89.6	10.7	92.78
p,p' DDE	88	11.5	103	7	93	11.5	89.6	12.2	93.4
Permethrin	95	11	99	9	101	11	88.5	11.7	95.76
Average	84.13		92.08		86.08		83.91		86.5
CV_{typ}		12.4		11.1		11.3		12.6	
Lsd	13.3		13.02		12.8		13.7		

Notes: see Table 3

Table 6 *Summary of recovery data in wheat*

	Fortification level [mg/kg]								
	0.02		0.3		1		2		
	Q	CV_A	Q	CV_A	Q	CV_A	Q	CV_A	Qa
Dimethoate	81.6	13.1	88	12	86	13.1	87.4	12.6	85.75
Methamidophos	84.5	13.1	91	12	88.9	13.1	90.4	12.6	88.7
Propargite	84.5	9.95	91	9	88.9	9.98	90.4	9.41	88.7
Chlorpyrifos	86.4	9.95	93	9	90.9	9.98	92.4	9.41	90.66
Ethion	86.4	9.95	93	9	90.9	9.98	92.4	9.41	90.66
Iprodione	87.3	13.1	94	12	91.9	13.1	93.4	12.6	91.65
Lindane	89.2	12.1	96	11	93.8	12	95.5	11.5	93.62
Azinphos-methyl	90.6	13.1	96	12	94.5	13.1	96.6	12.6	94.42
Endosulfan I	90.2	7.85	97	7	94.8	7.93	96.5	7.29	94.6
p,p' DDE	94.9	14.2	102	13	99.7	14.1	102	13.7	99.52
Diazinon	95.9	12.1	103	11	101	12	103	11.5	100.5
Permethrin	97.8	6.8	105	6	103	6.9	105	6.22	102.5
Average	89.08		95.75		93.61		95.3		93.4
CV_{typ}		11.8		10.7		11.9		11.2	
Lsd	13.4		13.01		14.2		13.6		

Notes: see Table 3

Table 7 *Summary of recovery (Q%) data in sunflower seed*

	Fortification level [mg/kg]								
	0.02		0.3		1		2		
	Q	CV_A	Q	CV_A	Q	CV_A	Q	CV_A	Qa
Chlorpyrifos	73.4	12.7	78	9	74.1	10.9	76.8	9.65	75.56
Azinphos-methyl	77	10.6	80	11	76	13.6	78.7	12.6	77.93
Ethion	73.4	11.1	83	7	78.9	9.83	81.6	9.06	79.22
Methamidophos	71.8	15.3	85	11	80.8	13.6	83.6	12.6	80.28
Lindane	75.8	10.6	86	12	81.7	13.6	84.6	12.1	82.02
p,p' DDE	80.7	12.1	85	12	80.8	14.7	83.6	13.7	82.5
Propargite	71.8	13.7	89	12	84.6	12.6	87.5	10.5	83.2
Dimethoate	69.4	14.8	90	11	85.5	13.6	88.5	12.6	83.33
Endosulfan I	76.6	10.6	88	9	83.6	9.88	86.5	8.08	83.68
Diazinon	81.5	11.6	89	14	84.6	14.7	87.5	12.7	85.62
Permethrin	83.1	11.6	90	6	85.5	7.7	88.5	6.41	86.76
Iprodione	74.2	13.7	96	9	91.2	12.5	94.3	12	88.93
Average	76		86.58		82.3		85.2	11	82.4
CV_{typ}		10.8		10.9		12.8		11.4	
	10.4		11.99		13.4		12.3		

Notes: see Table 3

In order to estimate the typical parameters a number of statistical tests should be performed if necessary.

(a) Testing of recovery data for outliers

Since all CV values were within the acceptable range, and there was no result which was suspected as outlier, the Dixon outlier test was not carried out for the recovery data.

> Note: Rejection of outliers should be considered ONLY in cases where either there is clear experimental evidence that something went wrong or the result is so clearly part of "another population" that there is no possibility that it could be an extreme result from the same population. The use of outlier tests on small sets of data (from which it is impossible to determine whether they are normally distributed) should be very carefully considered and fully justified. It should never be applied as a means for "cleaning up" the performance of methods. If a method produces the occasional extreme result (and they all do, even if it's because the analyst has approached the limits of ruggedness), it is likely to be true reflection of the performance of the method.

(b) The homogeneity of variances was tested with Cochran test[15].

> The Cochran test compares the largest variance observed to the sum of total variances. If the ratio is smaller than the tabulated value the variances observed can be considered to be derived from one population.

The Cochran tests revealed that the differences in the variances were not significant. For example, the calculated values are given in Table 8.

Table 8 *Summary of recoveries and their variation in samples of high water content at 0.3 mg/kg fortification level*

	Cabbage				Apple				Orange			
	Q	CV_A	S_A	V_A	Q	CV_A	S_A	V_A	Q	CV_A	S_A	V_A
Azinphos-methyl	85	10	8.5	72	95	8	7.6	58	92	12	11.04	122
Chlorpyrifos	87	14	10.4	109	78	6	4.68	22	88	9	7.92	63
p,p' DDE	95	6	5.7	32	100	12	12	144	103	7	7.21	52
Diazinon	90	14	12.6	159	97	9	8.73	76	95	13	12.35	153
Dimethoate	75	10	7.5	56	81	13	10.5	111	79	12	9.48	90
Endosulfan I	85	8	6.8	46	91	6	5.46	30	98	10	9.8	96
Chlorothalonil	90	13	11.7	137	93	10	9.3	86	93	13	12.09	146
Lindane	80	6	4.8	23	95	8	7.6	58	102	8	8.16	67
Iprodione	95	16	15.2	231	110	11	12.1	146	96	14	13.44	181
Methamidophos	72	14	10.1	102	74	12	8.88	79	75	11	8.25	68
Permethrin	105	7	7.35	54	103	10	10.3	106	99	9	8.91	79
Propargite	88	11	9.68	94	91	11	10	100	85	8	6.8	46
Sum				1115				1016				1162
Cochran 5/12=0.34			0.21				0.14				0.16	
CV ave(arithm.)		10.6				9.67				10.5		
CV_{typ}		11				10				11		

Table 9 *Recoveries % of representative analytes in cabbage at 0.3 mg/kg fortification level*

Analyte	Recoveries % in replicate analytical portions					Mean	SD	V	CV
Methamidophos	63.3	63.3	70.0	76.7	86.7	72.00	9.88827	97.78	14
Dimethoate	65.0	70.0	76.7	80.0	83.3	75.00	7.45356	55.56	10
Lindane	75.0	76.7	80.0	83.3	86.7	80.33	4.77261	22.78	6
Azinphos-methyl	76.7	78.3	82.7	91.7	95.0	84.87	8.11925	65.92	10
Endosulfan I	78.3	78.3	85.0	90.0	93.3	85.00	6.77003	45.83	8
Chlorpyrifos	76.7	73.3	86.7	100.0	100.0	87.33	12.5610	157.78	14
Propargite	75.0	83.3	86.7	93.3	100.0	87.67	9.54521	91.11	11
Diazinon	76.7	76.7	96.7	100.0	101.7	90.33	12.6051	158.89	14
Ethion	76.7	83.3	86.7	101.7	103.3	90.33	11.6905	136.67	13
p,p' DDE	88.3	90.0	95.0	96.7	103.3	94.67	5.93951	35.28	6
Iprodione	80.0	81.7	91.7	110.0	111.7	95.00	15.1383	229.17	16
Permethrin	96.7	96.7	110	110.0	110.0	104.67	7.30297	53.33	7

ANOVA

Source of Variation	SS	Df	MS	F	P-value	F crit
Between Groups	4394.044	11	399.4586	4.167941	0.000251	1.994579
Within Groups	4600.356	48	95.84074			

ANOVA

Source of Variation	SS	Df	MS	F	P-value	F crit
Total	8994.4	59				
least significant difference, Sa*sqrt(2/n)*t		12.4452				

(c) The significance of the differences of average recoveries

The recoveries of the representative compounds in one commodity and at one spiking level were tested with one way ANOVA.

The summary of ANOVA calculation, together with one set of raw data, are shown in Table 9. The results indicate that the average recoveries are not derived from the same population.

Note: Between group variance (one group consists of the recovery data obtained with a representative analyte) is significantly larger than the within group (variance of replicate analysis): $F_{calc} = 4.1679 > F_{critical}$.

The calculated least significant difference ($S_A*t*\sqrt(2/n} = 12.44\%$, $t_{v=48, 0.05} = 2.01$, n=5) confirmed that the average recoveries of methamidophos and dimethoate are significantly different from lindane, and the recovery of permethrin (105%) is significantly greater than for iprodion (95%). The recalculated ANOVA without these compounds indicated that the difference for the rest of the recoveries are not significant ($F= 1.079 < F_{critical}$).

Significantly different recoveries from apple and orange, indicated by shading in Tables 3-7, were obtained for methamidophos and dimethoate in 5 and 2 other cases, respectively. The average recovery data for permethrin was not significantly different from the remaining analytes in the other commodities. Consequently, the recovery data of methamidophos and dimethoate, but not that of permethrin, were considered separately from the data for other representative compounds in the case of commodities of high water content.

(d) Comparison of recoveries obtained at different spiking level

Assuming similar standard deviations, the differences between two mean values can be calculated with the t-test:

$$t = \frac{Q_1 - Q_2}{s\sqrt{\frac{1}{n_1} + \frac{1}{n_2}}}$$

eq. 10

From eq. 10 the difference in average recoveries (Q_1-Q_2) , which must be exceeded to obtain significant difference, can be calculated.

For example, the differences are 11.1% and 6.9%, respectively, for two sets of data consisting of 5 and 20 replicate measurements with a mean recovery of 90% and 12% CV ($s = 10.8$).

(e) Testing to determine if the average recovery is significantly different from 100%

The formula for comparing the average recovery (Q) to $\mu=100\%$

$$t = \frac{(Q-100)\sqrt{n}}{s}$$

eq. 11

Where s is the standard deviation of n replicate recovery tests giving an average recovery of Q.

Since the recovery can be smaller or larger than 100%, a two-tailed test is used at P=0.05. When the difference is significant, the results should be corrected for the average recovery[16].

Note however, that the measured residue values and the average recovery should be reported in every case regardless whether the average recovery was used for correction or not.

3.6 Estimation of typical performance characteristics of the method

The typical values of recovery and within laboratory reproducibility of the method are estimated from the recoveries obtained for the representative analyte sample matrices at various fortification levels. For the estimation of typical recovery or reproducibility those recovery values can be considered which are not different significantly.

3.6.1 Typical Recoveries. The average recoveries obtained for one analyte commodity matrix did not show any concentration relationship. The differences between their average recoveries obtained at various spiking levels were not significant either. Moreover, the Q_a values obtained for cabbage, apple and orange (commodities of high water content) did not differ significantly. Consequently, their average could be calculated as a typical value for the given matrix or group of commodities. The summary of the results is given in Table 10.

Table 10 *Typical recovery values estimated for commodity groups*

Representative analyte	Groups I, II & III.		Cereals	Oil seeds
Azinphos-methyl	87.7		94.4	77.9
Chlorpyrifos	83.5		90.7	75.6
p,p' DDE	92.7		99.5	82.5
Diazinon	89.7		100.5	85.6
Dimethoate	80.4		85.7	83.3
Endosulfan I	90.9		94.6	83.7
Ethion	88.5		90.7	79.2
Lindane	86.7		93.6	82.0
Iprodione	82.7		91.6	88.9
Permethrin	97.3		102.5	86.8
Propargite	85.3		88.7	83.2
Dimethoate		80.4	85.7	83.3
Methamidophos		76.1	88.7	80.3
Typical for the group	**89.6**		**93.4**	**82.4**

3.6.2 Typical within Laboratory Reproducibility. The typical reproducibility of the method may be calculated from the pooled variances of recoveries which are assumed to be derived from the same population.

Each set of replicate recovery tests, performed with a representative compound in a representative commodity, provides an estimate of the precision of the method. Since the estimation of variances based on small number of samples is very imprecise, the best estimate of the precision of the method is the average of the variances obtained.

The theoretically correct calculation of the typical CVs is:

$$CV_{Typ} = \frac{\sqrt{V_{ave}}}{Q_a}$$

eq. 12

Where the V_{ave} is the average variance of recoveries and Q_a is the average recovery. CV_{typ} may be calculated for various combinations. For example, recoveries of several representative analytes in one representative commodity at one fortification level; recoveries of one analyte-matrix combination at different fortification levels; recoveries of several analytes at several fortification levels in one sample, etc.

> Though theoretically not correct, if the differences of CVs obtained for different analytes in a commodity are in a relatively narrow range (e.g. 9-15%), their simple arithmetic mean can be calculated if the detailed records on the SD or variance of the results are not available. It is usually only slightly different than the correctly calculated value (Table 8).

The CV_{typ} values calculated with eq. 12 are listed in Tables 3-10. In view of the limited data on which the estimations are based, estimating typical CVs for a concentration interval does not contradict the average precision - concentration relationship described by the Horwitz equation[17]. The typical values calculated as the averages of the CV values are shown in Table 11.

Table 11 *Typical CV values of representative analytes for 0.02-2mg/kg residue levels*

	Cabbage	Apple	Orange	Wheat	Sunflower
Azinphos-methyl	11.8	9.7	11.1	12.7	12.0
Chlorpyrifos	12.5	10.6	10.7	9.6	10.6
Diazinon	15.1	11.0	12.2	11.6	13.4
Dimethoate	9.9	12.5	12.2	12.7	12.9
Endosulfan I	11.6	9.9	10.2	7.5	9.4
Ethion	12.1	12.4	12.6	9.6	9.2
Iprodione	15.4	13.0	12.3	12.7	12.7
Lindane	9.8	10.9	10.2	11.6	11.2
Methamidophos	10.7	12.3	12.4	12.7	13.1
p,p' DDE	10.2	12.3	10.5	13.7	13.2
Permethrin	10.6	11.9	10.7	6.5	8.1

	Cabbage	Apple	Orange	Wheat	Sunflower
Propargite	11.6	10.8	10.7	9.6	12.2
CV$_{typ}$	**11.5**	**11.1**	**11.1**	**11.1**	**11.6**

The results indicate that the typical CVs are around 10-15% with an overall CV$_{typ}$ of 11-12%, which characterises the within laboratory reproducibility of the method.

3.6.3 Limit of Quantification, Detection Capability. When samples are analysed, the factors affecting the recovery measurements (e.g. extraction, cleanup, evaporation) will contribute to the variability of the results. Thus, the S$_B$ in the second part of equation 7 can be replaced with the standard deviation of the recoveries performed at LD. Performing the recoveries at LD may not be practical, therefore the blank samples may be fortified 2-5 times higher concentrations than LD. Spiking with higher concentrations than 5LD may result in better precision and average recovery, which lead to underestimating the limit of determination, LOD.

> LOD is used for the calculated limit to distinguish it from the reported Limit of quantification which must be based on replicate recovery tests. The LOQ should only be reported as a performance criterion of the method.

To make the precision data independent of the analyte concentration, the relative standard deviation should be used in the calculation for estimating LOD:

$$C_{LOD} = C_B + t_{v,1-\alpha}S_B + t_{v,1-\beta}\frac{LD\ CV_{LOD}}{Q} \qquad \text{eq. 13}$$

Where
C$_B$ is the analyte concentration equivalent of the average blank response;
S$_B$ is the standard deviation of the analyte equivalents of blank responses;
LD is the limit of detection calculated with eq. 7;
CV$_{LOD}$ is the relative standard deviation of the recovered analyte concentrations at around LOD
Q is the average recovery expressed as average analyte recovered/analyte added.

The calculation of LOD is illustrated, as an example, with the calibration data obtained for dimethoate (Table 2). The residue concentration equivalent of the Decision limit of 15.7 pg is 0.007 mg/kg. Inserting in the eq. 13 the decision limit (C$_B$+t$_{v,\ 1-\alpha}$S$_B$) = 0.007 mg/kg, LD = 0.014 mg/kg, Q$_{typ}$ =90 % and CV$_{typ}$ = 12% , and t=1.64 (since the typical values were obtained from over 100 measurements) the LOD for commodities of high water content may be calculated as:

LOD [mg/kg] = 0.007+ 1.64*0.014*0.12/0.90 =0.01

Similarly calculating with the CCα and LD values estimated based on 3-level calibration, the estimated LOD = 0.014+1.64*0.028*0.12/0.9= 0.02.

Since the reporting limit is 0.05 mg/kg, the Limit of quantification, **LOQ**, was confirmed experimentally with replicate recovery studies performed at 0.02 mg/kg level. The tests resulted in recoveries and CV values in the acceptable range, therefore the <u>reported limit of quantification of the method is 0.02 mg/kg</u> for commodities of high water content.

> The calibration function varied during the method validation, and larger variation is likely to occur during the long term practical use of the method. This consequently leads to the variation of estimated LD or CCα and LOD which should only be used as guidance values and not as a constant parameter. Therefore the properly selected Lowest Calibrated Level may be equally well used as a practical alternative to LD for indicating the detectability of pesticide residues.

4 CHARACTERISATION OF THE METHOD

In conclusion, the method may be characterised as follows:

The method is applicable within the analytical range of 0.02 and 2 mg/kg for a wide range of GC amenable pesticides, the physico-chemical properties of which are within the ranges of representative compounds given in Table 1. The typical recoveries from plant commodities of high water content are about 90 % (except the highly water soluble compounds such as methamidophos), cereals 93% and oil seeds 82% with a typical CV of ≤12%.

The typical or average residue values and typical CV-s characterise the trueness and precision of the method in general. The particular values characterise the method for specific pesticide/commodity combinations. For the represented, but not tested, pesticide/ commodity combinations, the typical values for the group should be used as an initial guidance. Their performance characteristics should be checked during the extension of the method.

The daily performance of the method must be checked with appropriate internal quality control procedures during the routine use of the method. The performance characteristics established during method validation should be adjusted, if necessary, based on the results of internal quality control/performance verification tests.

References

1. W. J. Youden and E.H. Steiner, 'Statistical Manual of the AOAC', AOAC International, Gaithersburg, MD, 1975.
2. Ambrus, A., Solymosné, E.. and Korsós, I.; Estimation of uncertainty of sample preparation for the analysis of pesticide residues, J. Environ. Sci. Health, B31, 443-450 (1996).
3. B. Maestroni, A. Ghods, M. El Bidaoui, N. Rathor, O. Jarju and A. Ambrus, Testing the efficiency and uncertainty of sample preparation using 14C-labelled chlorpyrifos: PART II, Proceedings of International Workshop on Method Validation, Budapest, 1999.
4. AOAC/FAO/IAEA/IUPAC Expert Consultation, Guidelines for Single-laboratory Validation of Analytical Methods for Trace-level Concentrations of Organic Chemicals
5. A. Anderson and H. Palsheden, *Fresenius J. Anal Chem.*, 1991, **339**, 365.

6. A. de Kok, H. A. van der Schee, K. Vreeker, W. Verwaal, E. Besamusca, M. de Kroon and A. Toonen, Poster 8E-014, Presented at 9th IUPAC Congress of Pesticide Chemistry (IUPAC) London, 1998.

7. A. de Kok, H. A. van der Schee, K. Vreeker, W. Verwaal, E. Besamusca, M. de Kroon, Poster 8E-013, Presented at 9th IUPAC Congress of Pesticide Chemistry (IUPAC) London, 1998.

8. FAO, Pesticde residues in Food - Evaluations, FAO Rome (annual publication)

9. Tomlin, Clive, (ed) The Pesticide Manual, 10th ed. 1994

10. M. El-Bidaoui, O. Jarju, B. Maestroni, Y. Phakaiew and A. Ambrus, Testing the effect of sample processing on the stability of residues, Proceedings of International Workshop on Method Validation, Budapest, 1999.

11. E. Soboleva, N. Rathor, A. Mageto and A. Ambrus, Estimation of significance of "matrix-induced" chromatographic effects, Proceedings of International Workshop on Method Validation, Budapest, 1999.

12. L. A. Currie, *Chemometrics and Intelligent Laboratory Systems*, 1997, **37**, 151.

13. L. A. Currie, *Pure &Appl. Chem.*, 1995, **67**, 1699.

14. European Commission: Commission Decision laying down analytical methods to be used for detecting certain substances and residues thereof in live animals and animal products according to Council Directive 96/23/EC, Final Version: approved by CRLs

15. W. Horwitz, *Pure & Appl. Chem.*, 1988, **60**, 855

16. M. Thompson, S. L. R. Ellison, A. Fajgelj, P. Willetts and R. Wood, *Pure & Appl. Chem.*, 1999, **71**, 337.

17. W. Horwitz, L. R. Kamps and K. W. Boyer, *J. Assoc. Off. Anal. Chem.*, 1980, **63**, 1344.

EU Guidance Document on Residue Analytical Methods

Ralf Hänel[1], Johannes Siebers[1] and Karen Howard[2]

[1] FEDERAL BIOLOGICAL RESEARCH CENTRE FOR AGRICULTURE AND FORESTRY (BBA), CHEMISTRY DIVISION, BRAUNSCHWEIG, GERMANY
[2] PESTICIDES SAFETY DIRECTORATE (PSD), MINISTRY OF AGRICULTURE, FISHERIES AND FOOD, YORK, UK

1 INTRODUCTION

In the context of a harmonised registration of Plant Protection Products (PPP) within the EU (Directive 91/414/EEC)[1] applicants have to submit among other things validated residue analytical methods for post-registration control and monitoring purposes. The Guidance document on residue analytical methods (8064/VI/97-rev. 4)[2] – finalised and endorsed by the Standing Committee on Plant Health at the 1st December 1998 – is an interpretation of Annex II (96/46/EC)[3] of 91/414/EEC and provides guidance to applicants and Member States on the data requirements and assessment for these methods for Annex I-listing and national authorisation procedure of active ingredients. Depending on proposed use, analytical methods for crops, food of plant and animal origin, feedingstuff, soil, water and air must be provided. Additionally, for toxic active ingredients methods for body fluids and tissues are required.

2 REQUIREMENTS ON ANALYTICAL METHODS FOR PLANT MATERIALS

As far as practicable methods must employ the simplest approach, involve the minimum cost, and require **commonly available equipment**. The following methodologies are regarded as commonly available:

GC	PND, FPD, ECD, FID, MS-detector
HPLC	UV, DAD, Fluorescence detector, Electrochemical detector, normal phase, reversed phase, ion-exchange, ion-pair; column switching
AAS	

The list is valid until 31st December 2000 and will be regularly reviewed in order not to inhibit development. For example LC-MS-MS can be a powerful tool in residue analysis, but it cannot be regarded as commonly available at the moment within the EU.

Residue methods should be **standard multi-residue methods**. In the Guidance document a scheme showing constituent elements of multi-residue methods is given.

The components suggested for a multi-method for plant materials are present in figure 1.

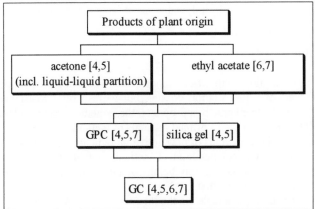

Figure 1 Constituent elements of multi-residue methods for food of plant origin

Where residues cannot be determined using a multi-residue method, an alternative method must be proposed. However, the methods must be suitable for the determination of all compounds included in the residue definition for compliance with the MRL.

The following factors must be taken into account for the validation of residue analytical methods:
- A full description of the method (incl. used reagent, apparatus, procedure, calculation, representative chromatograms) must be provided.
- The recovery rates for all fortification levels and commodities must be reported.
- A sample set of 5 determinations each at the LOQ and at 10 times LOQ or at the MRL (when the MRL is higher than 10 times LOQ) and 2 control samples should be performed.
- The limit of quantification (LOQ) is defined as the lowest acceptably validated fortification level.
- The mean recovery should be in the range of 70 - 110% [trueness, 96/46/EC] with a relative standard deviation (rsd) of less than 20% [repeatability, 96/46/EC]. Both must be reported.
- Blank values should not be higher than 30% of the LOQ [specifity, 96/46/EC].
- An additionally confirmatory technique is required to demonstrate specificity, if the proposed enforcement method is not highly specific [specifity, 96/46/EC].
- Plant commodities are listed in four crop matrix groups (dry, high water-, high fat- and high acid content). The method must be tested with representatives of the groups relevant to the intended uses.
- Additional validation is also required for matrices which are difficult to analyse (e.g. hops, tea, herbs, brassicas, bulb vegetables), where these are relevant to the intended uses.
- The proposed method(s) must be validated by an independent laboratory (ILV) [reproducibility, 96/46/EC].
- Hazardous reagents should be avoided, if possible.

In addition, the above mentioned Guidance document on residue analytical methods (8064/VI/97-rev. 4) gives details concerning the requirements for analytical methods for the determination of residues in soil, water, air, food of animal origin and body fluids.

3   PUBLICATION

In principle, Registration Authorities cannot make available methods which have been submitted as part of a Registration dossier, but we can offer assistance in obtaining relevant multi-residue methods. If appropriate, these methods can be adopted – with any further validation necessary – to official manuals of methods.

Additionally, residue methods of new active ingredients are reviewed by the BBA in a German journal. A UK compendium of methods suitable for monitoring, developed with government funding, is currently being compiled. It is the intention to eventually make this available via the internet.

Further information is available via the internet:

http://www.bba.de/analytik/analytik.htm
http://maffweb/aboutmaf/agency/psd/psdhome.htm

References

[1]   Council Directive of 15 July 1991 concerning the placing of plant protection
        products on the market (91/414/EEC), Official Journal of the European
        Communities No L 230/1, 19.08.1991.
[2]   Guidance document on residue analytical methods, 8064/VI/97-rev 4, 15.12.98,
        European Commission, Directorate General for Agriculture VI B II.1, finalised
        and endorsed by the Standing Committee on Plant Health at 1 December 1998
[3]   Commission Directive 96/46/EC of 16 July 1996 amending Council Directive
        91/414/EC concerning the placing of plant protection products on the market,
        Official Journal of the European Communities No L 214/18, 23.08.1996
[4]   *Manual of Pesticide Residue Analysis*, H.-P. Thier, H. Zeumer (Edit.), DFG
        Pesticides Commission, *Vol. I + II*, VCH Weinheim, 1987.
[5]   W. Specht, S. Pelz, W. Gilsbach, *Fresenius J. Anal. Chem.*, 1995, **353**, 183-190.
[6]   A. Anderson, H. Pålsheden, *Fresenius J. Anal. Chem.*, 1991, **399**, 365-367.
[7]   *Analytical Methods for Pesticide Residues in Foodstuffs*, Ministry of Public
        Health, Welfare and Sport, The Netherlands, *6. Edition*, 1996.

Guidelines for Single-Laboratory Validation of Analytical Methods for Trace-level Concentrations of Organic Chemicals

Overview

1   What is method validation?

An analytical method is the series of procedures from receipt of a sample to the production of the final result. Validation is the process of verifying that a method is fit for purpose.

2   Purpose of this document

Method validation guidelines for use in trace analysis have been proposed by various authors but there is little consistency in the recommended approaches. The general guidelines for methods of chemical analysis proposed by standards organisations are impractical for use in laboratories involved in monitoring of trace level organic chemicals.

The guidelines were elaborated by the AOAC/FAO/IAEA/IUPAC Expert consultation with the participation of L. Alder, A. Hill, P.T. Holland, J. Lantos, S. M. Lee, J. D. MacNeil, J. O'Rangers, P. van Zoonen, and A. Ambrus (Scientific Secretary) The background of the elaboration of the Guidelines is described in Annex 6.

The validity of methods has often been established on the basis of inter-laboratory studies of performance but that process can be uneconomic and too slow or restricted in scope. Validation in a single laboratory can provide a practical and cost-effective alternative (or intermediate) approach.

These Guidelines identify adequate minimum requirements for the validation of methods of analysis for "trace levels" of organic chemicals, in order to verify that a method is fit for purpose, according to the current general requirements for the reliability of analytical data.

This document aims to assist analysts to perform method validation in a scientifically sound and practical manner, and national authorities and accreditation bodies in assessing acceptability of the methods developed and applied in testing laboratories.

The requirements identified here are intended for guidance, to be applied pragmatically rather than prescriptively. The analyst should address only those requirements appropriate to the method and the purposes for which the method will be used. They may be modified or adapted, as necessary for the specific purpose.

3   Relationship between method validation, method development, quality control, proficiency testing and reporting of uncertainty

The method may be developed in-house, taken from the literature or otherwise obtained from a third party. The method may then be adapted or modified to match the requirements and capabilities of the laboratory and/or the purpose for which the method will be used.

Typically, validation follows completion of the development of a method and it is assumed that requirements such as calibration, system suitability, analyte stability, etc., have been established satisfactorily. When validating and using a method of analysis, measurements must be made within the calibrated range of the detection system used. In general, validation will precede practical application of the method to the analysis of samples but subsequent internal quality control (IQC, referred to in the present document as performance verification) is an important continuing aspect of the process. Requirements for performance verification data[1,2,3,4,5] have been identified as a subset of those required for method validation.

Proficiency testing[4,6,7,8] (or other inter-laboratory testing procedures[9,10,11]), where practicable, provides an important means for verifying the general accuracy of results generated by a method, and provides information on the between-laboratory variability of the results. However, proficiency testing generally does not address analyte stability or homogeneity and extractability of analytes in the processed sample.

Where uncertainty data must be reported to the client (the purchaser or user of results), this information should incorporate performance verification data and not rely solely on method validation data.

4 Fitness for purpose of methods and results

Analytical methods, and the validation required for them, must be appropriate to the purpose for which the results are likely to be used. In general, method performance and validation requirements should be agreed between the analyst and the client (the purchaser/user of results, as appropriate). Method performance requirements have been defined by governmental authorities[12,13,14,15] and by AOAC International[5,16].

General guidelines for method validation have been developed by various organisations.[4,5,,17,18,19,19,20,21] Specific guidelines have been developed for single-laboratory validation of methods for pesticide residues[22] and are under development for residues of veterinary drugs in food and animal feeds[23].

Methods and analytical results are often classified loosely as quantitative, semi-quantitative or qualitative (screening). These categories do not have well-defined or universally accepted boundaries. The method performance criteria given in these Guidelines indicate acceptability for quantitative data, but the boundaries to be applied in practice should be agreed between the analyst and the client.

5 Limits to the scope of validation

Extensions of methods to new analytes, matrices (eg., additional tissues or commodity groups), much lower concentrations, much smaller test portions, use of the method in other laboratories, etc., should be validated as indicated in this document. Minor changes in methods may be validated through performance verification but, where the change leads to unsatisfactory performance, the method may require modification and re-validation, as indicated here.

For multi-analyte and/or multi-matrix methods, it is likely to be impractical to validate a method for all combinations of analyte, concentration and type of sample matrix that may be encountered in subsequent use of the method. Initial validation should incorporate as many of the target analytes and matrices as practicable. However it is proposed that a method may be validated initially for "representative matrices" and, "representative analytes". This can limit the cost and time required before samples are analysed, while establishing the key performance characteristics of the method. Subsequently, performance validation data must be generated to establish method performance for all other analytes and matrices for which results will be reported.. Where the performance verification data indicate that method performance is not adequate, the method may be modified as appropriate and subsequently validated. If the modification is not successful, an alternative method is necessary.

6 Requirements for validation

Diagrams of the decision processes (Figures 1, 2) and parameters to be assessed (Table 1) in determining overall validation requirements are presented in this document.

The amount of effort allocated to the validation of a method may vary considerably. Where appropriate data are already available, it may not be necessary for the analyst to perform all the tests. However, all required information must be included or referred to in the validation report.

6.1. Parameters to be assessed should be restricted to those which are appropriate both to the method and to the purpose for which the particular method is to be applied. In many cases, performance characteristics with respect to several parameters may be obtained simultaneously using a single experiment. Test designs where different factors are changed at the same time under statistical control (factorial experiment designs), may help to minimise the resources required and provide additional information, such as the ruggedness of methods[24,25,26,27].

6.2. The criteria to be applied are likely to be specific to the purpose for which the method is to be used. Examples are given in Annexes 1 and 2, describing validation of methods for the determination of residues of pesticides and veterinary drugs, respectively. Different criteria may be defined according to the purpose. If the desired criteria cannot be met, either expectations about the quality of data must be revised or an improved method must be sought.

7. General criteria for multi-analyte methods for pesticide and veterinary drug residues

7.1   The general performance criteria required for analysis of pesticide and veterinary drug residues are summarised in Table 2. Deviation from these criteria may be acceptable for certain analytes and matrices when supported with appropriate justification.

The analytical techniques or their combination listed in Table 3 are generally suitable for the confirmation of analytes in test samples. National authorities may set particular requirements for specified compounds. Analysts should be aware of such requirements when conducting method validations to demonstrate "fitness for purpose".

Table 1. Summary of parameters to be assessed for method validation

Parameters to be tested	Existing analytical method, for which previous tests of the parameter have shown that it is valid for one or more analyte/matrix combinations					Modification of an existing method	New method, not yet validated	Experiment types which may be combined
	Performance verification*	Additional matrix	Additional analyte	Much lower concentration of analyte	Another laboratory			
Specificity (show that the detected signal is due to the analyte, not another compound)	No (provided criteria for matrix blanks and confirmation of analyte are met)	Yes, if interference from matrix is apparent in QC	Yes	Yes, if interference from matrix is apparent in QC	Rigorous checks not necessary if the performance of the determination system is similar or better	Yes or No. Rigorous checks may be necessary if the determination system is fundamentally different or where the extent of interferences from the matrix is uncertain	Yes. Rigorous checks may be necessary if the determination system is different or where the extent of interferences from the matrices are uncertain, compared with existing methods	
Analytical Range, Recovery through extraction, clean-up, derivatisation and measurement	Yes	Yes	Yes	Yes	Yes	Yes	Yes	Calibration range – analytical range - LOD/LOQ - matrix effect.
Calibration range for determination of analyte	No	No	Yes	Yes	Yes, for representative analytes	Yes, for representative analytes	Yes, for representative analytes	Linearity, reproducibility and signal/noise
LOD and LOQ	No	Yes, (partial if matrix is from a represented class)	Yes, partial for represented analytes	Yes	Yes	Yes	Yes	Lowest calibrated level, and low level spike recovery data
Reporting Limit, LCL	Yes	No	No	No	No	No	No	
Analyte stability in sample extracts*†	No	Yes, unless matrix is from a represented class	Yes, unless the analyte is represented	Yes	No	No, unless extraction/final solvent is different, or the clean-up is less	Yes, if extraction/final solvent is different from that used in an existing method, or the	

Parameters to be tested	Existing analytical method, for which previous tests of the parameter have shown that it is valid for one or more analyte/matrix combinations					Modification of an existing method	New method, not yet validated	Experiment types which may be combined
	Performance verification*	Additional matrix	Additional analyte	Much lower concentration of analyte	Another laboratory			
Analyte stability during sample storage*°	Yes	Yes	Yes,	Ideally	No	No	clean-up is less stringent, compared with existing methods used.	
Extraction efficiency*♦	No	Ideally	Ideally	Ideally	No	No, unless different extraction conditions employed	Yes, unless previously tested extraction procedure is used.	
Homogeneity* of analytical samples	Yes†	No, unless the matrix is substantially different	No	No	No, unless the equipment is changed	No, unless the equipment is changed	Yes, unless a previously tested sample processing procedure is used	See below
Analyte stability in sample processing*	No	Yes, unless a represented matrix	Yes, unless a represented analyte	Ideally	No	No, unless procedure involves higher temperature, longer time, coarser comminution, etc.	No, unless procedure involves higher temperature, longer time, finer comminution, etc. than validated procedures.	Repeatability, re-producibility

* On-going quality control

‡ If relevant information is not available

† Representative analytes may be chosen on the basis of hydrolysis, oxidation and photolysis characteristics

° Stability data in/on representative commodities should provide sufficient information. Additional tests are required, for example, where:

(a) samples are stored beyond the time period tested (eg. stability tested up to 4 weeks and measurable analyte loss occurs during this period, samples not analyzed until 6 weeks),

(b) stability tests were performed at ≤ -18 °C, but the samples are stored in the laboratory at ≤ 5 °C;

(c) samples are normally stored at ≤ -15°C, but storage temperature rises to +5°C).

♦ Information on efficiency of extraction may be available from the manufacturer or company which is registering the compound.

✶ Occasionally with repeated analysis of test portions of positive samples.

Table 2. Within Laboratory Method Validation Criteria for Analysis of pesticide residues and veterinary drugs[1]

Concentration	Repeatability		Reproducibility		Trueness[2,]
	$CV_A\%$ [3]	$CV_L\%$ [4]	$CV_A\%$ [3]	$CV_L\%$ [4]	Range of mean % recovery
≤1 μg/kg	35	36	53	54	50–120
> 1 μg/kg ≤ 0.01 mg/kg	30	32	45	46	60–120
> 0.01 mg/kg ≤ 0.1 mg/kg	20	22	32	34	70–120
> 0.1 mg/kg ≤ 1 mg/kg	15	18	23	25	70–110
> 1 mg/kg	10	14	16	19	70–110

1. With multi-residue methods, there may be certain analytes where these quantitative performance criteria cannot be strictly met. The acceptability of data produced under these conditions will depend on the purpose of the analyses e.g. when checking for MRL compliance the indicated criteria should be fulfilled as far as technically possible, while any data well below the MRL may be acceptable with the higher uncertainty.
2. These recovery ranges are appropriate for multi-residue methods. Stricter criteria may be necessary for some purposes e.g. methods for single analytes or veterinary drug residues (see Codex V3, 1996).
3. CV_A: Coefficient of variation for analysis excluding sample processing. The parameter can be estimated from tests performed with reference materials or analytical portions spiked before extraction. A reference material prepared in the laboratory may be used in the absence of a certified reference material.
4. CV_L: Overall coefficient of variation of a laboratory result, allowing up to 10% variability of sample processing.

Table 3. Examples of detection methods suitable for the confirmatory analysis of substances

Detection method	Criterion
LC or GC and Mass spectrometry	if sufficient number of fragment ions are monitored
LC-DAD	if the UV spectrum is characteristic
LC – fluorescence	in combination with other techniques
2-D TLC – (spectrophotometry)	in combination with other techniques
GC-ECD, NPD, FPD	only if combined with two or more separation techniques[1]
Derivatisation	if it was not the first choice method
LC-immunogram	in combination with other techniques
LC-UV/VIS (single wavelength)	in combination with other techniques

1. Other chromatographic systems (applying stationary and/or mobile phases of different selectivity) or other techniques.

8. References

1. *Codex Alimentarius*, Food and Agriculture Organisation of the United Nations, Rome, 1993, vol. 2, pp. 405-415.
2. *Annual Report of the Working Party on Pesticide Residues: 1992*, Supplement to The Pesticides Register, HM Stationery Office, London, 1993, pp. 12-28.
3. Codex Alimentarius Volume 3 - 1995. Residues of Veterinary Drugs in foods.
4. Thompson, M., & Wood, R., Pure & Appl. Chem., 67, 1995, 649.
5. AOAC Official Methods Program, *Quick & Easy Associate Referees Manual on Development, Study, Review and Approval Process*, 1995, AOAC International, 481
6. Pocklington, W. D., Pure & Appl. Chem., Vol. 62, No. 1, pp. 149 - 162, 1990.
7. Thompson, M., & Wood, R., Pure & Appl. Chem., 65, 1993, 2123.
8. Thompson, M., & Wood, R., J. of AOAC International, 76, 1993, 926-940, 1993.
9. Horwitz, W., *Pure & Appl. Chem.*, **60**, 1988, 855-864.
10. Horwitz, W., *Pure & Appl. Chem.*, **66**, 1994, 1903.
11. Horwitz, W., *Pure & Appl. Chem.*, **67**,1995, 331.
12. *Anonymous, Council Directive 96/23/EC of 29 April 1996, OJ No. L 125 23.5.96, p 10 – 32*
13. Hill, A.R.C., *Quality control procedures for pesticide residues analysis - guidelines for residues monitoring in the European Union*, Document *7826/VI/97*, European Commission, Brussels, 1997.
14. European Commission, 1999, Guidance Document on Residue Analytical methods, VI B II.1 8064/VI/97-rev 4 15.12.98.
15. US Environmental Protection Agency Office of Pesticide Programme, Tolerence Method Validation SOP: # TMV-1
16. AOAC Peer Verified Methods Policies and Procedures, 1993, AOAC International, 2200 Wilson Blvd, Suite 400, Arlington, Virginia 22201-3301 USA North Frederick Avenue, Suite 500, Gaithersburg, MD 20877-2417 USA
17. AOAC Peer-Verified Methods Programme Manual on Policies and Procedures, Checklist for Method Development, Characterization and Rugedness Testing for Peer-Verified Method Submission - Methods for Chemical Residues in Foods
18. EURACHEM, The fitness for Purpose of Analytical Methods - A Laboratory Guide to Method Validation and Related Topics Guide 3.0, prepared by David Holcomb.
19. Wood, R., Harmonised GL for in-house method validation (IUPAC, in preparation)
20. Bravenboer J, Putten A. v.d., Verwaal W., van Bavel-Salm M., Validation of methods, Inspectorate for Health Protection, Report No. 95-001, NL, 1995
21. Nordic Committee on Food Analysis, NMKL, Validation of chemical analytical methods, NMKL Procedure No. 4., 1996 issued in February 1997.
22. Hill A. R.C., Reynolds S.L., 1999, Guidelines for In-House Validation of Analytical Methods for Pesticide Residues in Food and Animal Feeds, Analyst, 1999, 124, 953-958.
23. European Commission: Commission Decision laying down analytical methods to be used for detecting certain substances and residues thereof in live animals and animal products according to Council Directive 96/23/EC, Final Version: approved by CRLs
24. Jülicher, B., Gowik, P. and Uhlig, S., 1999, Analyst, 124, 537 - 545.
25. Gowik, P., Jülicher, B. and Uhlig, S., J. 1998, Chromatogr. B, 716, 221 - 232.
26. Jülicher, B., Gowik, P. and Uhlig, S., 1998, Analyst, 123, 173 - 179.
27. Youden & Steiner (1975) Statistical Manual of the AOAC, AOAC International, Gaithersburg, MD.

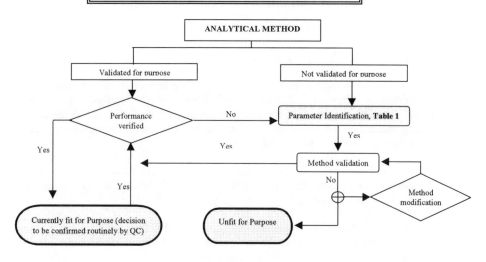

Figure 1. Overview of Method Validation

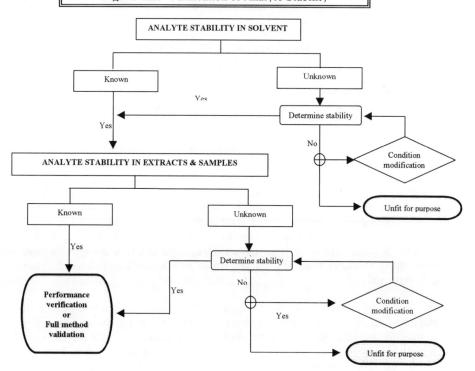

Figure II.2. Verification of Analyte Stability

Method Validation for Pesticide Residues

1   *Validation of methods for pesticide residues*

1.1   Successful participation in proficiency test programmes does not replace the establishment of within laboratory performance of the method. It provides complementary evidence of method and laboratory performance, as does the demonstration of the repeatability of the method in other laboratories by peer verification or collaborative studies

1.2   Before validation of a method commences, the method must be optimised, its robustness tested and the analyst(s) performing the validation should be experienced with the method.

1.3   Consider all residue components included in the definition of residues for the specified purposes. Note that the definitions of residues of a pesticide may be different for enforcement and dietary intake purposes.

1.4   Clearly written detailed instructions (preferably a standard operating procedure) for the method, including accurate descriptions of equipment and reagents to be used, must be available. These instructions should be closely adhered to during the validation process.

1.5   Détailed records on method validation experiments must be maintained where required according to the principles of GLP or ISO 17025.

1.6   Based on the validation data generated, a QC scheme should be designed for the procedure, including appropriate limits, frequency of checks and system suitability tests for equipment.

1.7   Not all procedures can be validated. Sample preparation - as distinct from sample processing - may be impossible or unnecessary to validate. For example, preparation procedures such as gentle rinsing or brushing to remove soil, or the removal of sample parts (such as taking the outer leaves from cabbages) which are not to be analysed, cannot be validated. Similarly, sub-sampling without sample processing (e.g. reducing the size of grain samples), may be required for certain purposes but the procedure is unlikely to be validated as part of a quantitative method. Standard operating procedures (SOPs) must describe these procedures with sufficient clarity that they are performed in a scientifically defensible and consistent manner.

1.8   Inefficiency of extraction can be the major source of bias. Rigorous validation of extraction efficiency of organic analytes can only be performed with samples containing analyte(s) incurred by the route through which the trace levels would normally be expected to arise. Recovery of analytes from samples spiked shortly before extraction does not necessarily reveal correct information on the extractability of incurred residues or native analytes. Suitable certified reference materials containing incurred residues of pesticides, drugs and other contaminants are rarely available, so that validation normally requires identification and quantification of "field-applied" radio-labelled analyte(s), including resulting metabolites and all other degradation products. This is beyond the capability of most laboratories involved in routine trace-level monitoring. Alternative approaches involve

(i)   comparison of the results obtained with extraction of samples containing incurred residues with the procedure tested and with another procedure which has previously been validated rigorously;

(ii)   comparison of the results obtained from extraction of samples containing incurred residues by a very different extraction technique; for example, using a very different solvent (especially if coupled with a more effective sample disintegration technique); or

(iii)   analysis of proficiency test material containing incurred analyte, where the consensus analyte level has been determined by a number of laboratories using either a rigorously validated extraction technique or several different extraction techniques.

1.9   For method validation purposes, commodities should be differentiated sufficiently but not unnecessarily. For example, some products are available in a wide range of minor manufactured variants, or cultivated varieties, or breeds, etc. Generally, though not invariably, a single variant of a particular commodity may be considered to represent others of the same commodity but, for example, a single fruit or vegetable species must not be taken to represent all fruit or vegetables. The analyst must consider each case on its merits but where particular variants within a commodity are known to differ from others in their effects on method performance, they should be the subject of ruggedness testing. Considerable differences in quantitation may occur from species to species.

Some examples of how the validation results may be extended to other commodities in the case of the analysis of pesticide residues:

cereals, validation for whole grains cannot be taken to apply to bran or bread but validation for wheat grain may apply to barley grain;

animal products, validation for muscle should not be taken to apply to fat or offal but validation for chicken fat may apply to cattle fat;

fruit and vegetables, validation for a whole fresh product cannot be taken to apply to the dried product but validation on cabbages may apply to Brussels sprouts.

1.10   Where experience shows similar performance of extraction and cleanup between broadly similar commodities/sample matrices, a simplified approach may be adopted for performance validation. Representative commodities/matrices may be selected for pesticide residue analysis from Table 4 for each group of common properties, and used for validation of the procedure or method. In Table 4, the commodities are classified according to the Codex Classification[1].

1.11 Similarly representative analytes may be used for characterisation of the method. Select compounds to cover physical and chemical properties of analytes that are intended to be determined by the method.

1.11.1 The representative analytes selected should:

   (i) possess sufficiently wide range of physico-chemical properties to include those of represented analytes

   (ii) be those which will be analysed regularly, or for which critical decisions shall be made based on the results.

1.12 The concentration of the analytes used to characterise a method should be selected to cover the AL-s of all analytes planned to be represented in all commodities. Therefore the selected representative analytes should include, among others, those which have high and low AL-s as far as practical. Consequently, the fortification levels used in performance testing with representative analytes/representative commodities may not necessarily correspond to the actual AL-s.

1.13 Individual methods should be fully validated with all analyte(s) and sample materials specified for the purpose, or using sample matrices representative of those to be tested by the laboratory.

1.14 Group specific methods (GSM) should be validated initially with one or more representative commodities and a minimum of two representative analytes selected from the group (see Table 5).

1.15 MRMs may be validated with representative commodities and representative analytes As far as practicable, all analytes should be included in the initial validation process which will have to be tested regularly and which can be determined simultaneously by the determination system used.

1.15.1 The full method validation shall be performed in all matrices and for all compounds specified, if required by relevant legislation.

1.16 The applicability of a method should be characterised by performing the tests described in Table 5.

1.17 Several performance characteristics may be determined simultaneously. To minimise workload it is advisable to combine experiments, leading to experimental plans which limit the workload as far as possible. Sufficient replication, performed over days, analysts and equipment, is essential to provide statistically supportable validation.

---

[1] Codex Alimentarius, Volume 2, 2nd ed., Pesticide Residues in Food, pp. 147-365, FAO, 1993

2.   Extension of the method to a new analyte or matrix

2.1 The applicability of a MRM or GSM for a new analyte or commodity shall be tested according to section 2 of Table 5.

2.1.1   The applicability of the method for the additional analytes and commodities shall be verified as part of the internal quality control programme. Note that ruggedness tests carried out with representative analytes and matrices may not apply to all other analytes or commodities.   Therefore, the performance verification should be carried out more frequently at the beginning of the application of an extended method than during the regular use of a well-established method.

Table 4. Representative commodities/samples for validation of analytical procedures for pesticide residues

Group	Common properties	Commodity group	Representative species
Plant products			
I.	High water and chlorophyll content	Leafy vegetables Brassica leafy vegetables Legume vegetables	spinach or lettuce broccoli, cabbage, kale green beans
II.	High water and low or no chlorophyll content	Pome fruits Stone fruits Berries Small fruits Fruiting vegetables Root vegetables	apple, pear peach, cherry strawberry grape, tomato, bell pepper, melon, mushroom potato, carrot, parsley
III.	High acid content	Citrus fruits	orange, lemon
IV.	High sugar content		raisins, dates
V.	High oil or fat	Oil seeds Nuts	avocado, sunflower seed, walnut, pecan nut, pistachios
VI.	Dry materials	Cereals Cereal products	wheat, rice or maize grains wheat bran, wheat floor
	Commodities requiring individual test		e.g. garlic, hops, tea, spices, cranberry
Products of animal origin			
		Meats	Cattle meat, chicken meat
		Edible offals	Liver, kidney
		Fat	Fat of meat
		Milk	Cow milk
		Eggs	Chicken egg

Note: The method should be validated with representative analytes for each commodity, which is difficult to analyse and requires individual tests (e.g. garlic)

2.1.2   All reported data for a specific pesticide matrix combination should be supported with either validation or performance verification performed on that particular combination.

2.2   The extension of a method may be carried out as a planned activity before commencing the analyses, or as a result of the detection of an unexpected compound.

2.3   When the performance characteristics of the method with a new analyte or matrix match the typical performance characteristics of the method, the method is considered validated for the new analyte.

2.4   When the performance characteristics do not satisfy the criteria set in Section 1 of Table 5, the method is not applicable for that analyte.

3.   *Adaptation of a method in another laboratory*

3.1   The method must have been fully validated in one laboratory, and must be:

   (i)   described in sufficient detail, including the results of performance tests carried out with representative analyte - sample matrix combinations, and providing information on all conditions which prevailed during validation;

   (ii)   used in the adapting laboratory, without any change affecting its performance, with the equipment and conditions in the laboratory being considered equivalent to those of the validating laboratory.

3.2   Missing information must be acquired in the adapting laboratory with appropriate tests described in section 1 of Table 5 by using analytes, matrices/commodities expected to be analysed in the laboratory adapting the particular method.

3.3   When a method is successfully adapted in a laboratory, and its performance tests, carried out according to Section 3 of Table 5, meet the performance parameters established by the validating laboratory, the method is considered applicable in the adapting laboratory for all analytes and sample matrices which were included in the original validation.

3.4   Compounds and commodities which can be considered as represented but were not included in the validated method shall be considered as new compounds and commodities, and tested according to section 2 of Table 5, respectively.

4.   *Internal quality control – performance verification*

The conditions of the application of a trace analytical method continually vary in practice. For instance, the composition of samples and the consequent interfering co-extracts or partition properties may be different depending on the variety and maturity of sample matrix , while the inertness and separation power of chromatographic columns gradually change depending on the sample loading and duration of use. Even the most rigorous robustness testing procedure cannot take into account such variability. Therefore, the performance of the method shall be regularly verified during its use as part of the internal quality control programme of the laboratory. In addition, each analyst using the method for the first time should complete the tests specified in sections 4.5.5 of Table 5 to demonstrate that they can use the method within the expected performance parameters established during method validation prior to applying the method for analysis of samples.

4.1   The major purposes of internal quality control /performance verification are to:

- monitor the performance of the method under the actual conditions prevailing during its use,

- take into account the effect of inevitable variations caused by, for instance, the composition of samples, performance of instruments, quality of chemicals, varying performance of analysts and laboratory environmental conditions,

- demonstrate that the performance characteristics of the method are similar to the accepted performance parameters, therefore the application of the method is under "statistical control", and the accuracy and uncertainty of the results are comparable to the performance characteristics established. (Note: The accepted performance parameters are initially those which were obtained during method validation, they may be refined with data collected from performance verification during the regular use of the method.

4.2   The results of internal quality control provide essential information for the confirmation, refinement and extension of performance characteristics established during the initial validation of the method.

4.3   The essential performance characteristics to be tested and the appropriate test procedures are described in Section 4 of Table 5.

4.4   For the most effective internal quality control, analyse samples concurrently with quality control check samples. The number of quality control tests depends on the frequency of use of the method and the number of samples which can be analysed in one batch.

4.5   Construction and use of control charts.

4.5.1   The initial control charts is constructed with the average recovery (Q) of representative analytes in representative matrices and the typical within laboratory reproducibility coefficient of variation (CV_{typ}) of analysis for checking acceptability of individual recovery results. The warning and action limits are $Q \pm 2*CV_{Atyp}*Q$ and $Q \pm 3*CV_{Atyp}*Q$, respectively.

4.5.2   At the time of the use of the method, the recoveries obtained for individual analyte/sample matrices are plotted in the chart. The average recovery and reproducibility standard deviation are calculated for the first 5 recovery studies carried out with a particular analyte/sample matrix. The average recovery and within laboratory reproducibility of the specific analyte/matrix combination should be within the ranges specified in Table 2.

4.5.3   If during the regular use of the method the average of the first 5 recovery tests for a particular analyte /sample matrix is significantly different (P=0.05) from the average recovery obtained for the representative analyte/sample matrices, refine recovery and precision data and construct a new control chart for the particular analyte/sample matrix applying the new average recovery and the typical CV values.

4.6   Based on the results of internal quality control tests, refine the control charts at regular intervals if necessary. The results are acceptable and the method is considered applicable as long as the criteria specified in Table 2 are fulfilled.

4.7   If the analyte content measured in the quality control check samples is outside the action limits, the analytical batch (or at least the analysis of critical samples in which residues found are ≥ 0.7 AL and 0.5 AL for regularly and occasionally detected analytes, respectively) may have to be repeated.

4.8   When the results of quality control check samples fall repeatedly outside the warning limits (1 in 20 measurements outside the limit is acceptable), the application conditions of the method have to be checked, the sources of error(s) have to be identified, and the necessary corrective actions have to be taken before the use of the method is continued.

4.9   Re-analyse regularly analytical portions of positive samples. The differences of the replicate measurements can be used to calculate the overall within laboratory reproducibility of the method (CV_{Ltyp}) in general or specifically for a particular analyte/sample matrix. The CV_{Ltyp} will also include the uncertainty of sample processing, but will not indicate if the analyte is lost during the process.

5.   *Changes in the implementation of the method*

5.1   When the conditions of the method are changed the affected part of the method shall be revalidated (see Table 1).

5.1.1   Replacement of basic equipment with new one does not require revalidation.

5.1.2   Replacement of major instruments or parts should always be followed by system suitability test(s); revalidation of detection conditions may also be necessary.

5.2   Some frequently occurring situations and the parameters to be tested are given in section 5.4 in Table 5.

5.3   If the performance characteristics achieved using the original conditions are not met, new performance characteristics shall be established, and adopted for the future application of the method. The new performance characteristics shall meet the performance criteria defined in Table 5.

Table 5. Summary of parameters and criteria for development, adaptation and validation of pesticide multi residue analytical procedures[2]

Parameter	Level(s)	No. of analyses or type of test required	Criteria — Quantitative method	Criteria — Screening method	Comments
1. Within-Laboratory (single laboratory) performance of optimised method					
1.1 Analyte stability in extracts and standard solutions	At ≤AL, or with well detectable residues	≥5 replicates at each appropriate point in time (including zero) and for each representative analyte/commodity. Fortify blank sample extracts to test stability of residues. Compare analyte concentration in stored and freshly made standard solutions.	No significant change in analyte concentration in stored extracts and analytical standards (P = 0.05)	At the end of the storage period, residues added at LCL are detectable	The test of stability in extracts is required if the analytical method is suspended during the determination process, and the material will likely be stored longer than during determination of precision, or if low recoveries were obtained during optimisation of the method. During method optimisation, recovery should be measured against both "old" and "freshly prepared" calibration standards, if the recovery extracts are stored. Storage time should encompass the longest period likely to be required to complete the analysis.
1.2 Calibration function Matrix effect	LCL to 2 (3) times AL	Test the response functions of all analytes included in the method with ≥2 replicates at ≥3 analyte levels plus blank sample. For non-linear response, determine response curve at ≥7 levels and ≥3 replicates. Test the matrix effect with all representative analytes and matrices. Apply the standards prepared in solvent and sample extracts randomly.	For linear calibration: regression coefficient for analytical standard solutions (r) ≥ 0.99. When weighed linear regression is used the SD of residuals (S_{yx}) ≤ 0.1 For polynomial function (r) ≥ 0.98. The matrix effect is confirmed if the difference is significant based on paired t-test at P = 0.05.	For linear calibration: regression coefficient (r) ≥ 0.98. SD of residuals (r) ≤ 0.2 For polynomial function (r) ≥ 0.95	Calibration parameters may be established during optimisation of the procedure, determination of precision or detection capability. Prepare calibration solutions of different concentrations For MRM perform calibration with mixtures of analytes ("standard mixture"), which can be properly separated by the chromatographic system. Use matrix matched analytical standards for further tests if matrix effect is significant. The method validation may not give definite information for the matrix effect, because matrix effects change with time, with sample (sometimes), with column, etc.
1.3 Analytical range, accuracy, trueness, precision, limit of detection (LD), limit of quantitation (LOQ)	LCL to 2 (3) times AL*	Analyse representative analyte matrix combinations: ≥ 5 analytical portions spiked at zero, LCL, AL and ≥3 replicates at 2–3 AL level. The recovery tests should be divided among the analysts, who will use the method, and instruments which will be involved in the analysis.	LOQ should be fit for purpose. Mean recovery and CV_A see Table 2. Reference materials, mean residue* is not significantly different from consensus value (P = 0.05).	All recoveries are detectable at LCL	The analysts should demonstrate that the method is suitable for determining the presence of the analyte at the appropriate AL with the maximum (false negative and false positive) errors specified. For MRM, the fortification level of blank samples should cover the AL's of analytes represented. Consequently they may not correspond with the actual AL for the representative analytes. Fortify analytical portions with standard mixtures. The accuracy and precision ranges determined for representative analyte/matrix combinations can be considered typical for the method, and will be used as applicability criteria for extension to new analytes and commodities, as well as initial guidance for internal quality control of the

[2] Unless otherwise specified replace representative analytes and representative commodities with individual analyte and sample matrices for single analyte methods, and tests for banned substances.

Parameter	Level(s)	No. of analyses or type of test required	Criteria — Quantitative method	Criteria — Screening method	Comments
					method. Report uncorrected results, mean recovery and CV_A of replicates. CV_A is equivalent to the within laboratory reproducibility of analysis of samples. * Correct the results for mean recovery if it is significantly different from 100 %. Where the method does not permit recovery to be estimated, accuracy and precision are those of calibration.
1.4 Specificity and selectivity of analyte detection	At lowest calibration level (LCL)	Identify by mass spectrometry, by a similarly specific technique, or by the appropriate combination of separation and detection techniques available. Analyse ≥5 blanks of each representative commodity obtained preferably from different sources. Report analyte equivalent of blank response. Determine and report selectivity (δ) of detector and relative response factors of representative analytes (RRF) with specific detectors used.	Measured response is solely due to the analyte. Residues measured on two different columns should be within the critical range of replicate chromatographic determinations.	The rate of false negative samples (β error) at AL should typically be < 5‰.	Applies only to a specific combination of separation and detection technique. Samples of known treatment history may be used instead of untreated samples, for analytes other than that applied during treatment. Maturity of sample matrices may significantly affect the blank sample response. Blank values shall also be regularly checked during performance verification (see Section 5 below). Report typical peaks present in the extracts of blank samples. The LCL should preferably be ≤ 0.3AL, except when the AL is set at or about the limit of quantitation. The test may be performed in combination with the determination of decision limit and detection capability and will also provide information for the RRTs and RRFs of compounds. Alter chromatographic conditions if blank sample response interfere with the analyte or use an alternative detection system. Suitable combination of selective detectors increases specificity, because the amount of information about the analyte is increased.
1.5 Selectivity of separation	At AL	Determine RRt values for all analytes to be tested by the method (not only the reference compounds). When chromatographic techniques are used without spectrometric detection, apply different separation principles and/or determine RRt-s on columns of different polarity. Determine and report resolution (R_s) and tailing factors (T_f) of critical peaks.	The nearest peak maximum should be separated from the designated analyte peak by at least one full width at 10% of the peak height, or more selective detection of all analytes is required.	Tentative identification of all analytes tested. (Not all analytes need to be separated)	Unless the chromatographic separation and spectrometric detection is used in combination, report RRt values on columns of different polarity, which enable the separation (minimum R ≥ 1.2) of all analytes tested. The test may be combined with the determination of calibration function and matrix effect (see. 1.7)
1.6 Homogeneity of analyte in analytical sample	At about AL or well detectable residues	Analyse ≥5 replicate test sample portions of one representative commodity from each group (Table 4), post-processing. Determine CV_{Sp} with analysis of variance. The analyte homogeneity should be checked with analytes known to be stable.	$CV_{Sp} \leq 10\%$.	$CV_{Sp} \leq 15\%$ For screening methods it may be desirable to take a portion in which residues can be expected to be highest (e.g. citrus peel) and achievement of homogeneity may	Use preferably commodities with incurred stable surface residues or treat the surface of a small portion of the natural units (<20%) of laboratory sample before cutting or chopping to represent worst scenario of sample processing. Processing validated for use with any subsequent procedure. Validation applicable to other commodities with similar

Parameter	Level(s)	No. of analyses or type of test required	Criteria		Comments
			Quantitative method	Screening method	
				be unnecessary.	physical properties, and it is independent of the analyte. The test may be combined with testing stability of analyte (see Section 1.7 of this Table) Determine the sampling constant[3,4]. to calculate the size of analytical portion required to satisfy quality criteria of CV_{Sp} ≤ 10% specified. The CV_{Sp} may not need to be determined separately if the CV_L of the incurred residues are within the limits specified in Table 2.
1.7 Analyte stability during sample processing	About AL	Fortify interested commodities with known amounts of analytes before processing the sample. Analyse ≥5 replicates of each commodity, post-processing. Apply a notionally stable marker compound together with the analytes tested For MRM and GSM, several analytes, which can be well separated, can be tested together.	The stability of the analyte need not be specified if the average overall recovery of analyte added before sample processing (including procedural recovery) and CV_A are within the ranges specified in Table 2 Quantify stability if the overall recovery and the procedural recovery is significantly different (P=0.05).	Analyte added at LCL remains detectable after processing	The temperature of the sample during processing may be critical. Processing validated for use with any subsequent procedure. Validation may be specific to analyte and/or sample matrix. For testing stability determine the mean recovery and CV_L of labile and stable marker compounds. Use these compounds for internal QA tests (see section 5). Express the ratio of average concentration of labile and stable compounds to indicate stability of residues. CV's of stable compounds will indicate the within laboratory repeatability as well.
1.8 Extraction efficiency	About AL or readily measurable residues	Analyse ≥5 replicate portions of samples or reference material with incurred residues. Compare the reference (or different) procedure with that under test. For MRM the analytes tested should preferably have a wide range of Pow values. Only be determined using incurred residues.	For samples with incurred residues, the mean result obtained with the reference procedure and the tested procedure should not differ significantly at P=0.05 level applying CV_L in the calculation. Or: the consensus value of reference material and the mean residue should not differ significantly at P=0.05 level when calculated with CV_A of the method tested. When the CV_A of the method is larger than 10%, the number of replicate analyses has to be increased to keep the relative standard error of the mean < 5%. Otherwise quantify and report the efficiency of extraction (excluding the recovery of analytical	The mean incurred residues, known to be present at or about the LOQ or LCL, are actually detectable in the samples.	Temperature of the extract, speed of blender or Ultra Turrax, time of extraction and solvent/water/matrix ratio may significantly effect the efficiency of extraction. The effect of these parameters can be checked with ruggedness test. The optimised conditions should be kept constant as far as possible. Validation is generally applicable for commodities within one group and represented analytes of similar physical and chemical properties. Validation is independent from subsequent procedures in the method.. The average recovery of each method shall be determined from spiked analytical portions. Correct results with average recovery of analysis if its is significantly different from 100%. According to some regulations the ability of screening kits should be tested to detect a positive at 95% confidence.

[3] Wallace, D. and Kratochvil, B., Analytical Chemistry, **59**, 1987, 226.
[4] Ambrus, A., Solymosné, E.M. and Korsós, I., J. Environ. Sci. and Health, **B31**, 1996, 443.

Parameter	Level(s)	No. of analyses or type of test required	Quantitative method	Screening method	Comments
				phase).	
1.9 Analyte stability during sample storage	About AL	Analyse freshly homogenised samples containing incurred residues, or homogenise and spike blank samples (time 0), and then analyse samples stored according to normal procedures of the laboratory (usually at ≤ -18 °C). The storage time should be ≥ than the longest interval foreseen between sampling and analysis. ≥5 replicates at each time point. When the stored portions are analysed ≥ 4 occasions, test ≥2 spiked portions, and ≥ 1 blank portion spiked at the time of analysis.	No significant loss of analyte during storage (P = 0.05)	Analyte added at lowest calibration level, LCL, remains detectable after storage	Storage is validated for use with any subsequent procedure. Validation is specific to analyte. However, generally storage stability data obtained with representative sample matrices can be considered valid for similar matrices. The matrices shall be selected taking into account the chemical stability (e.g. hydrolysis) of the analyte and the intended use of the substance. Useful information can be obtained on stability during storage from the JMPR evaluations[5] or from dossiers submitted for registration Report the initial residue concentration, the remaining residue concentration and the procedural recovery of the analyte. Unnecessary sample storage can be avoid by a careful planning for sampling and consequent analysis through administrative arrangement, which is not a part of analytical method.

2. Extension of the validated method

Parameter	Level(s)	No. of analyses or type of test required	Quantitative method	Screening method	Comments
2.1 Analyte stability during sample storage, processing, and in extracts and standard solutions.	See 1.1, 1.2 & 1.9				Only if information on stability under the processing conditions and on the representative matrix is not already available
2.2 Calibration function, matrix effect	LCL to 2 (3) AL:	Three point calibration embracing AL with and without matrix matched analytical standards	For linear calibration: regression coefficient for analytical standard solutions (r) ≥ 0.99. SD of residuals ($S_{y/x}$) ≤ 0.1 For polynomial function (r) ≥ 0.98.	For linear calibration: regression coefficient (r) ≥ 0.98. SD of residuals ≤ 0.2 For polynomial function (r) ≥ 0.95.	The method validation may not give definite information for the matrix effect, because matrix effects change with time, with sample (sometimes), with column, etc.
2.3 Accuracy, precision, LOD, LOQ	at AL	Planned in advance: (a) Analyse 3 analytical portions of representative sample matrices of interest fortified at AL Unexpectedly found: Fortify 2 preferably 3 additional portions of analytical sample approximately at the level of the new analyte. Calculate the recovery of added analyte. Use similar	The residues recovered should be within the repeatability limits of the method: Three portions: C_{max}- C_{min} ≤ 3.3CV_{Atyp}Q Two portions: C_{max}- C_{min} ≤ 2.8*CV_{Atyp}Q Q =average recovery of the new	Analytes added to blank samples at target reporting level should be measurable in all tests.	Use CV_{Atyp} established during method validation. The method should only be tested with commodities representing the intended use (possible misuse) of the analyte.

[5] FAO, Pesticide Residues in Food – Evaluations; published annually in the series of FAO Plant Production and protection Papers

Parameter	Level(s)	No. of analyses or type of test required	Criteria Quantitative method	Criteria Screening method	Comments
		sample matrix for recovery test if appropriate amount of analytical sample is not available.	analyte, and it shall comply with Table 2.		
2.4 Specificity and selectivity of analyte detection	At LCL	Identify by mass spectrometry, or by the appropriate combination of separation and detection techniques available. Planned in advance: (a) Analyse one representative blank sample from each commodity group of interest (in which the new analyte is likely to be present). Analyse new matrix with representative compounds. Unexpectedly found: (b) Check response of blank sample (if available), or demonstrate that the response measured corresponds solely to the analyte, using the best technique available in the laboratory. Check δ and RRF of detection and RRt-s of representative analytes. Compare RRt and response of new analyte with other analytes tested during method validation and with blank responses obtained during extension of the method and the prior validation of the method.	Measured response is solely due to the analyte. The detection system used should have equal or better detector performance than those applied during method validation. Residues measured on two different columns should be within the critical range of replicate chromatographic determinations. RRts of representative analytes obtained during method validation and measured should be within 2 % for GLC and 5 % for HPLC determinations.	The rate of false negative samples (β error) at AL should be < 5‰.	When the extension for a new analyte is planned, the applicability of the method shall be checked for all representative sample matrices in which the analyte may occur. When an analyte is unexpectedly detected, the performance check may be carried out for the actual matrix alone. See also 1.4. The responses of blank sample(s) should not interfere with the analytes, which are likely to be measured in the sample. Report typical peaks present in blank extracts. The background noise of a new matrix extract should be within the range obtained for representative commodities/sample matrices. If the selectivity of detection does not eliminate the matrix response, use appropriate combination of chromatographic columns which enables the separation of analytes from the matrix peaks. See other options in Table 3.
2.5 Selectivity of separation	See 1.5	See 1.5	See 1.5	See 1.5	See 1.5 Only if information is not available
2.6 Extraction Efficiency	See 1.8	See 1.8	See 1.8	See 1.8	See 1.8 Only if information is not available
3. Adaptation of the validated method in another laboratory					
3.1 Purity and suitability of chemicals, reagents and ad(ab)sorbents	See 1.10	Test reagent blank, applicability of ad(ab)sorbents and reagents. Perform derivatization without and with sample.	No interfering response above 0.3 LCL.	No interfering response above 0.5 AL	Some of the most common problems in method transfer involve differences in selection of reagents, solvents and chromatographic media, or in equipment capabilities. Whenever possible, try to confirm actual materials and equipment used by the method developer, if that information is not provided with the method or publication, as received. Substitutions can be tried after the method is working within your laboratory.
3.2 Analyte stability in extracts and stan-	See 1.1	See 1.1	See 1.1	See 1.1	This testing may be omitted if full information on analyte stability is provided with the method or if the method is re-placing a previously used method for the analyte and the

Parameter	Level(s)	No. of analyses or type of test required	Criteria		Comments
			Quantitative method	Screening method	
dard solutions					stability information has been previously generated for the previous method.
3.3 Calibration function Matrix effect	LCL to 2 (3) times AL	Test the response functions of representative analytes included in the method at ≥3 analyte levels plus blank. For non-linear response, determine response curve at ≥7 levels and ≥3 replicates. Test the matrix effect with representative analytes and matrices.	For linear calibration: regression coefficient for analytical standard solutions (r) ≥ 0.99. For weighed regression SD of residuals (S_{yx}) ≤ 0.1 For polynomial function (r) ≥ 0.98.	For linear calibration: regression coefficient (r) ≥ 0.98. For weighed regression SD of residuals ≤ 0.2 For polynomial function (r) ≥ 0.95.	Sees: 1.2
3.4 Analytical range Accuracy and precision, limit of detection, limit of quantitation	Blank extract and or AL	Analyse representative analyte/matrix combinations : ≥ 5 analytical portions each of blank samples spiked at 0 and AL, and 3 portions spiked at 2 AL. The recovery tests should be divided among the analysts, who will use the method, and instruments which will be involved in the analysis.	Average recovery and CV_A should be within the ranges given in Table 2.	All recoveries detectable at LCL. Reference materials at AL: analyte detected.	See comments in 1.3.
3.5 Specificity and selectivity of analyte detection	At AL	Check performance characteristics of detectors used and compare them with those specified in the method. Check response of one blank of each representative commodity, otherwise perform test as described in section 1.4.	Measured response is solely due to the analyte. The detector performance (sensitivity and selectivity) should be equal or better than specified in the method. See section 1.4	The rate of false negative samples (β error) at AL should typically be < 5%.	The relative response of specific detectors can substantially vary from model to model. Proper checking of specificity of detection is critical for obtaining reliable results. Compare blank response observed with typical peaks reported in blank extracts See other comments under section 1.4.
3.6 Analyte "homogeneity"	At about AL or well detectable residues	Test two representative commodities of different nature	CV_{Sp} <10%	CV_{Sp} <15% For screening methods it may be desirable to take a portion in which residues can be expected to be highest (e.g. citrus peel) and achievement of homogeneity may be unnecessary.	The tests are performed to confirm similarity of application conditions and applicability of parameters obtained by the laboratory validating the method. When the test results in similar CV_{Sp} as reported, the conditions of sample processing may be considered similar and further tests are not required for the validation of the method.
3.7 Analyte stability in extracts and standard solutions	See 1.1	See 1.1	See 1.1	See 1.1	This testing may be omitted if full information on analyte stability is provided with the method or if the method is replacing a previously used method for the analyte and the stability information has been previously generated for the previous method.

4. Quality control (performance verification)

4.1 Methods used regularly

Parameter	Level(s)	No. of analyses or type of test required	Criteria		Comments
			Quantitative method	Screening method	
4.1.1 Suitability of chemicals, adsorbents and reagents		For each new batch: Test reagent blank, applicability of ad(ab)sorbents and reagents Perform derivatization without sample.	No interfering response ≥0.3 LCL.	No interfering response ≥ 0.5AL.	Alternately, if the sample blank, calibration and the recovery are satisfactory then the suitability of reagents etc are confirmed.

Parameter	Level(s)	No. of analyses or type of test required	Criteria		Comments
			Quantitative method	Screening method	
4.1.2 Calibration and analytical range		Single point calibration may be used with standard mixtures, if the intercept of calibration function is close to 0. Apply multi point calibration (3x2) for quantitative confirmation.	The analytical batch may be considered to be under statistical control if the analytical standards and sample extracts are injected alternately, and the calculated $S_{y/x}$ is ≤0.1.	Analyte is detected at LCL.	Standard solution and samples should be injected alternately. Bracketing with appropriate standard injections may provide a time saving alternative to multi point calibration especially if auto sampler is not available. As system response often changes multi point calibration shall be performed regularly to confirm that the intercept is close to zero. Multi point calibration is not necessary for quantitative confirmation if the calibrant is very close in concentration to that of the sample.
4.1.3 Accuracy and precision	Within analytical range	Include in each analytical batch ≥1 sample either: fortified with standard mixture, or the reanalysis of a replicate portion of a positive sample,	The performance of detector and chromatographic column shall be equal or better than specified in the method. Preferably all recoveries should be within the warning limit of control chart constructed with the specific or typical CV_A of analytes. On a long run one of every 20 or 100 samples may be outside the warning and action limits, respectively. The analytical batch should be repeated if any of the recoveries falls outside the action limits, or the results of the replicate analyses of the positive sample exceeds the critical range. C_{max} - C_{min} > 2.8*$CV_{L,y}Q$ Q is the average residue obtained from the replicate measurements, the $CV_{L,y}$ is the measure of within laboratory reproducibility which includes the combined uncertainty of sample processing and analysis.		Fortify analytical portion with standard mixture(s). Alter standard mixtures in different batches to obtain recoveries for all analytes of interest at regular intervals. Perform alternately recovery studies at AL as well as at LCL and 2 times AL, as appropriate, to confirm applicability of the method within the analytical range. The frequency of recovery studies at AL should be 2 to 3 times higher then those at other levels. Repeated analysis of positive samples may replace the recovery test in a particular batch. For MRM prepare commodity/sample specific standard mixtures from the analytes which may occur in a particular sample. The selection of analytes for one mixture should assure selective separation/detection without any problem. For tentative identification: prepare analytical batches containing the appropriate detection test mixture, and samples. For quantitative determination/confirmation include in the analytical batch the detection test mixture, appropriate number of calibration mixtures, fortified blank sample(s), or one repeated positive sample and the new positive samples alternately.
4.1.4 Selectivity of separation, Specificity of detection Performance of detectors		Include appropriate detection test mixture in each chromatography batch. Include untreated commodity (if available) in analytical batch. Use standard addition if no untreated sample (similar to those analysed in the batch) is available Confirm identity and quantity of each analyte present ≥0.7 AL level.	R_s, T_r of test compounds, and RRF and δ of the detection should be within the specified range. RRt-s should be within 2 % for GLC and 5 % for HPLC determinations Detector performance should be within specified range. Sample co-extractives interfering with the analyte should not be present ≥ 0.3 LCL. The recovery of added standard should be	Detector performance should be within specified range. Analyte should be seen above LCL or CCα for banned compounds.	This is also sometimes referred to as a "system suitability" test. Prepare detection test mixture for each method of detection. Select the components of the mixture in order to indicate the characteristic parameters of chromatographic separation and detection. Adjust relative retention data base for the compounds of detection test mixture and analytes used for calibration. Define the RRF specific for the detection system. Perform quantitative confirmation with analytical standards prepared in blank matrix extract if matrix effect is significant.

Parameter	Level(s)	No. of analyses or type of test required	Criteria — Quantitative method	Criteria — Screening method	Comments
			within the acceptable recovery range of the analyte.		
4.1.5 Analyte homogeneity in processed sample	At well detectable analyte concentration.	Select a positive sample randomly. Repeat analysis of another one or two analytical portions.	The residues measured on two different days should be within the reproducibility limit of replicate analytical portions: $C_{max} - C_{min} \leq 2.8*CV_{L,rp}Q$ Q is the combined uncertainty of sample processing and analysis obtained during method validation.		Perform test alternately to cover each commodity analysed. Test homogeneity at the beginning of growing season, or at the start of the analysis of the given type of samples. The acceptable results of the test also confirm that the reproducibility of the analyses (CV_A) was appropriate.
4.1.6 Extraction efficiency					The efficiency of the extraction cannot be controlled during the analysis. To ensure appropriate efficiency, the validated extraction procedure should be carried out without any change.
4.1.7 Duration of analysis			The samples, extracts etc. should not be stored longer than the period for which the storage stability was tested during method validation. Storage conditions should be regularly monitored and recorded.		Examples for the need of additional storage stability tests are given under Table 2.
4.2 Analyte detected occasionally					
Follow tests described in 4.1 with the following exceptions					
4.2.1 Accuracy and precision	At around AL	Reanalyse another analytical portion; Use standard addition at the measured level of analyte.	The residues measured on two different days should be within the critical range: $C_{max}-C_{min} \leq 2.8*CV_{L,rp}Q$ Q is the average residue obtained from the replicate measurements, the $CV_{L,rp}$ is obtained during method validation. The recovery following standard addition shall be within action limits.		Check accuracy if residue found at $\geq 0.5AL$.
4.3 Methods used at irregular intervals					
Follow tests described in 4.1 with the following exceptions					
4.3.1 Accuracy and precision (repeatability)	At AL and LCL	Include one fortified sample at LCL and two samples at AL in each analytical batch. Use standard addition if untreated sample (similar to those analysed in the batch) is not available. Perform analysis with ≥2 analytical portions.	Minimum two recoveries shall be within warning limit, one may be within action limit. The residues measured in replicate portions should be within the critical range: $C_{max}- C_{min} \leq 2.8*CV_{L,rp}Q$ or $C_{max}- C_{min} \leq f_{(n)}*CV_{L,rp}Q$ Q is the average residue obtained from the replicate measurements, the $CV_{L,rp}$ is obtained during method validation.		The acceptable results also prove the suitability of chemicals, adsorbents and reagents used. Confirm residues above 0.5AL. If performance criteria were not satisfied, the method shall be practised and its performance characteristics (Q, CV_{App-}, $CV_{L,rp}$) re-established during partial revalidation of the method.
4.4 Changes in implementation of the method					
For test methods and acceptability criteria see the appropriate sections of Appendix 1.					
Change	Parameters to be tested				
4.4.1 Chromatographic column	Test selectivity of separation, resolution, inertness, RRt values		Performance characteristics should not be affected		Apply appropriate test mixtures to obtain information on the performance of the column.
4.4.2 Equipment for sample processing	Homogeneity of processed sample; Stability of analytes		Test described in 1.6 and 1.7 shall be performed and they should give results conforming with the relevant criteria.		Homogeneity test is only necessary if the degree of comminution and/or mixing is inferior to that of the original equipment. The stability of analytes need to be tested if the processing time and temperature are significantly increased.
4.4.3 Equipment for extraction	Compare field incurred residue levels detected with the old and new equipment in ≥ 5 replicates		The mean residues should not be significantly different at p=0.05 level.		Test is necessary if a new type of equipment is used

Parameter	Level(s)	No. of analyses or type of test required	Criteria		Comments
			Quantitative method	Screening method	
4.4.4 Detection	Test selectivity of separation and selectivity and sensitivity of detection		Performance characteristics should be the same or better specified in the description of the method.		Test also detectability separately with new detection reagents.
4.4.5 Analyst	≥5 recovery tests at each level (LCL, AL and 2 (3) AL), re-analysis of one blank sample and two positive samples (unknown to the analyst)		All results should be within the warning limits specified for the method in the laboratory. Replicate sample analysis shall be within the critical range.		This is a minimum requirement. Laboratories in some areas of residue work use a more detailed protocol which includes: (1) generation of standard curve within acceptability criteria; (2) minimum of 2 analytical runs for each matrix, containing representative analytes fortified by the analyst at a minimum of 3 levels in duplicate; (3) minimum of 1 analytical run containing fortified or incurred samples, 3 levels in duplicate, provided as unknowns to the analyst. All results must meet acceptability criteria, or be repeated.
4.4.6 Laboratory	Accuracy and precision ≥3 recovery tests at each level (LCL, AL and 2 (3) AL) by (different) analyst(s) on different days.		All results should be within the warning limits specified for the method in the laboratory.		The reproducibility of the method under the new conditions must be established and it has to be done by more than one analyst if available.

METHOD VALIDATION FOR RESIDUES OF VETERINARY DRUGS

1. *Validation of methods for veterinary drug residues*

   1.1 See Sections 1.1 – 1.9 of Annex 1. The general instructions provided there are also relevant to the validation of methods of analysis for veterinary drug residues.

   1.2 Ideally, a method of analysis for veterinary drug residues should be validated for the analysis of the four tissues generally classed as "edible tissues", which are fat, liver, kidney and muscle. Local dietary preferences may require validation of methods for other tissues which are normally consumed in a country or region. In addition, there may be a regulatory requirement to analyse urine or other body fluids for residues, particularly if live animal testing is part of a regulatory program. From a practical approach, the usual minimum requirement is that an analytical method should be validated for what is normally termed as "target tissue", which is the tissue from a treated animal in which the highest and most persistent concentrations of the drug residue are expected to be found. This would usually be the tissue collected for a national residue monitoring program. In addition, there is a requirement to test the "tissue in trade" when products are shipped between countries. This is most commonly muscle tissue, but may include other tissues. General guidance as to the selection of suitable target tissues and the expected "tissue in trade" is provided in Table 6. A knowledge of the metabolism and tissue distribution should ideally be gained for each drug residue before a final selection of appropriate tissues for validation is made.

   1.3 It is also essential to determine that the method is validated for the appropriate marker residue, which may be either the parent compound, a metabolite or a representative compound formed during the analysis which includes both parent compound and metabolites. Guidance on the selection of the marker residue is provided in the reports of the Joint FAO/WHO Expert Committee on Food Additives[1].

   1.4 The concentration of the analytes used to characterise a method should be selected to cover the AL-s of all analytes planned to be represented in all matrices to be tested. Therefore the selected representative analytes should include, among others, those which have high and low AL-s as far as practical. Consequently, the fortification levels used in performance testing with representative analytes/representative commodities may not necessarily correspond to the actual AL-s.

   1.5 Characterise the applicability of a method by performing the tests described in Table 7 for veterinary drug residues. See also Annex 1, Sections 1.14 and 1.16 (apply to Table 7).

---

[1] *Residues of some veterinary drugs in animals and foods*, FAO Food & Nutrition Paper, Series 41, Food and Agriculture Organization of the United Nations, Rome.

Table 6. Practical guidance on selection of appropriate test matrix for examination for residues of veterinary drugs in foods.

Species/Commodity for method validation	Usual target tissue or matrix for method validation	
	Water-soluble	Fat-soluble
Ruminant (e.g. cattle, sheep)[*]	Liver or kidney, muscle[**]	Fat, muscle
Non-ruminant (e.g. pig)[*]	Liver or kidney, muscle[**]	Fat, muscle
Poultry (e.g. chicken, turkey)*	Liver, muscle	Fat ,or muscle with adhering skin in normal proportions[**]
Fish	Muscle with adhering skin in normal proportions	Muscle with adhering skin in normal proportions
Shellfish/Crustacean (e.g. prawn)	Muscle	Muscle
Milk (usually cows' milk)	Whole milk	Whole milk
Honey	Honey	Honey

* Validation should be conducted for all major species from which samples will be collected for routine testing. For minor use applications, it may be acceptable to conduct a performance verification for the new species if full validation has previously been conducted for another species from the group (e.g., ruminant).

** Residues of water-soluble compounds are usually found at highest concentrations in either liver or kidney, with the choice of tissue being made based on distribution studies provided by the drug sponsor at the time of registration by a national authority. Fat-soluble compounds are usually present as residues at highest concentrations in fat, so in such instances the selection of test matrices is typically fat and muscle. However, in the case of poultry and fin-fish, where food preparation and consumption frequently include both the muscle and skin with fat, a suitable guideline may be "muscle with adhering skin in normal proportions", reflecting the combined muscle tissue, fat and skin which may be consumed. Such requirements should be clearly established with the client (the purchaser or user of results) before beginning method validation. National or regional authorities, or purpose of testing may require validation for different or additional matrices.

2. *Extension of the method to a new analyte or matrix*

    See Annex 1 and refer to Table 7.

3. *Adaptation of a method in another laboratory*

    See Annex 1 and refer to Table 7.

4. *Internal quality control – performance verification*

    See Annex 1 and refer to Table 7.

5. *Construction and use of control charts*

    See Annex 1.

6. *Changes in the implementation of the method*

    See Annex 1 and refer to Table 7.

Table 7. Summary of parameters and criteria for development adaptation and validation of single analyte, group specific and multi-residue analytical procedures for veterinary drug residues.

Parameter	Level(s)	No. of analyses or type of test required	Codex Level I method (Confirmatory/Quantitative)	Codex Level II method (Quantitative)	Codex Level III method (Screening – qualitative or semi-quantitative)	Comments
				Criteria		
1. Within-laboratory performance of optimised method						
1.1 Analyte stability during sample storage	About AL	Analyse representative samples (time 0) and samples stored according to normal procedures of the laboratory (e.g. at ≤ -18°C). The storage time should be ≥ than the longest interval foreseen between sampling and sample disposal. Repeat at -70 °C if analyte stability does not meet criteria at ≤-18°C. ≥5 replicates at each time point.	No significant loss of analyte during storage (P = 0.05)	No significant loss of analyte during storage (P = 0.05)	No false negatives at after storage.	Storage stability should be assessed using incurred tissues, when available. Otherwise, prepare fortified test materials using different pools of blank tissue to reflect the expected variability of the samples to which the method is to be applied. Storage is validated for use with any subsequent procedure. Validation may be specific to analyte. However, generally storage stability data obtained with representative sample matrices can be considered valid for similar matrices. The matrices shall be selected taking into account the chemical stability of the analyte. Useful information can be obtained on stability during storage from the JECFA evaluations[2], or from dossiers submitted for registration.
1.2 Analyte stability during sample processing	About AL	Treat representative tissue matrices with known amount of analyte(s). Analyse ≥5 replicates of each representative commodity, post-processing,	No significant loss of analyte during processing (P = 0.05)	No significant loss of analyte during processing (P = 0.05)	No false negatives at AL after processing.	Factors such as exposure to light, the temperature of the sample during processing and the extent of sample processing (e.g. homogenization time) may be critical. Processing validated for use with any subsequent procedure. Validation may be specific to analyte and/or sample matrix. For testing stability determine the mean recovery and CV of representative marker compounds. Use these compounds for internal QA tests (see section 5). CV of each compound will indicate the within laboratory repeatability as well.
1.3 Analyte stability in extracts and standard	At AL, with well detectable residues	≥5 replicates at each appropriate point in time (including zero) and for each representative analyte/matrix.	At the end of the storage period, the recoveries should be within the range specified in Table 2. No significant change in analyte	At the end of the storage period, the recoveries should be within the range	At the end of the storage period, all recoveries detectable at AL.	The test of stability in extracts is required if the semi-processed material will likely be stored longer than during determination of precision, or low recoveries were obtained during optimization of the method. Storage time

[2] *Residues of some veterinary drugs in animals and foods*, FAO Food & Nutrition Paper, Series 41, Food and Agriculture Organization of the United Nations, Rome.

Parameter	Level(s)	No. of analyses or type of test required	Criteria			Comments
			Codex Level I method (Confirmatory/Quantitative)	Codex Level II method (Quantitative)	Codex Level III method (Screening – qualitative or semi-quantitative)	
solutions		Fortify blank extracts to test stability of residues. Compare analyte concentration in stored and freshly made standard solutions.	concentration in stored analytical standards (P = 0.05)	specified in Table 2. No significant change in analyte concentration in stored analytical standards (P = 0.05)		should encompass the longest period likely to be required to complete the analysis, including any subsequent confirmation using the extract.
1.4 Extraction efficiency	About AL	Analyse ≥5 replicate portions of samples or reference material with incurred residues. Compare the reference (or different) procedure with that under test.	For samples with incurred residues, the mean result obtained with the reference procedure and the tested procedure should not differ significantly at P=0.05 level. If using a reference material, the mean concentration of the residue should not differ significantly at P=0.05 level, calculated with CV_A of the method tested, from the consensus value for the residue in the reference material. When the CV_A of the method is larger than 10%, the number of replicate analyses has to be increased to keep the relative standard error of the mean < 5%. Otherwise quantify and report the efficiency of extraction (excluding the recovery of analytical phase).	For samples the mean residues obtained with the reference procedure and the tested procedure should not differ significantly at P=0.01 level applying CV_L in the calculation. Or, the consensus value of reference material and the mean residue, calculated with CV_A of the method tested, should not differ significantly at P=0.01 level. Otherwise quantify and report the efficiency of extraction (average recovery of extraction excluding the recovery of analytical phase).	No false negatives at AL	Some residues may be conjugated or otherwise bound to the tissue matrix and sample pretreatment (e.g. glucuronidase) may be required to release such residues and thereby improve analyte recovery. Temperature of the extract, speed and duration of blending or homogenizing, time of extraction and volumes and ratios of extracting solvents may significantly effect the efficiency of extraction. The effect of these parameters can be checked with a ruggedness test. The optimised conditions should be kept constant as far as possible and may be generally applicable for similar matrices and analytes of similar physical and chemical properties.
1.5 Selectivity of separation	About AL	Determine RRt values for all analytes to be tested by the method (not only the reference compounds). When chromatographic techniques are used without spectrometric detection, apply different separation principles and/or	Peaks should be baseline resolved or sufficiently separated to permit accurate identification and quantitation. The nearest peak maximum should be separated from the designated analyte peak by at least one full width at 10% of	Peaks should be baseline resolved or sufficiently separated to permit accurate quantitation. The nearest peak maximum should be separated from the designated	For chromatographic methods, peaks should be sufficiently resolved to permit tentative identification of all analytes tested at AL. Other types of screening methods, such as ELISA,	Use information obtained from these experiemnts in establishing system suitability criteria for the analysis. System suitability involves injection of analytes to demonstrate adequate performance of the chromatographic system (i.e. peak resolution as specified by method or client requirement).

Parameter	Level(s)	No. of analyses or type of test required	Criteria			Comments
			Codex Level I method (Confirmatory/Quantitative)	Codex Level II method (Quantitative)	Codex Level III method (Screening – qualitative or semi-quantitative)	
		determine RR-s on columns of different polarity. Determine and report resolution (Rs) and tailing factors (Tf) of critical peaks.	the peak height, or more selective detection of all analytes is required.	analyte peak by at least one full width at 10% of the peak height, or more selective detection of all analytes is required.	should detect analytes at the AL.	
1.6 Specificity and selectivity of analyte detection	About AL	Identify by mass spectrometry, or by the appropriate combination of separation and detection techniques available. Analyse ≥5 blanks of each representative commodity obtained preferably from different sources, Report analyte equivalent of blank response. Determine and report selectivity (δ) of detector and relative response factors of representative analytes (RRF) with specific detectors used.	Analyte may be identified and, if necessary, quantified, by mass spectrometry or other suitable technique. Analyse ≥5 blanks of each representative commodity obtained preferably from different sources, Report analyte equivalent of blank response. Determine and report selectivity (δ) of detector and relative response factors of representative analytes (RRF) with specific detectors used.	Analyte peak sufficiently resolved from other peaks in chromatogram for quantitative determination. Evidence of no co-eluting compounds should be provided.	False negatives (β-error) ≤ 5%; false positives (α-error) ≤10%. (see Codex Alimentarius, Volume 3).[3]	Applies only to a specific combination of separation and detection technique. Samples of known treatment history may be used instead of untreated samples. Maturity of sample matrices may significantly affect the blank response and consequently the selectivity of detection. Blank values shall also be regularly checked during performance validation (see Section 5 below). Report typical peaks present in blank extracts. The LCL should preferably be ≤ 0.5AL. Alter chromatographic conditions if blank response interfere with the analyte. The targets for false positive and false negative rates for screening tests are based on Codex Alimentarius Volume 3.
1.7 Calibration function Matrix effect	About AL	Test the response functions of all analytes included in the method on a minimum of 2 occasions with ≥2 replicates at ≥3 analyte levels plus blank.	For linear calibration: regression coefficient for analytical standard solutions (r) ≥ 0.99. SD of residuals $(S_{y/x}) ≤ 0.1$	For linear calibration: regression coefficient (r) ≥ 0.99. SD of residuals $(S_{y/x}) ≤ 0.1$	Not applicable.	Calibration parameters may be established during optimization of the procedure, determination of precision or detection capability. Prepare calibration solutions of different concentrations independently from the stock solution. For MRM perform calibration with mixtures of analytes ("standard mixture"), which can be properly separated by the chromatographic system to take into account the "multi component effect".
1.8 Analytical range, accuracy, precision, limit of detection (LD), limit of	About AL	Analyse ≥5 blank samples and analytical portions spiked LCL, plus ≥3 analytical portions spiked at each of ≥ .5, 1 and 2 times AL. Where practical, method	Method must positively confirm presence of analyte at AL and, used quantitatively, must meet performance criteria in Table 2. LOQ must be fit for purpose.	Method must meet performance criteria in Table 2.	False negatives (β-error) ≤ 5%; false positives (α-error) ≤10%. (see Codex Alimentarius, Volume 3).	The analysts should demonstrate that the method is suitable for determining the presence of the analyte at the appropriate AL with the maximum errors specified. See Annex 4 for calculation. The confidence interval around the calculated mean

Parameter	Level(s)	No. of analyses or type of test required	Criteria			Comments
			Codex Level I method (Confirmatory/Quantitative)	Codex Level II method (Quantitative)	Codex Level III method (Screening – qualitative or semi-quantitative)	
quantitation (LOQ).		performance tests should be divided among the analysts, who will use the method, and instruments which will be used in the analysis.				depends on the number of data points used for the calculation. The decision limit and detection capability for specified analyte/matrix combinations can be determined according to Section 1.1 and 1.2 of Annex 3 by analysing ≥ 5 blank samples and analytical portions spiked at AL and 0.5 and 2 times the AL, or by applying ISO Standard 11843. Estimates of method accuracy, precision and recovery should be available to users of data generated with the method.
2. Extension of the method to new analyte and matrices having similar properties to those of representative analytes and matrices						
2.1 Analyte stability during sample storage, processing, and in extracts and standard solutions	See. 1.1, 1.2, and 1.3	See 1.1, 1.2, 1.3	See. 1.1, 1.2, 1.3	See. 1.1, 1.2, 1.3	See. 1.1, 1.2, 1.3	See. 1.1, 1.2, 1.3
2.2 Extraction Efficiency	About AL	See 1.4	See 1.4	See 1.4	See 1.4	See 1.4
2.3 Selectivity of separation	About AL	See 1.5	See 1.5	See 1.5	See 1.5	See 1.5
2.4 Specificity and selectivity of analyte detection	About AL	Check response of ≥ 3 different (if available) blank samples.	See 1.6.	See 1.6.	See 1.6	Some authorities now recommend that 6 or more representative blanks be used for each new matrix. If the selectivity of detection does not eliminate the matrix response, use appropriate combination of chromatographic columns which enables the separation of analytes from the matrix peaks. Report typical peaks present in blank extracts. See 1.6
2.5 Calibration function, matrix effect	About AL	See 1.7	See 1.7	See 1.7	See 1.7	See 1.7
2.6 Analytical range, accuracy, precision, limit	About AL	Fortify blank analytical portions with relevant representative analytes at 3 concentrations, in	Method must positively confirm presence of analyte at AL and, used quantitatively, must meet	Meets performance specifications in Table 2.	Analytes added to blank samples at AL should be detectable in all tests.	Relevant representative analyte: analyte which may occur in a particular sample. See 1.8

Parameter	Level(s)	No. of analyses or type of test required	Codex Level I method (Confirmatory/Quantitative)	Criteria		Comments
				Codex Level II method (Quantitative)	Codex Level III method (Screening – qualitative or semi-quantitative)	
of detection (LD), limit of quantitation (LOQ).		duplicate. See 1.8	performance criteria in Table 2. See 1.8	See 1.8	See 1.8	
2.7 Analyte homogeneity	See 1.3.	See 1.3.	See 1.3.	See 1.3.	See 1.3.	Biological variability may result in differences in analyte homogeneity in, for example, liver from different species.
2.8 Matrix effect	About AL	Test the matrix effect using blanks in combination with 3.4.	Method must positively confirm presence of analyte at AL, and, used quantitatively, must meet performance criteria in Table 2.	Meets performance specifications in Table 2. No matrix effect observed.	Analytes added to blank samples at AL should be detectable in all tests.	If method performance criteria are not met due to matrix effects, method requires revision to be applied to the new matrix.

3. Adaptation of the method in another laboratory

Parameter	Level(s)	No. of analyses or type of test required	Codex Level I method (Confirmatory/Quantitative)	Codex Level II method (Quantitative)	Codex Level III method (Screening – qualitative or semi-quantitative)	Comments
3.1 Purity and suitability of chemicals, reagents and ad(ab)sorbents		Test reagent blank, applicability of ad(ab)sorbents and reagents. Perform derivatization without and with sample.	No interfering response.	No interfering response.	Verify screening test performs within manufacturer's specifications.	
3.2 Analyte "homogeneity"						No test required unless evidence of heterogeneity is found through quality control procedures during method application.
3.3 Selectivity of separation	About AL	Verify system suitability.	Specified separation achieved.	Specified separation achieved.	Analytes added to blank samples at AL should be detectable in all tests.	System suitability samples are usually prepared by dissolving the analyte(s) in the solvent used in the final extract of the method. They are injected prior to running samples to ensure that the chromatographic separation achieved is within the requirements of the method.
3.4 Calibration function, matrix effect	About AL	Test the response functions of representative analytes included in the method on a minimum of 2 occasions at ≥3 analyte levels plus blank, in duplicate on each occasion. Test the matrix effect with representative analytes and matrices.	Method must positively confirm presence of analyte at AL, and, used quantitatively, must meet performance criteria in Table 2. No matrix effects observed.	Meet requirements of Table 2. No matrix effects observed.	Analytes added to blank samples at AL should be detectable in all tests.	Calibration parameters may be established during optimization of the procedure, determination of precision or detection capability. Prepare calibration solutions independently from the stock solution. For MRM perform calibration with mixtures of analytes ("standard mixture"), which can be properly separated by the chromatographic system to take into account the "multi component effect". Use matrix matched analytical standards for quantitative tests if matrix effect is significant.

Parameter	Level(s)	No. of analyses or type of test required	Codex Level I method (Confirmatory/Quantitative)	Codex Level II method (Quantitative)	Codex Level III method (Screening – qualitative or semi-quantitative)	Comments
3.5 Specificity of analyte detection	About AL	Check performance characteristics of detectors used and compare them with those specified in the method. Check response of one blank of each representative commodity, otherwise perform test as described in section 1.6.	Measured response is solely due to the analyte. The detector performance (sensitivity and selectivity) should be equal or better than specified in the method. Response of blank sample should not interfere with those of the analytes.	False negatives (β-error) ≤5%; false positives (α-error) ≤ 10% (see Codex Alimentarius, Volume 3).	The relative response of specific detectors can substantially vary from model to model. Proper checking of specificity of detection is critical for obtaining reliable results. Compare blank response observed with typical peaks reported in blank extracts. See other comments under 1.6.	
3.6 Analytical range, accuracy, precision, decision limit, detection capability,	About AL	See 1.8	See 1.8	See 1.8	Establish that original performance characteristics of method are met or exceeded, or document performance achieved. If method is fit for purpose, establish QC criteria based on the within-laboratory performance achieved during validation	**See comments in 1.8.**
3.7 Analyte stability in extracts and standard solutions		No test, unless problems arise during evaluation of performance.				See 1.9 if problems arise

4. Quality control (performance validation)
4.1 Methods used regularly

Parameter	Level(s)	No. of analyses or type of test required	Codex Level I method	Codex Level II method	Codex Level III method	Comments
4.1.1 Suitability of chemicals, adsorbents and reagents		For each new batch: Test reagent blank, applicability of ad(ab)sorbents and reagents. Perform derivatization without sample.	No interfering response ≥LCL.	No interfering response ≥ LCL.	No interfering response at minimum concentration specified.	

Parameter	Level(s)	No. of analyses or type of test required	Criteria			Comments
			Codex Level I method (Confirmatory/Quantitative)	Codex Level II method (Quantitative)	Codex Level III method (Screening – qualitative or semi-quantitative)	
4.1.2 Analyte stability during sample processing and analysis	About AL	Fortify blank sample matrix of known origin with appropriate test compounds (See 1.2) and analyse them together with other samples in the analytical batch.	Recoveries of the test compounds should be within the action limits of control chart, if method used for quantitation; otherwise, analyte should be confirmed at lowest concentration specified by requirement.	Recoveries of the test compounds should be within the specified limits (usually 2σ) of control chart.	Analyte added at lowest concentration specified by requirement remains detectable after storage.	Test stability during period when seasonal changes may result in fluctuations in laboratory environment (temperature, humidity, etc.).
4.1.3 Analyte homogeneity in processed sample	About AL	Select a positive sample randomly. Repeat analysis of another one or more analytical portions.	The replicates should be within the reproducibility limit of Table II.2, if method includes quantitation. For confirmation only, results should confirm within method criteria (e.g. ion ratios).	The replicates should be within the reproducibility limit of Table II.2.	All results should be positive at or above the minimum detection requirements specified.	Perform test alternately to cover each commodity analysed. Test homogeneity at the start of the analysis of the given type of samples. The acceptable results of the test also confirm that the reproducibility of the analyses (CV_A) was appropriate.
4.1.4 Extraction efficiency						The efficiency of the extraction cannot be controlled during the analysis. To ensure appropriate efficiency, the extraction should be carried out without any change.
4.1.5 Selectivity of separation, performance of detectors	About AL	Include appropriate detection test mixture (system suitability) in each chromatography batch.	System suitability demonstrated.	System suitability demonstrated.	Not applicable for most screening tests (test kits) – see 5.1.1	Prepare detection test mixture for each method of detection. Select the components of the mixture to indicate the characteristic parameters of chromatographic separation and detection. Adjust chromatographic conditions to obtain required separation, if required.
4.1.6 Specificity of analyte detection	About AL	Include blank matrix in analytical batch. Use standard addition if no untreated sample (similar to those analysed in the batch) is available. Confirm identity and quantity of each analyte present AL.	Sample co-extracts interfering with the analyte should not be present.	Sample co-extracts interfering with the analyte should not be present.	False negatives (β-error) $\leq 5\%$, false positives (α-error) $\leq 10\%$. (see Codex Alimentarius, Volume 3).	Include appropriate detection test mixture in each chromatography batch (system suitability). Perform quantitative analysis with analytical standards prepared in blank matrix extract if matrix effect is significant.
4.1.7 Calibration and analytical range	About AL	Usually prepared at a minimum of 0.5, 1 and 2 times the AL of each analytical run.	Ion ratios for peaks used for mass spectral confirmation must be within limits specified in method. Usually $r = 0.98$ or better for each calibration curve used in quantitation.	Usually $r = 0.98$ or better for each calibration curve.	Usually not applicable. Include appropriate standards to verify test performance at minimum level specified by requirement.	
4.1.8 Accuracy	Within	Include in each analytical batch ≥ 1	The performance of detector and	The performance of	Usually not applicable.	Fortify analytical portion with standard mixture(s) within

Parameter	Level(s)	No. of analyses or type of test required	Criteria			Comments
			Codex Level I method (Confirmatory/Quantitative)	Codex Level II method (Quantitative)	Codex Level III method (Screening – qualitative or semi-quantitative)	
and precision	analytical range	blank sample either: fortified with standard mixture, replicate portion of a positive sample, or a re-analysis of a positive sample. Certified reference materials, if available, may also be used.	chromatographic column shall be equal or better than specified in the method. For quantitative methods, preferably all recoveries should be within the warning limit of control chart constructed with the specific or typical CV_A of analytes. On a long run one of 20 or 100 samples may be outside the specified limits for the control chart. The analytical batch should be repeated if any of the recoveries falls outside the action limits, or the results of the replicate analyses of the positive sample exceeds the critical range.	detector and chromatographic column shall be equal or better than specified in the method. For quantitative methods, preferably all recoveries should be within the warning limit of control chart constructed with the specific or typical CV_A of analytes. Occasionally one sample may be outside the specified limits for the control chart. The analytical batch should be repeated if any of the recoveries falls outside the action limits, or the results of the replicate analyses of the positive sample exceeds the critical range.	Apply criteria for false positives and false negatives.	the analytical range of interest, particularly at concentrations near an AL to be detected.
4.1.9 Duration of analysis			The samples, extracts etc. should not be stored longer than the period for which the storage stability was tasted during method validation. Storage conditions should be regularly monitored and recorded.			
4.2 Analyte detected occasionally						
Follow tests described in 4.1 with the following exceptions						
4.2.1 Accuracy and precision	About AL	Reanalyse another analytical portion; or use standard addition	Replicate analyses should agree within confirmatory criteria of	Replicates should agree within the	Replicates should be in agreement.	Check accuracy if residue found at ≥0.5 AL.

Parameter	Level(s)	No. of analyses or type of test required	Criteria: Codex Level I method (Confirmatory/Quantitative)	Criteria: Codex Level II method (Quantitative)	Criteria: Codex Level III method (Screening – qualitative or semi-quantitative)	Comments
		at the measured concentration of analyte.	method. For quantitative purposes, replicates should agree within the specifications in Table 2.	specifications in Table 2.		

4.3 Methods used at irregular intervals

Follow tests described in 4.1 with the following exceptions

Parameter	Level(s)	No. of analyses or type of test required	Codex Level I method (Confirmatory/Quantitative)	Codex Level II method (Quantitative)	Codex Level III method (Screening – qualitative or semi-quantitative)	Comments
4.3.1 Accuracy and precision (repeatability)	About AL	Include fortified samples at 0.5, 1 and 2 times AL in each analytical batch. Use standard addition if untreated sample (similar to those analysed in the batch) is not available. Perform analysis with ≥2 analytical portions.	Replicate analyses should agree within confirmatory criteria of method. For quantitative purposes, replicates should agree within the specifications in Table 2.	Replicates should agree within the specifications in Table 2.	Replicates should be in agreement.	The acceptable results also prove the suitability of chemicals, adsorbents and reagents used. If performance criteria were not satisfied, the method shall be practised and it performance characteristics re-established during partial revalidation of the method.

4.4 Changes in implementation of the method

Change		Parameters to be tested	For test methods and acceptability criteria see the appropriate sections of Appendix 1.			Comments
4.4.1 Reagent/ new materials: different supplier or quality		Test blank value and perform derivatization without sample. Test recoveries in two replicates at 0.5, 1 and 2 times AL.				Method performance characteristics should not be changed. Modify method protocol to include acceptable change in specified quality or supplier.
4.4.2 Chromatographic column		Test selectivity of separation, resolution, inertness, RRt values (system suitability).				Method performance characteristics should not be changed significantly. Some adjustment or modification of chromatographic conditions may be required and should be documented accordingly. Modify method protocol to include acceptable change in specified item or supplier.
4.4.3 Equipment for sample processing		Homogeneity of processed sample; Stability of analytes				Method performance characteristics should not be changed. Modify method protocol to include acceptable change in item or supplier.
4.4.4		Compare results using incurred samples or	The mean residues should not	Method performance		

Parameter	Level(s)	No. of analyses or type of test required	Criteria			Comments
			Codex Level I method (Confirmatory/Quantitative)	Codex Level II method (Quantitative)	Codex Level III method (Screening – qualitative or semi-quantitative)	
Equipment for extraction		suitable surrogates (fortified matrix blanks when these have been shown previously to reflect extraction for incurred samples) detected after extraction with the old and new equipment in ≥ 5 replicates	be significantly different at p=0.05 level.	characteristics should not be changed. Modify method protocol to include acceptable change in item or supplier.		
4.4.5 Detection		Test selectivity of separation and selectivity and sensitivity of detection				Test also detectability separately with new detection reagents. Method performance characteristics should not be changed. Modify method protocol to include acceptable change in item or supplier.
5.4.6 Analyst		≥5 recovery tests at each level (LCL, AL and 2 (3) AL), re-analysis of one blank sample and two positive samples (unknown for the analyst).	All results should be within the warning limits specified for the method in the laboratory. Replicate sample analysis shall be within the critical range.	Document analyst familiarization with method (i.e. analyst is "ready to perform" tests on samples). This is a minimum requirement. Some veterinary drug residue laboratories use a more detailed protocol which includes (1) generation of standard curve within acceptability criteria; (2) minimum of 2 analytical runs for each matrix, containing representative samples fortified by the analyst at a minimum of 3 concentrations, in duplicate; and (3) minimum of 1 analytical run containing fortified or incurred residues, at 3 concentrations in duplicate, provided as unknowns to the analyst. All results at each stage must meet acceptability		

Parameter	Level(s)	No. of analyses or type of test required	Criteria			Comments
			Codex Level I method (Confirmatory/Quantitative)	Codex Level II method (Quantitative)	Codex Level III method (Screening – qualitative or semi-quantitative)	
5.4.7 Laboratory		Accuracy and precision ≥3 recovery tests at each concentration, 0.5, 1 and 2 times AL, preferably by (different) analyst(s) on different days.	All results should be within the warning limits specified for the method in the laboratory. criteria, or be repeated.			

Worked example for validation of a Multi Residue Method for Pesticide Residue Analysis

1 INTRODUCTION

The aim of this worked example is to show a possible way to apply the principles described in the Guidelines for Single-laboratory Validation of Analytical Methods for Trace-level Concentrations of Organic Chemicals (referred to further on as Guidelines)

Naturally, other options are also possible, and the practical application of the principles depends on the particular situation, available information and results of the tests.

The robustness test performed according to the Youden procedure is not included in this worked example as it is considered part of the method development/optimisation procedure, and it is very well explained elsewhere[1].

The tests recommended for determining the homogeneity of analyte and the "well mixed" condition of processed samples are discussed in elsewhere[2,3]. However it is pointed out, that those tests are independent from the analyte. Once they are properly carried out, the results can be used for any other analytical method based on the same sample processing.

The abbreviations used are specified in the Glossary of the Guidelines[4].

2 ASSUMPTIONS

The laboratory has to determine the residues of some 150 pesticide active ingredients amenable to direct GC analysis, and has to report residues present in samples at ≥ 0.05 mg/kg concentration. The laboratory had used a method for several years based on Ethyl Acetate extraction and GPC cleanup with additional cleanup on silicagel for ECD determination of residues if necessary. The final cleaned extract contains 2.5 g sample equivalent, of which 2.5 mg is injected to the GC.

The method was recently optimised and its robustness was tested in the laboratory with the participation of 2 technicians who will also apply the method in the future. The optimisation work was documented according to GLP and all records (e.g. sample blanks, calibration curves, recovery data, relative retention data of analytes of interest on 3 columns of different polarity) are available.

In addition, information on stability of residues during sample processing and analysis is available from previous analysis of samples, storage stability data, and results of comparative extraction of incurred residues and stability of analytical standards in stock and intermediate solutions are available for a large number of pesticide from the JMPR evaluations, submissions for registration and literature references[5,6,7], respectively.

The laboratory had to validate the method to characterise and record its performance parameters. (Availability of validated methods is also a requirement for formal accreditation according to either ISO 17025 and/or GLP.)

3. VALIDATION OF THE METHOD

The work was conducted by two technicians both *being familiar with the method*. Each technician analysed all representative compounds in each sample matrix. The validation was performed on several days. Thus the precision of the data reflects the within laboratory reproducibility.

3.1 Selection of representative analytes and representative compounds

The laboratory selected 12 representative pesticides (Table 1) and 5 representative commodities (Tables 3 - 7) for validating the method and to determine its typical characteristics.

The analysts chose representative compounds with wide a range of physico-chemical properties: water solubility from 0.065 mg/l to 200 g/l; $logP_{ow}$: from - 0.8 to 6.1; vapour pressure: from $5x10^{-4}$ to 12 mPa; and hydrolytic stability from 1 to 800 days at pH 7.

Notes: Naturally other or more compounds could also be selected to represent the analytes intended to be determined with the method (represented analytes), taking into account three important criteria:

1. As many analytes as it is possible to quantify in one mixture should be included in the method validation;

2. The analytes critical for the purpose of the study for which the method will be used should always be included;

3. The physico-chemical properties of selected compounds should represent a sufficiently wide range to include the properties of the "represented analytes";

Table 1. Summary of physico-chemical properties[8,9] of selected representative compounds

Active ingredient	Water solubility		$LogP_{ow}$		Vapour pressure		Hydrolysis	
	mg/l	pH; °C		°C	mPa	°C	DT50 [day]	pH· °C
DDE-p,p	0.065	24						
Permethrin	0.2	30	6.1	20	0.045	25	>720	4, 50
Endosulfan a	0.32	22	4.74	PH 5	0.83	25ª		
Chlorothalonil	0.81		2.89		0.076	25		
Chlorpyrifos	1.4	20	4.7		2.7	25	water, 1.5	8, 25
Lindane	7.3	25			0.051	25	191	7, 22
Iprodione	13	20	3	PH 3, 5	0.0005	25	1 to7	7
Dimethoate	23.3	5, 20	0.704		1.1	25	12	9, -
Azinphos-Methyl	28	20	2.96		0.18	20	87	4, 22
Diazinon	60	20	3.3		12	25	0.49 185	3.1, 20 7.4, 20
Progargite	632	25	3.73		0.006	25	800	7, -
Methamidophos	200,000	20	-0.8	20	2.3	20	657	4, 22

Notes: (a) 2:1 mixture of a and b isomers

3.2 Analytical range, calibration

The method was validated at 4 residue concentrations within the range of 0.02 mg/kg - 2 mg/kg.

> Note: for MRM the concentrations of representative analytes do not necessarily relate to their actual Accepted Limits (ALs).

Calibration mixtures were prepared for ECD and NPD detection at 7 concentration levels both in solvent and in cleaned sample extract containing the equivalent of 2.5 g blank sample. The standards were injected into on-column injector in duplicate (number of injections in one detection system k =2*7=14). The GC was prepared according to the daily maintenance routine and its operating conditions, checked with appropriate test mixtures, were within the acceptable range. No special procedure was applied before the validation. In order to demonstrate the effect of the number of calibration points, a three-level calibration (k=2*3=6) is simulated in this example with the omission of 4 levels from the calculations.

As an example, the injected amount and average response (Y) relationship (rounded to whole number) is given hereunder for dimethoate and chlorothalonil. The shaded values indicate the three-level calibration. The standard solution prepared in iso-octane containing 2.5 mg/ml matrix equivalent is indicated by m2.5 mg.

Injected [pg]	9.2	36.8	184	368	552	827	1103
Dimethoate	385	3337	20006	40669	59065	82929	109285
Dimethoate m2.5mg	898	3739.33	18834	37550	55553	79397	100381
Y_{matr}-Y_{solv} = d_I	512.66	402.333	-1171.67	-3118.67	-3512	-3532	-8904.67

Mean of differences ($\bar{d}$): -2760, standard deviation of differences (S$_D$): 3321.8

Injected [pg]	16.3	65.2	326	652	979	1468	1958
Chlorthalonil	0	61.3	2318	4653	7007	10215	13521
Chlorthalonil m2.5	40.66	476	2434	5010	7336	10665	14164
Y_{matr}-Y_{solv} = d_I	40.66	414.7	116	357	328	450	643

Mean of differences ($\bar{d}$): 335.9, standard deviation of differences (S$_D$): 203.8

The linear regression equations for the dimethoate and chlorothalonil are given in Table 2.

The significance of matrix effect was tested for all compounds with paired t-test.

The t value is calculated with equation:

$$t = \left| \frac{\bar{d}\sqrt{n}}{S_D} \right|$$

eq. 1

where $\bar{d}$ is the average and S$_D$ is the standard deviation of differences (d_i):

$$S_D = \sqrt{[\Sigma(d_i - \bar{d})^2/n-1]}$$

eq. 2

with degrees of freedom of $v=n-1$, where n is the number of data pairs.

Table 2 Summary of calibration parameters and calculated decision limits at $\alpha=0.05$ *

Analyte	Level of Calibr.	Regression equation	R^2	S_b	Decision limit CCα [pg]
Dimethoate	7	Y= 99.5x + 1318	0.9975	878	15.7
Diethoate in m.	7	Y= 92.2x + 1798	0.9967	906	17.5
Dimethoate	3	Y= 98.6x+ 1459	0.9987	1612	34.8
Diethoate in m.	3	Y= 90.1x + 1815	0.9979	1412	33.4
Chlorothalonil	7	Y= 7.03x - 97	0.9987	150.5	38.1
Chlorothalonil in m.	7	Y= 7.26x + 67.8	0.9993	128.4	31.5
Chlorothalonil	3	Y= 6.95x-17.7	0.9980	179.3	55
Chlorothalonil in m.	3	Y= 7.24x + 69.7	0.9950	310.1	91

Surprisingly, the test did not indicate significant matrix effect for dimethoate (tabulated value for $t_{v=6,\ 0.05}$ is 2.45) (t_{calc}: 2.267), while it did for chlorothalonil (t_{calc}: 4.36). The matrix effect can also be seen from the substantially increased intercept compared with standards without matrix (Table 2) in case of chlorothalonil, and a lesser extent for dimethoate.

Notes: 1Chlorothalonil is used here to illustrate the matrix effect. It was not included in the method validation programme as it undergoes significant degradation during sample processing[10].

2. Although the difference in slope is larger in case of dimethoate, the paired t-test shows no matrix effect (Figures 1 & 2). Other tests performed under varying conditions indicated that the **paired t-test** is not a sensitive indicator of matrix effect and should not be used and relied on alone. The authors provide alternative possibilities for testing the matrix effect[11].

No significant matrix effect was observed with the paired t-test for the other representative compounds which is in accordance with the practical experience gained with on-column injection, which generally induce smaller matrix effect than split/splitless injectors. Since the average recoveries were within the acceptable range (70-110%) for all compounds, the quantification of the chromatograms was carried out with analytical standard mixtures prepared without matrix according to the guidance given by the Guidelines.

Note: If the recoveries measured with standards prepared in pure solvent were outside the acceptable range, then matrix matched standards would be used for quantification. Practical experience indicates that the matrix effect may vary from day to day during the regular use of the method depending on the condition of the gas chromatographic system. Consequently, the extent of matrix effects found during method validation is only for information. It must be checked regularly as part of internal quality control. As suggested in the Guidelines, when residues measured during the application of the method for analysis of samples are between 0.7 AL (0.5AL) and 2AL the confirmation of the analyte concentration should always be carried out with matrix matched standard solutions. Alternatively, standard addition calibration may be used, which will also dispel any doubts that the blank matrix extracts available produce effects identical to those of the sample extract.

*See also Figures 1–4.

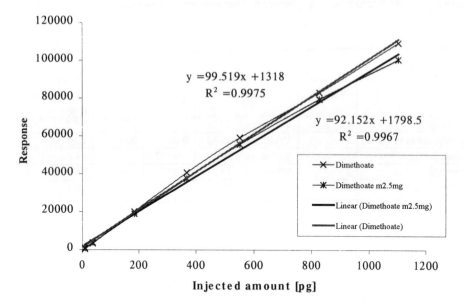

Figure 1. Calibration chart for dimethoate in isooctane and matrix solution containing 2.5 g sample equivalent/ml.

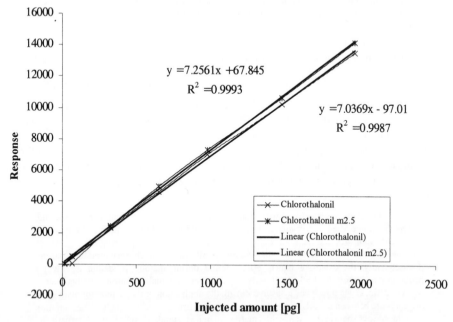

Figure 2. Calibration chart for chlorothalonil in isooctane and matrix solution containing 2.5 g sample equivalent/ml.

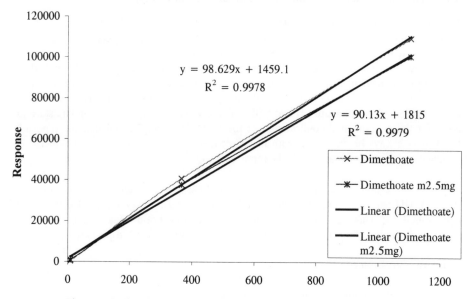

Figure 3 Calibration curve of dimethoate based on three level calibration

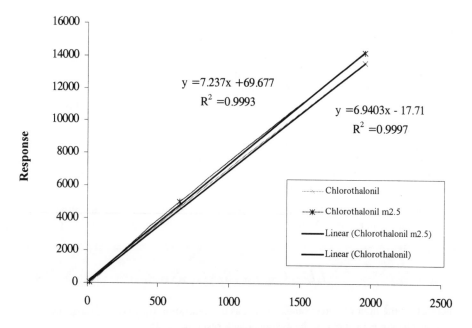

Figure 4 Calibration curve of chlorothalonil based on three level calibration

3.3 Calculation of Decision Limit and Limit of Detection (LD)

The terms limit of detection (**LD**), limit of determination (LOD) and limit of quantification (LOQ) are often used as synonyms, and there has been very little agreement in their calculation in the practice of residue analysts. Therefore the analysts should always report how the various quoted parameters were determined or calculated. Though they may be the subject of justified criticism, and more precise procedures are available and used in other areas of chemometrics[12,13] the procedures described hereunder may be used for estimation of approximate LD and LOD. These values provide an appropriate starting point for the experimental confirmation of the LOQ. It is emphasised that the reported LOQ must always be supported with mean recovery and its relative standard deviation obtained with spiking of samples at the reported LOQ.

As the blank response was often negligible and not easily measurable, the decision limits, ($CC\alpha$[14]), (IUPAC recommends[13] the term *Detection decision L_C)* were calculated with equations 3-5. This was possible because a reasonably good linear relationship was observed at the low concentration range and the intercepts were >0, especially with NPD.

Linear regression equation:
$$Y = ax + b$$
eq. 3

The upper 95% confidence level of the intercept:
$$Y_{x=0,0.95} = b + t_{0.05} * S_b$$
eq. 4

The corresponding analyte concentration:
$$CC\alpha = L_C = (Y_{x=0,\,0.95}-b)/a = t_{0.05}S_b/a$$
eq. 5

where $Y_{x=0,\,0.95}$ is the response at the upper 95% confidence level, $t_{0.05}$ is the tabulated value of one tail Student distribution with degree of freedom $v=k-2$ ($t_{v:12}= 1.782$; $t_{v:4}=2.132$), S_b is the standard deviation of the intercept (b). For testing the presence of banned substances $\alpha=0.01$ may be chosen, allowing only 1 percent probability for false positive detection.

The regression equations for 7 and 3- level calibrations were reasonably similar, but the calculated $CC\alpha$ is strongly influenced by the number of calibration points and the consequent degree of freedom, which affects the *t* value. The calculated values for dimethoate and chlorothalonil are shown in Table 2.

Assuming normally distributed peak areas/heights and similar standard deviations at the Decision Limit and at the LD, the LD can be approximately calculated with the following equation:

$$LD = C_B + t_{v,1-\alpha}S_B + t_{v,1-\beta}S_B$$
eq. 7

where LD is the limit of detection expressed as concentration (or mass) of the analyte;
C_B is the analyte equivalent of the average blank response
S_B is the standard deviation of the analyte equivalent of blank response.

$t_{v, 1-\alpha}$ *and* $t_{v, 1-\beta}$ are the <u>one tailed</u> Student *t* values corresponding to a specified degree of freedom (v) and probabilities (α indicating the probability of false positive detection or β indicating the probability of false negative detection).

A number of points should be considered in applying the above equations:

In computing the detection and quantification limits, the effects of analyte concentration, matrix effect and matrix interference on the variance of the estimated quantity, as well as the degrees of freedom, and the acceptable probability of false positive and negative detection must be taken into account.

The response of the analyte is most sensitively effected by the matrix and detection conditions at the low concentration range. The linear calibration function is often distorted at the lower end. Consequently **the best estimate of the CCα and LD may be obtained from the replicate analysis of different blank samples** (e.g. different variety and/or stage of maturity of apples).

The other alternative may be to estimate the LD from the calibration points:

$$Y_{LD}= Y_{0, 0.95} + t_{v, 1-\alpha}S_b + t_{v, 1-\beta}S_b \qquad \text{eq. 8}$$

$$LD = (Y_{LD}\text{-}b)/a=\delta_{\alpha,\beta}S_b/a \qquad \text{eq. 9}$$

Where $\delta_{\alpha,\beta}= t_{v, 1-\alpha} + t_{v, 1-\beta}$

The estimation of the true standard deviation, and consequently the LD, based on a few measurements is very imprecise. The number of measurements is partly taken into account by the increasing *t* value, and reflected by the nearly doubled CCα values obtained from the 3-level (6 points) calibration compared with those calculated based on a 7-level calibration. Further on, the calculated LD is uncertain by the ratio σ/s[13].

The calculation of LD based on equations 7 - 9 provides the necessary flexibility for the analysts to estimate the limit of detection to fit best for the purpose of the analysis. (e.g. selecting α and β) and experimental conditions applied.

The estimation of Limit of Detection for dimethoate based on the 7 level calibration curve with equation 9 at $\alpha=\beta=0.05$ probability level ($S_b = 878$, $t_{12, 0.05} = 1.78$, a=99.51) resulted in 31.4 pg. Based on 3 level calibration, the estimated LD was about 70 pg. The estimated LD values correspond to 0.014-0.028 mg/kg residue concentration. Based on 7-level calibration, the range of estimated LD-s for the other compounds was similar.

3.4 Accuracy and precision of the method

Taking into account the estimated LD values and prior experience with the method, the accuracy and precision of the method were determined by the analyses of 5 replicate analytical portions of each representative commodity spiked at about 0, 0.02, 0.3, 1 and 2 mg/kg. The spiking solutions contained all of the twelve representative compounds which were determined together.

The summary of the validation results obtained is given in Tables 3-9. The results indicate that the performance of the method with the representative commodities and representative pesticides meets the requirements of a quantitative method (mean recovery 70-100%, reproducibility CV$_A$ 23-32%) specified in the Guidelines.

Table 3. Summary of recovery (Q%) data in cabbage

| | Fortification level [mg/kg] | | | | | | | | |
| | 0.02 | | 0.3 | | 1 | | 2 | | |
	Q	CV_A	Q	CV_A	Q	CV_A	Q	CV_A	Qa
Methamidophos	73	15	72	14	78	12	77	11	75.0
Dimethoate	77	15	75	10	78	9	76	11	76.5
Lindane	78	10	80	6	85	12	90	10	83.3
Chlorpyrifos	86	13	87	12	80	13	85	12	84.5
Propargite	80	14	88	11	86	12	84	9	84.5
Azinphos-methyl	87	11	85	10	84	12	83	14	84.8
Ethion	82	9	90	13	88	14	90	12	87.5
Endosulfan I	88	12	85	8	92	11	89	14	88.5
Iprodione	85	12	95	16	92	9	82	14	88.5
Diazinon	82	10	90	14	93	14	90	13	88.8
p,p' DDE	88	12	95	6	90	11	92	10	91.3
Permethrin	95	12	105	7	99	10	95	13	98.5
Average	83.4	12.1	87.3	10.7	87.1	11.6	86.1	11.9	86.0
Cvtyp		12.6		11		12		12.4	
Lsd	13.5		11.78		13.4		13.7		

Notes: Q is the average of replicate recovery tests at one level.
Qa is the average of recoveries at all fortification levels.
Lsd: Least significant difference (P=0.05)

Table 4. Summary of recovery (Q%) data in apple

| | Fortification level [mg/kg] | | | | | | | | |
| | 0.02 | | 0.3 | | 1 | | 2 | | |
	Q	CV_A	Q	CV_A	Q	CV_A	Q	CV_A	Qa
Methamidophos	72	14	74	12	75	11	82	12	75.8
Chlorpyrifos	80	11	78	6	82	9	84	14	81.0
Dimethoate	80	13	81	13	86	13	87	11	83.5
Propargite	84	12	91	11	88	9	91	11	88.5
Ethion	84	12	93	10	93	15	89	12	89.8
Azinphos-methyl	83	9	95	8	95	12	90	9	90.8
Lindane	87	10	95	8	91	13	92	12	91.3
Endosulfan I	92	8	91	6	93	11	90	13	91.5
Diazinon	85	12	97	9	92	12	93	11	91.8
p,p' DDE	88	11	100	12	94	14	92	12	93.5
Permethrin	95	10	103	10	100	14	92	13	97.5
Iprodione	92	14	110	11	89	12	99	15	97.5
Average	85.2		92.3		83.8		90.1		92.4
CV_{typ}		12.7		11		12		13	
Lsd	13		12.32		14.3		14.3		

Notes: see Table 3

Table 5. Summary of recovery (Q%) data in orange

| | Fortification level [mg/kg] | | | | | | | | |
| | 0.02 | | 0.3 | | 1 | | 2 | | |
	Q	CV$_A$	Q	CV$_A$	Q	CV$_A$	Q	CV$_A$	Qa
Methamidophos	72.5	14.5	75	11	77.5	10.5	85.2	13.2	77.55
Dimethoate	78.5	14	79	12	83	10	84.8	12.7	81.33
Propargite	80	13	85	8	84	9.5	83.5	12	83.12
Chlorpyrifos	83	12	88	9	82	10	86.8	11.7	84.96
Azinphos-methyl	85	10	92	12	90.5	11	82.6	11.2	87.52
Ethion	83	10.5	93	13	91.5	13.5	84.9	12.7	88.11
Diazinon	83.5	11	95	13	93.5	12	82.3	12.2	88.57
Iprodione	88.5	13	96	14	91.5	9.5	82.2	12	89.54
Lindane	82.5	10	102	8	89	11.5	85.9	11.5	89.84
Endosulfan I	90	10	98	10	93.5	10	89.6	10.7	92.78
p,p' DDE	88	11.5	103	7	93	11.5	89.6	12.2	93.4
Permethrin	95	11	99	9	101	11	88.5	11.7	95.76
Average	84.13		92.08		86.08		83.91		86.5
CV$_{typ}$		12.4		11.1		11.3		12.6	
Lsd	13.3		13.02		12.8		13.7		

Notes: see Table 3

Table 6. Summary of recovery data in wheat

| | Fortification level [mg/kg] | | | | | | | | |
| | 0.02 | | 0.3 | | 1 | | 2 | | |
	Q	CV$_A$	Q	CV$_A$	Q	CV$_A$	Q	CV$_A$	Qa
Dimethoate	81.6	13.1	88	12	86	13.1	87.4	12.6	85.75
Methamidophos	84.5	13.1	91	12	88.9	13.1	90.4	12.6	88.7
Propargite	84.5	9.95	91	9	88.9	9.98	90.4	9.41	88.7
Chlorpyrifos	86.4	9.95	93	9	90.9	9.98	92.4	9.41	90.66
Ethion	86.4	9.95	93	9	90.9	9.98	92.4	9.41	90.66
Iprodione	87.3	13.1	94	12	91.9	13.1	93.4	12.6	91.65
Lindane	89.2	12.1	96	11	93.8	12	95.5	11.5	93.62
Azinphos-methyl	90.6	13.1	96	12	94.5	13.1	96.6	12.6	94.42
Endosulfan I	90.2	7.85	97	7	94.8	7.93	96.5	7.29	94.6
p,p' DDE	94.9	14.2	102	13	99.7	14.1	102	13.7	99.52
Diazinon	95.9	12.1	103	11	101	12	103	11.5	100.5
Permethrin	97.8	6.8	105	6	103	6.9	105	6.22	102.5
Average	89.08		95.75		93.61		95.3		93.4
CV$_{typ}$		11.8		10.7		11.9		11.2	
Lsd	13.4		13.01		14.2		13.6		

Notes: see Table 3

Table 7. Summary of recovery (Q%) data in sunflower seed

	Fortification level [mg/kg]								
	0.02		0.3		1		2		
	Q	CV_A	Q	CV_A	Q	CV_A	Q	CV_A	Qa
Chlorpyrifos	73.4	12.7	78	9	74.1	10.9	76.8	9.65	75.56
Azinphos-methyl	77	10.6	80	11	76	13.6	78.7	12.6	77.93
Ethion	73.4	11.1	83	7	78.9	9.83	81.6	9.06	79.22
Methamidophos	71.8	15.3	85	11	80.8	13.6	83.6	12.6	80.28
Lindane	75.8	10.6	86	12	81.7	13.6	84.6	12.1	82.02
p,p' DDE	80.7	12.1	85	12	80.8	14.7	83.6	13.7	82.5
Propargite	71.8	13.7	89	12	84.6	12.6	87.5	10.5	83.2
Dimethoate	69.4	14.8	90	11	85.5	13.6	88.5	12.6	83.33
Endosulfan I	76.6	10.6	88	9	83.6	9.88	86.5	8.08	83.68
Diazinon	81.5	11.6	89	14	84.6	14.7	87.5	12.7	85.62
Permethrin	83.1	11.6	90	6	85.5	7.7	88.5	6.41	86.76
Iprodione	74.2	13.7	96	9	91.2	12.5	94.3	12	88.93
Average	76		86.58		82.3		85.2	11	82.4
CV_{typ}		10.8		10.9		12.8		11.4	
	10.4		11.99		13.4		12.3		

Notes: see Table 3

3.5 Statistical tests for the evaluation of recovery data

The average recoveries, standard deviation, coefficient of variation and variances of recoveries were calculated for each analyte/sample matrix for each fortification level independently. Detailed results for cabbage fortified at 0.3 mg/kg level are given in Tables 8 and 9.

In order to estimate the typical parameters a number of statistical tests should be performed if necessary.
(a) Testing of recovery data for outliers

Since all CV values were within the acceptable range, and there was no result which was suspected as outlier, the Dixon outlier test was not carried out for the recovery data.

> Note: Rejection of outliers should be considered ONLY in cases where either there is clear experimental evidence that something went wrong or the result is so clearly part of "another population" that there is no possibility that it could be an extreme result from the same population. The use of outlier tests on small sets of data (from which it is impossible to determine whether they are normally distributed) should be very carefully considered and fully justified. It should never be applied as a means for "cleaning up" the performance of methods. If a method produces the occasional extreme result (and they all do, even if it's because the analyst has approached the limits of ruggedness), it is likely to be true reflection of the performance of the method.

(b) The homogeneity of variances was tested with Cochran test[15].

> The Cochran test compares the largest variance observed to the sum of total variances. If the ratio is smaller than the tabulated value the variances observed can be considered to be derived from one population.

The Cochran tests revealed that the differences in the variances were not significant. For example, the calculated values are given in Table 8.

Table 8. Summary of recoveries and their variation in samples of high water content at 0.3 mg/kg fortification level

	Cabbage				Apple				Orange			
	Q	CV_A	S_A	V_A	Q	CV_A	S_A	V_A	Q	CV_A	S_A	V_A
Azinphos-methyl	85	10	8.5	72	95	8	7.6	58	92	12	11.04	122
Chlorpyrifos	87	14	10.4	109	78	6	4.68	22	88	9	7.92	63
p,p' DDE	95	6	5.7	32	100	12	12	144	103	7	7.21	52
Diazinon	90	14	12.6	159	97	9	8.73	76	95	13	12.35	153
Dimethoate	75	10	7.5	56	81	13	10.5	111	79	12	9.48	90
Endosulfan I	85	8	6.8	46	91	6	5.46	30	98	10	9.8	96
Chlorothalonil	90	13	11.7	137	93	10	9.3	86	93	13	12.09	146
Lindane	80	6	4.8	23	95	8	7.6	58	102	8	8.16	67
Iprodione	95	16	15.2	231	110	11	12.1	146	96	14	13.44	181
Methamidophos	72	14	10.1	102	74	12	8.88	79	75	11	8.25	68
Permethrin	105	7	7.35	54	103	10	10.3	106	99	9	8.91	79
Propargite	88	11	9.68	94	91	11	10	100	85	8	6.8	46
Sum				1115				1016				1162
Cochran 5/12=0.34				0.21				0.14				0.16
CV ave(arithm.)		10.6				9.67				10.5		
CV_{typ}		11				10				11		

(c) The significance of the differences of average recoveries

The recoveries of the representative compounds in one commodity and at one spiking level were tested with one way ANOVA.

The summary of ANOVA calculation, together with one set of raw data, is shown in Table 9. The results indicate that the average recoveries are not derived from the same population.

Note: Between group variance (one group consists of the recovery data obtained with a representative analyte) is significantly larger than the within group (variance of replicate analysis): $F_{calc} = 4.1679 > F_{critical}$.

The calculated least significant difference ($S_A*t*\sqrt{(2/n} = 12.44\%$, $t_{v=48, 0.05} = 2.01$, n=5) confirmed that the average recoveries of methamidophos and dimethoate are significantly different from lindane, and the recovery of permethrin (105%) is significantly greater than for iprodion (95%). The recalculated ANOVA without these compounds indicated that the difference for the rest of the recoveries are not significant ($F = 1.079 < F_{critical}$).

Significantly different recoveries from apple and orange, indicated by shading in Tables 3-7, were obtained for methamidophos and dimethoate in 5 and 2 other cases, respectively. The average recovery data for permethrin was not significantly different from the remaining analytes in the other commodities. Consequently, the recovery data of methamidophos and dimethoate, but not that of permethrin, were considered separately from the data for other representative compounds in the case of commodities of high water content.

Table 9. Recoveries % of representative analytes in cabbage at 0.3 mg/kg fortification level

Analyte	Recoveries % in replicate analytical portions					Mean	SD	V	CV
Methamidophos	63	63.3	70.0	76.7	86.7	72.00	9.88827	97.78	14
Dimethoate	65	70.0	76.7	80.0	83.3	75.00	7.45356	55.56	10
Lindane	75	76.7	80.0	83.3	86.7	80.33	4.77261	22.78	6
Azinphos-methyl	77	78.3	82.7	91.7	95.0	84.87	8.11925	65.92	10
Endosulfan I	78	78.3	85.0	90.0	93.3	85.00	6.77003	45.83	8
Chlorpyrifos	77	73.3	86.7	100.0	100.0	87.33	12.5610	157.78	14
Propargite	75	83.3	86.7	93.3	100.0	87.67	9.54521	91.11	11
Diazinon	76.7	76.7	96.7	100.0	101.7	90.33	12.6051	158.89	14
Ethion	76.7	83.3	86.7	101.7	103.3	90.33	11.6905	136.67	13
p,p' DDE	88.3	90.0	95.0	96.7	103.3	94.67	5.93951	35.28	6
Iprodione	80	81.7	91.7	110.0	111.7	95.00	15.1383	229.17	16
Permethrin	96.7	96.7	110	110.0	110.0	104.67	7.30297	53.33	7

ANOVA

Source of Variation	SS	Df	MS	F	P-value	F crit
Between Groups	4394.044	11	399.4586	4.167941	0.000251	1.994579
Within Groups	4600.356	48	95.84074			
Total	8994.4	59				

least significant difference, Sa*sqrt(2/n)*t   12.4452

(d) Comparison of recoveries obtained at different spiking level

Assuming similar standard deviations, the differences between two mean values can be calculated with the t-test:

$$t = \frac{Q_1 - Q_2}{s\sqrt{\dfrac{1}{n_1} + \dfrac{1}{n_2}}}$$

eq. 10

From eq. 10 the difference in average recoveries (Q_1-Q_2) , which must be exceeded to obtain significant difference, can be calculated.

For example, the differences are 11.1% and 6.9%, respectively, for two sets of data consisting of 5 and 20 replicate measurements with a mean recovery of 90% and 12% CV (s = 10.8).

(e) Testing to determine if the average recovery is significantly different from 100%

The formula for comparing the average recovery (Q) to μ=100%

$$t = \frac{(Q - 100)\sqrt{n}}{s}$$

eq. 11

Where s is the standard deviation of n replicate recovery tests giving an average recovery of Q.

Since the recovery can be smaller or larger than 100%, a two-tailed test is used at P=0.05. When the difference is significant, the results should be corrected for the average recovery[16].

Note however, that the measured residue values and the average recovery should be reported in every case regardless whether the average recovery was used for correction or not.

3.6 Estimation of typical performance characteristics of the method

The typical values of recovery and within laboratory reproducibility of the method are estimated from the recoveries obtained for the representative analyte sample matrices at various fortification levels. For the estimation of typical recovery or reproducibility those recovery values can be considered which are not different significantly.

3.6.1 Typical Recoveries. The average recoveries obtained for one analyte commodity matrix did not show any concentration relationship. The differences between their average recoveries obtained at various spiking levels were not significant either. Moreover, the Q_a values obtained for cabbage, apple and orange (commodities of high water content) did not differ significantly. Consequently, their average could be calculated as typical a value for the given matrix or group of commodities. The summary of the results is given in Table 10.

Table 10 Typical recovery values estimated for commodity groups

Representative analyte	Groups I, II & III.		Cereals	Oil seeds
Azinphos-methyl	87.7		94.4	77.9
Chlorpyrifos	83.5		90.7	75.6
p,p' DDE	92.7		99.5	82.5
Diazinon	89.7		100.5	85.6
Dimethoate	80.4		85.7	83.3
Endosulfan I	90.9		94.6	83.7
Ethion	88.5		90.7	79.2
Lindane	86.7		93.6	82.0
Iprodione	82.7		91.6	88.9
Permethrin	97.3		102.5	86.8
Propargite	85.3		88.7	83.2
Dimethoate		80.4	85.7	83.3
Methamidophos		76.1	88.7	80.3
Typical for the group	**89.6**		**93.4**	**82.4**

3.6.2 Typical within Laboratory Reproducibility. The typical reproducibility of the method may be calculated from the pooled variances of recoveries which are assumed to be derived from the same population.

Each set of replicate recovery tests, performed with a representative compound in a representative commodity, provides an estimate of the precision of the method. Since the estimation of variances based on small number of samples is very imprecise, the best estimate of the precision of the method is the average of the variances obtained.

The theoretically correct calculation of the typical CVs is:

$$CV_{Typ} = \frac{\sqrt{V_{ave}}}{Q_a}$$

eq. 12

Where the V_{ave} is the average variance of recoveries and Q_a is the average recovery. CV_{typ} may be calculated for various combinations. For example, recoveries of several representative analytes in one representative commodity at one fortification level; recoveries of one analyte-matrix combination at different fortification levels; recoveries of several analytes at several fortification levels in one sample, etc.

> Though theoretically not correct, if the differences of CVs obtained for different analytes in a commodity are in a relatively narrow range (e.g. 9-15%), their simple arithmetic mean can be calculated if the detailed records on the SD or variance of the results are not available. It is usually only slightly different than the correctly calculated value (Table 8).

The CV_{typ} values calculated with eq. 12 are listed in Tables 3-10. In view of the limited data on which the estimations are based, estimating typical CVs for a concentration interval does not contradict the average precision - concentration relationship described by the Horwitz equation[17]. The typical values calculated as the averages of the CV values are shown in Table 11.

Table 11. Typical CV values of representative analytes for 0.02-2mg/kg residue levels

	Cabbage	Apple	Orange	Wheat	Sunflower
Azinphos-methyl	11.8	9.7	11.1	12.7	12.0
Chlorpyrifos	12.5	10.6	10.7	9.6	10.6
Diazinon	15.1	11.0	12.2	11.6	13.4
Dimethoate	9.9	12.5	12.2	12.7	12.9
Endosulfan I	11.6	9.9	10.2	7.5	9.4
Ethion	12.1	12.4	12.6	9.6	9.2
Iprodione	15.4	13.0	12.3	12.7	12.7
Lindane	9.8	10.9	10.2	11.6	11.2
Methamidophos	10.7	12.3	12.4	12.7	13.1
p,p' DDE	10.2	12.3	10.5	13.7	13.2
Permethrin	10.6	11.9	10.7	6.5	8.1
Propargite	11.6	10.8	10.7	9.6	12.2
CV_{typ}	11.5	11.1	11.1	11.1	11.6

The results indicate that the typical CVs are around 10-15% with an overall CV_{typ} of 11-12%, which characterises the within laboratory reproducibility of the method.

3.6.3 Limit of Quantification, Detection Capability. When samples are analysed, the factors affecting the recovery measurements (e.g. extraction, cleanup, evaporation) will contribute to the variability of the results. Thus, the S_B in the second part of equation 7 can be replaced with the standard deviation of the recoveries performed at LD. Performing the recoveries at LD may not be practical, therefore the blank samples may be fortified 2-5 times higher concentrations than LD. Spiking with higher concentrations than 5LD may result in better precision and average recovery, which lead to underestimating the limit of determination, LOD.

> LOD is used for the calculated limit to distinguish it from the reported Limit of quantification which must be based on replicate recovery tests. The LOQ should only be reported as a performance criterion of the method.

To make the precision data independent of the analyte concentration, the relative standard deviation should be used in the calculation for estimating LOD:

$$C_{LOD} = C_B + t_{v,1-\alpha} S_B + t_{v,1-\beta} \frac{LD \; CV_{LOD}}{Q} \qquad \text{eq. 13}$$

Where
C_B is the analyte concentration equivalent of the average blank response;
S_B is the standard deviation of the analyte equivalents of blank responses;
LD is the limit of detection calculated with eq. 7;
CV_{LOD} is the relative standard deviation of the recovered analyte concentrations at around LOD
Q is the average recovery expressed as average analyte recovered/analyte added.

The calculation of LOD is illustrated, as an example, with the calibration data obtained for dimethoate (Table 2). The residue concentration equivalent of the Decision limit of 15.7 pg is 0.007 mg/kg. Inserting in the eq. 13 the decision limit $(C_B + t_{v, 1-\alpha} S_B) = 0.007$ mg/kg, LD = 0.014 mg/kg, Q_{typ} =90 % and CV_{typ} = 12% , and t=1.64 (since the typical values were obtained from over 100 measurements) the LOD for commodities of high water content may be calculated as

LOD [mg/kg] = 0.007+ 1.64*0.014*0.12/0.90 =0.01

Similarly calculating with the CCα and LD values estimated based on 3-level calibration, the estimated LOD = 0.014+1.64*0.028*0.12/0.9= 0.02.

Since the reporting limit is 0.05 mg/kg, the Limit of quantification, **LOQ**,was confirmed experimentally with replicate recovery studies performed at 0.02 mg/kg level. The tests resulted in recoveries and CV values in the acceptable range, therefore the <u>reported limit of quantification of the method is 0.02 mg/kg</u> for commodities of high water content.

> The calibration function varied during the method validation, and larger variation is likely to occur during the long term practical use of the method. This consequently leads to the variation of estimated LD or CCα and LOD which should only be used as guidance values and not as a constant parameter. Therefore the properly selected Lowest Calibrated Level may be equally well used as a practical alternative to LD for indicating the detectability of pesticide residues.

4 CHARACTERISATION OF THE METHOD

In conclusion, the method may be characterised as:

The method is applicable within the analytical range of 0.02 and 2 mg/kg for a wide range of GC amenable pesticides, the physico-chemical properties of which are within the ranges of representative compounds given in Table 1. The typical recoveries from plant commodities of high water content are about 90 % (except the highly water soluble compounds such as methamidophos), cereals 93% and oil seeds 82% with a typical CV of ≤12%.

The typical or average residue values and typical CV-s characterise the trueness and precision of the method in general. The particular values characterise the method for specific pesticide/commodity combinations. For the represented, but not tested, pesticide/ commodity combinations, the typical values for the group should be used as an initial guidance. Their performance characteristics should be checked during the extension of the method.

The daily performance of the method must be checked with appropriate internal quality control procedures during the routine use of the method. The performance characteristics established during method validation should be adjusted, if necessary, based on the results of internal quality control/performance verification tests.

5. REFERENCES

1. W. J. Youden and E.H. Steiner, 'Statistical Manual of the AOAC', AOAC International, Gaithersburg, MD, 1975.

2. Ambrus, A., Solymosné, E.. and Korsós, I.; Estimation of uncertainty of sample preparation for the analysis of pesticide residues, J. Environ. Sci. Health, B31, 443-450 (1996).

3. B. Maestroni, A. Ghods, M. El Bidaoui, N. Rathor, O. Jarju and A. Ambrus, Testing the efficiency and uncertainty of sample preparation using 14C-labelled chlorpyrifos: PART II, Proceedings of International Workshop on Method Validation, Budapest, 1999.

4. AOAC/FAO/IAEA/IUPAC Expert Consultation, Guidelines for Single-laboratory Validation of Analytical Methods for Trace-level Concentrations of Organic Chemicals

5. A. Anderson and H. Palsheden, *Fresenius J. Anal Chem.*, 1991, **339**, 365.

6. A. de Kok, H. A. van der Schee, K. Vreeker, W. Verwaal, E. Besamusca, M. de Kroon and A. Toonen, Poster 8E-014, Presented at 9th IUPAC Congress of Pesticide Chemistry (IUPAC) London, 1998.

7. A. de Kok, H. A. van der Schee, K. Vreeker, W. Verwaal, E. Besamusca, M. de Kroon, Poster 8E-013, Presented at 9th IUPAC Congress of Pesticide Chemistry (IUPAC) London, 1998.

8. FAO, Pesticde residues in Food - Evaluations, FAO Rome (annual publication)

9. Tomlin, Clive, (ed) The Pesticide Manual, 10th ed. 1994

10. M. El-Bidaoui, O. Jarju, B. Maestroni, Y. Phakaiew and A. Ambrus, Testing the effect of sample processing on the stability of residues, Proceedings of International Workshop on Method Validation, Budapest, 1999.

11. E. Soboleva, N. Rathor, A. Mageto and A. Ambrus, Estimation of significance of "matrix-induced" chromatographic effects, Proceedings of International Workshop on Method Validation, Budapest, 1999.

12. L. A. Currie, *Chemometrics and Intelligent Laboratory Systems*, 1997, **37**, 151.

13. L. A. Currie, *Pure &Appl. Chem.*, 1995, **67**, 1699.

14. European Commission: Commission Decision laying down analytical methods to be used for detecting certain substances and residues thereof in live animals and animal products according to Council Directive 96/23/EC, Final Version: approved by CRLs

15. W. Horwitz, *Pure & Appl. Chem.*, 1988, **60**, 855

16. M. Thompson, S. L. R. Ellison, A. Fajgelj, P. Willetts and R. Wood, *Pure & Appl. Chem.*, 1999, **71**, 337.

17. W. Horwitz, L. R. Kamps and K. W. Boyer, *J. Assoc. Off. Anal. Chem.*, 1980, **63**, 1344.

Statistical Procedures

1. Propagation of random errors within a laboratory

Within-laboratory precision of an analytical method is determined by errors arising in the sample processing and analysis steps.

$$S_L = \sqrt{S_{Sp}^2 + S_A^2}$$

eq. 1

S_L overall standard deviation of laboratory results for a sample (If the tests are performed on different days by different operators, S_L gives an estimate of the within laboratory reproducibility. Note, however, that the "real" within laboratory reproducibility should also include the maximum effect of all variables occuring during routine use of the method, which can only be estimated through the on-going performance verification)
S_{Sp} standard deviation of sample processing (i.e. heterogeneity of the analytical sample)
S_A standard deviation of sample analysis

The within laboratory reproducibility is very often estimated from the results of two different sets of experiments carried out at different analyte concentrations. For instance, the reproducibility of withdrawing analytical samples (S_{Sp}) was established at 0.5 mg/kg analyte concetration and the analysis is performed at 0.02 mg/kg level. In such cases the within laboratory reproducibility must be calculated by inserting the estimated coefficient of variation of sample processing and analysis in eq.1.

$$CV_L = \sqrt{CV_{Sp}^2 + CV_A^2}$$

eq. 2

CV_A can be determined by analysis of replicate analytical portions, spiked before the extraction. It reflects only the random variation of analysis from the point of extraction. The contribution of sample processing; CV_{Sp}, may be estimated by analysing replicate test portions of samples containing incurred residues or processing natural units of the samples which were treated on their surface before processing. When S_L and S_A are significantly different, the S_{Sp} component can be separated from the S_L with eq. 1.

Although in principle a sample may be made perfectly homogeneous if it is, or is converted to, a liquid, in practice some heterogeneity is inevitable. Thus the variability of residues between test portions depends on the mass of the test portions. For example, an apple sample thoroughly homogenised to a visibly uniform purée, was associated with a CV_{Sp} of incurred residues of 18% or 56%, depending on whether 50 g or 5g test portions were taken[1]. CV_A values depend on residue levels and the "difficulty" presented by the analyte but may be on average in the range of 53% to 16% for concentrations in the range

---

[1] Ambrus Á., 1999, Quality of residue data, in Brooks G.T. & Roberts, T.R. eds Pesticide Chemistry and Bioscience - The Food and Environment Challenge, Royal Society of Chemistry, pp 339-350.

of 1 µg/kg to 1 mg/kg (see Table 1). The contributions of these two components to the variability of the results of analysis are illustrated in Table 2.

Table 1   Examples for the contributions of the variability arising from sample processing and analysis to the precision of an analytical method

$CV_{Sp}\%$	$CV_A\%$	$CV_L\%$
56	53	77
56	16	58
25	53	59
18	32	37
18	16	24
10	16	19

2. Within laboratory repeatability and reproducibility

The within laboratory reproducibility (S_L) should be determined for the analyte/sample matrix/concentration tested by analysing replicate analytical portions of samples containing incurred residues on different days by different operators. The reproducibility of analytical phase (S_A) can be determined by analysing replicate blank samples spiked before the extraction.

The estimation of mean value and especially the standard deviation cannot be exact when based on a small number of tests. At the time of method validation more reliable estimates can be obtained for the mean recovery and repeatability/ reproducibility of the method by combining the appropriate results obtained. The calculated typical average recoveries and CVs can be used to characterise the MRM and GSM. Using the residue concentrations determined during the recovery studies with representative analytes and representative matrices, perform the following calculations (see also Annex 3):

2.1   Remove outliers based on Dixon's test from recovery data obtained for individual analyte/sample matrix combinations.

> Note: The use of outlier tests on small sets of data (from which it is impossible to determine whether they are normally distributed) should be very carefully considered and fully justified. Rejection of Outliers should be rejected only in cases where either there is clear experimental evidence that something went wrong or the result is so clearly part of "another population" that there is no possibility that it could be an extreme result from the same population.

2.2   Check similarity of recoveries and reproducibility of the determination of the representative analytes in various commodity/sample matrices.

2.2.1   Calculate the average recoveries as percentages and variances of recoveries, in the same percentage units (not percentage CVs), for each analyte/sample matrix. Perform the calculations independently for each fortification level.

2.2.2   Check with Cochran test if any of the the variances obtained at one spiking level for all representative analytes are significantly different in one representative commodity. Perform the tests separately for each commodity and fortification level.

2.2.3   If the tests performed under 2.2.2 show the homogeneity of variances at one spiking level in one representative commodity, perform the test of homogeneity of the variances of recoveries of representative analytes obtained at various fortification levels in each commodity and/or between different representative commodities.

2.2.4   Check with one way ANOVA if the average recoveries obtained for all representative analytes in one representative commodity and fortification level, (which forms one set of recovery data) are not significantly different. Similarly check if the differences of average recoveries obtained for each analyte sample matrix combination at various fortification levels are significant

2.3   Where the variances and the minimum and maximum mean recovery values do not differ significantly:
          * calculate the typical average recovery from all recovery results ($Q_{typ,}$),
          * calculate the average of variances of each set of recovery data and the corresponding standard deviation obtained with a given analyte/commodity matrix,
          * calculate the CV_{Atyp} from the pooled variances, $\overline{V}$ and typical average recovery.

$$CV_{Atyp} = \frac{\sqrt{\overline{V}}}{Q_{typ}} \qquad \text{eq. 3}$$

If some of the recoveries or variances are significantly different from the others, then they should be removed from the group and specific average recovery and CV parameters can be calculated for the remaining analyte/matrix combinations.
The typical values obtained during method validation should be regularly refined with the additional data collected during the performance verification, which provides more realistic estimate of the method characteristics than those deriving only from method validation.

3. Critical Range

3.1   The critical range test

This should be used when the differences between two, or a few, results obtained by analysis of single analytical portions are being compared to the distributions of data that are characteristic of the method used to produce the results. The comparison of the difference between the minimum and maximum values of the test results is made to the appropriate Critical Range. The critical range of a normal distribution is that multiple of the standard deviation for the whole population of results, within which the absolute difference between the highest and lowest value in a random set of 2 or more results is expected to lie, with a 95% probability. Where the critical range is exceeded, the set of results has less than 5% probability of conforming to the original population distribution (i.e. the data from validation or performance verification).

The reproducibility of the method during regular application, and the applicability of the method for new analytes and matrices in respect to variability of the results, and the agreement between initial and confirmatory tests may be judged on this basis. When the absolute difference exceeds the appropriate limit (critical range), as given in the following sections, then all measurements that have given rise to this difference should be considered as potentially atypical of the method and requiring further investigation.

When the difference between the maximum and minimum test results for the sample are smaller than the critical range the difference is not significant.

When examining two independent test results for an analyte in a single sample, obtained under repeatability or reproducibility conditions, the comparison of the difference shall be made with the critical ranges for repeatability (S_r) or reproducibility (S_R) , which are:

$$r = 2.8 * S_r \quad \text{or} \quad R = 2.8 * S_R \qquad \text{eq.4}$$

In case of three or more measurements the Critical Range is:

$$r = f_{(n)} * S_r \quad \text{or} \quad R = f_{(n)} * S_R \qquad \text{eq. 5}$$

When typical precision values are used the equations modified as:

$$r = f_{(n)} * CV_{rtyp} * X \quad \text{or} \quad R = f_{(n)} * CV_{Rtyp} * X \quad \text{eq. 6}$$

Where X is the average of the test results compared, CV_{typ} is the typical coefficient of variation characterising either the reproducibility or repeatabilty of the measurements
and $f_{(n)}$ is the critical range factor (see Table 2).

Table 2   Critical range factors[a] for multiple determinations

N	$f_{(n)}$
2	2.8
3	3.3
4	3.6
5	3.9
6	4.0
7	4.2

(a) The critical range factor $f_{(n)}$ is the 95% quantile of the distribution of $(x_{max} - x_{min})/\sigma$, where x_{max} and x_{min} are the extreme values in a sample of size n from a normal distribution with standard deviation of σ

3.2. Comparison of results of chromatographic determinations

The critical range test should be used to establish whether or not a difference between test results obtained for the same extract analysed using two different determination systems, e.g. for confirmatory purposes, is actually significant. For such comparisons multi-point calibration (minimum three concentration levels injected in two replicates) shall be used to estimate the standard deviation (S_{x0}) of the predicted concentrations on each column. Each calibration should satisfy the acceptable performance criteria specified.
Check with F-tets if the S_{x0} values are not significantly different at 90% level of confidence, as assuming that they are coming from the same population when they are actually not would lead to wrong conclusion. Calculate the pooled $\overline{S}_{x0}$ from the mean of

variances of S_{x0} values obtained in the two chromatographic systems. Use $\overline{S}_{x0}$ to calculate the critical range and determine whether the difference in test results falls within this range.

The S_{x0} values are calculated from the calibration regression equations using the following approximate formula[2]:

$$S_{X0} = \frac{S_{y/x}}{b}\left\{\frac{1}{m}+\frac{1}{n}+\frac{(y_0-\overline{y})^2}{b^2\sum_i(x_i-\overline{x})^2}\right\}^{0.5}$$

eq. 7

$$S_{y/x}=\left[\frac{\sum_i(y_i-\hat{y}_i)^2}{n-2}\right]^{0.5}$$

eq. 8

Where:

y_i is the response obtained from injecting x_i analytical standard.

$\hat{y}_i$ is the point corresponding with x_i on the regression line.

$\overline{y}$ is the arithmetic mean of the y_i responses for all standard injections.

n is the total number of standard injections e.g. when the calibration is made at three level with duplicate injections, then **n** is equal to 6.

$\overline{x}$ is the arithmetic mean of the x_i concentrations of analytical standards.

b is the slope obtained from the linear regression.

y_0 is the mean response (**m** replicate injections) used to calculate the concentration value X_0

If a weighted regression is used for the calibration, then S_{X0} is calculated as:

$$S_{X0w} = \frac{S_{(y/x)w}}{b}\left\{\frac{1}{w_0 m}+\frac{1}{n}+\frac{(y_0-\overline{y}_w)^2}{b^2(\sum w_i x_i^2-n\overline{x}_w^2)}\right\}^{0.5}$$

eq. 9

$$S_{(y/x)w}=\sqrt{\frac{\sum_i w_i(y_i-\hat{y}_i)^2}{n-2}}$$

eq. 10

---

[2] Miller J.C. and Miller J.N., Statistics for Analytical Chemistry 3rd ed. pp. 110-128, Ellis Horwood PTR Prentice Hall, 1993

Where (additional symbols):

w_0 is the weighting appropriate to value of y_0

w_i is the weighting appropriate to value of y_i

$\overline{y}_w$ is the arithmetic mean of the weighted y_{iw} responses from all standard injections.

$\overline{x}_w$ is the arithmetic mean of the weighted x_{iw} concentrations of standards.

4. Limits of Detection and Limits of Quantitation

The detection and quantification capabilities of a method are important for planning measurements, and for selecting or developing methods that can meet specified needs. The terms of limit of detection **(LD)**, limit of determination (LOD) and limit of quantitation (LOQ) are often used nterchangeably, with insufficient regard to their real meanings, and there has been very little agreement about appropriate methods for estimating them in practice. Therefore the analysts should always report or check the definition used to obtain the quoted value.

Although they are not constant characteristics, the LOQ and LD values should be determined during method validation and reported, even where the initial purpose of the method may not require measurements at these lower limits.
However there are some fundemental objections to using LOD and LOQ values as reporting limits during routine application of trace methods when they have been derived from method validation:

1. The limits will change with time due to changes in method performance, particularly in the determinative step and due to differences in matrices. These changes can be tracked by use of appropriate QC samples but good estimates for the true LOD and LOQ may still not be practical without a large and continuing effort.
2. The estimate for LD is generally based on an extrapolation of the calibration line. In organic trace analysis, low level calibrations frequently exhibit curvature due, for example, to non-proportional losses in the analytical system. This phenomenon is prevalent in gas chromatographic determinations where irreproducible effects can occur in the injector and column. Under these conditions the LD and any derived LOQ may be seriously underestimated. These problems are compounded with multi-residue methods.

It is recognised that the LOD and LOQ can have relevance to trace analytical problems, in that for some methods and matrices they are stable and reproducible parameters, and because their reporting with test data may be a regulatory requirement. However the following conservative recommendations are made in regard to setting reporting limits for most multi-residue methods.

1. All reported test data must fall within the calibrated range of the method.
2. The lowest calibrated level (LCL) will represent the lower reporting limit for test data.

3. The LCL should or must be supported by some recovery data which demonstrate adequate accuracy at concentrations in this vicinity (factor of two of LCL).
4. Supporting precision estimates can be provided using recovery data derived from multiple determinations near the LCL and generated under within-laboratory reproducibility conditions.

5. Decision Limit and Detection Capability

In the veterinary drug residue area there has been extensive research into statistical techniques for assessing the performance of methods intended to establish the presence or absence of banned substances in foods of animal origin.   This has included sophisticated approaches to estimating the concentration cut-offs for decision making. Related techniques may also be applied for determining whether set ALs have been exceeded with specified probability[3].

The application of the procedures described in the EU Comission Document[3] is not currently a standard practice in other areas of residue analysis, but can provide some useful additional information on method performance. Though the methods recommended for the determination of CCα and CCβ remain the subject of some debate, and more precise procedures are available and used in other areas of chemometrics,[4,5] the concepts have definite merits and may provide useful additional information. They provide the necessary flexibility for the analysts to define the detection limit, taking into account the number of replicate measurements (degree of freedom), with selected probabilities of false positive and false negative detection to fit best for the purpose of the analysis.   Until the utility and limitations of these techniques have been tested thoroughly for the particular analytical field, it is recommended that analysts should adopt them with due caution.

---

[3] European Commission: Commission Decision laying down analytical methods to be used for detecting certain substances and residues thereof in live animals and animal products according to Council Directive 96/23/EC, (Final Version prepared in 1998: approved by CRL.)
[4] L. A. Currie, Chemometrics and Intelligent Laboratory Systems, 1997, **37**, 151.
[5] L. A. Currie, *Pure &Appl. Chem.*, 1995, **67**, 1699.

Glossary of terms

Accepted Limit (AL)	Concentration value for an analyte corresponding to a regulatory limit or guideline value which forms the purpose for the analysis, e.g. MRL, MPL; trading standard, target concentration limit (dietary exposure assessment), acceptance level (environment) etc. For a substance without an MRL or for a banned substance there may be no AL (effectively it may be zero or there may be no limit) or it may be the target concentration above which detected residues should be confirmed (action limit or administrative limit).
Accuracy	Closeness of agreement between a test result and the accepted reference value.
Alpha (α) Error	Probability that the true concentration of analyte in the laboratory sample is less than a particular value (e.g. the AL) when measurements made on one or more analytical/test portions indicate that the concentration exceeds that value (false positive). Accepted values for this probability are usually in the range 1 to 5%.
Analyte	The chemical substance sought or determined in a sample.
Analyte Homogeneity (in sample)	Uniformity of dispersion of the analyte in matrix. The variability in analytical results arising from sample processing depends on the size of analytical portion. The sampling constant[1] describes the relationship between analytical portion size and the expected variation in a well mixed analytical sample: $K_S = w\ (CV_{Sp})^2$, where w is the mass of analytical portion and CV_{Sp} is the coefficient of variation of the analyte concentration in replicate analytical portions of w (g)which are withdrawn from the analytical sample
Analytical portion	A representative quantity of material removed from the analytical sample, of proper size for measurement of the residue concentration.
Analytical sample	The material prepared for analysis from the laboratory sample, by separation of the portion of the product to be analysed and then by mixing, grinding, fine chopping, etc., for the removal of analytical portions with minimal sampling error.
Applicability	The analytes, matrices and concentrations for which a method of analysis has been shown to be satisfactory.
Beta (β) Error	Probability that the true concentration of analyte in the laboratory sample is greater than a particular value (e.g. the AL) when measurements made on one or more analytical portions indicate that the concentration does not exceed that value (false negative). Accepted values for this probability are usually in the range 1 to 5%.
Bias	Difference between the mean value measured for an analyte and an accepted reference value for the sample. Bias is the total systematic error as contrasted to random error. There may be one or more systematic error components contributing to the bias. A larger systematic difference from the accepted reference value is reflected by a larger bias value.

---

[1] Wallace, D. and Kratochvil, B., Analytical Chemistry, 59, 226-232, 1987
[2] Ambrus, A., Solymosné, E.. and Korsós, I. J. Environ. Sci. Health, B31, (3) 1996

Collaborative Study (analytical method)	Inter-laboratory study in which all laboratories follow the same written protocol for the test method and use similar equipment to quantify the analyte(s) in sets of identical test samples. The reported results are used to estimate the performance characteristics of the method. Usually these characteristics are within-laboratory and between-laboratory precision, and when necessary and possible, other pertinent characteristics such as systematic error, recovery, internal quality control parameters, sensitivity, limit of determination, and applicability. Also called method-performance study. (See "Protocol for the Design, Conduct and Interpretation of Method Performance Studies", Pure & Appl. Chem. 67 (1995) 331-343).
Commodity Group	Group of foods or animal feeds sharing sufficient chemical characteristics as to make them similar for the purposes of analysis by a method. The characteristics may be based on major constituents (e.g. water, fat, sugar, and acid content) or biological relationships, and may be defined by regulations.
Confirmatory Method	Methods that provide complete or complementary information enabling the analyte to be identified with an acceptable degree of certainty [at the Accepted Limit or level of interest]. As far as possible, confirmatory methods provide information on the chemical character of the analyte, preferably using spectrometric techniques. If a single technique lacks sufficient specificity, then confirmation may be achieved by additional procedures consisting of suitable combinations of clean-up, chromatographic separation(s) and selective detection. Bioassays can also provide some confirmatory data.
	In addition to the confirmation of the identity of an analyte, its concentration shall also be confirmed. This may be accomplished by analysis of a second test portion and/or re-analysis of the initial test portion with an appropriate alternative method (e.g. different column and/or detector). The qualitative and quantitative confirmation may also be carried out by the same method, when appropriate.
Critical Difference	Value below which the absolute difference between two single test results may be expected to lie with a specified probability. May be determined under conditions of repeatability or reproducibility. ISO 5725-1986 and Thompson and Wood PAC 67. 649-669, 1995
Decision Limit (CCα)	Limit at which it can be decided that the concentration of the analyte present in a sample truly exceeds that limit with an error probability of α (false positive). In the case of substances with zero AL, the CCα is the lowest concentration level, at which a method can discriminate with a statistical probability of $1 - \alpha$ whether the identified analyte is present. The CCα is equivalent to the limit of detection (LOD) under some definitions (usually for $\alpha = 1\%$).
	In the case of substances with an established AL, the CCα is the measured concentration, above which it can be decided with a statistical probability of $1 - \alpha$ that the identified analyte content is truly above the AL.

Detection Capability (CCß)	Smallest true concentration of the analyte that may be detected, identified and quantified in a sample with a beta error (false negative). In the case of banned substances the CCβ is the lowest concentration at which a method is able to determine the analyte in contaminated samples with a statistical probability of $1 - ß$. In the case of substances with an established MRL, CCβ is the concentration at which the method is able to detect samples that exceed this MRL with a statistical probability of $1 - ß$. When it is applied at the lowest detectable concentration, this parameter is intended to provide equivalent information to the Limit of Quantitation (LOQ), but CCβ is always associated with a specified statistical probability of detection, and therefore it is preferred over LOQ..
Detection Test Mixture	Mixture of analytical standards which are suitable to check the conditions of chromatographic separation and detection. The detection test mixture should contain analytes which provide information for the selectivity and response factors for the detectors, and the inertness (e.g. characterised by the tailing factor Tf) and separation power (e.g. resolution Rs) of column, and the reproducibility of RRt. The detection test mixture may have to be column and detector specific.
Incurred Residue	Residues of an analyte in a matrix arising by the route through which the trace levels would normally be expected, as opposed to residues from laboratory fortification of samples. Also weathered residue.
Individual Method	Method which is suitable for determination of one or more specified compounds. A separate individual method may be needed, for instance to determine some metabolite included in the residue definition of an individual pesticide or veterinary drug.
Laboratory Sample	The sample as received at the laboratory (not including the packaging).
Limit of Detection (LD)	Smallest concentration where the analyte can be identified. Commonly defined as the minimum concentration of analyte in the test sample that can be measured with a stated probability that the analyte is present at a concentration above that in the blank sample. IUPAC and ISO have recommended the abbreviation LD. See also Decision Limit.
Limit of Quantitation (LOQ)	Smallest concentration of the analyte that can be quantified. Commonly defined as the minimum concentration of analyte in the test sample that can be determined with acceptable precision (repeatability) and accuracy under the stated conditions of the test. See also Detection Capability.
Lowest Calibrated Level (LCL)	Lowest concentration of analyte detected and measured in calibration of the detection system. It may be expressed as a solution concentration or as a mass ratio in the test sample and must not include the contribution from the blank
Matrix	Material or component sampled for analytical studies, excluding the analyte.
Matrix Blank	Sample material containing no detectable level of the analytes of interest.

Matrix-matched Calibration	Calibration using standards prepared in an extract of the commodity analysed (or of a representative commodity). The objective is to compensate for the effects of co-extractives on the determination system. Such effects are often unpredictable, but matrix-matching may be unnecessary where co-extractives prove to be of insignificant effect.
Method	The series of procedures from receipt of a sample for analysis through to the production of the final result.
Method Validation	Process of verifying that a method is fit for purpose.
Multi residue Method, MRM	Method which is suitable for the identification and quantitation of a range of analytes, usually in a number of different matrices.
Negative Result	A result indicating that the analyte is not present at or above the lowest calibrated level. (see also Decision Limit and Limit of Detection)
Performance Verification	Sets of quality control data generated during the analysis of batches of samples to support the validity of on-going analyses. The data can be used to refine the performance parameters of the method.
Positive Result	A result indicating the presence of the analyte with a concentration at or above the lowest calibrated level. (see also Detection Capability)
Precision	Closeness of agreement between independent test results obtained under stipulated conditions.
Recovery	Fraction or percentage of an analyte recovered following extraction and analysis of a blank sample to which the analyte has been added at a known concentration (spiked sample or reference material).
Quantitative Method	A method capable of producing results, expressed as numerical values in appropriate units, with accuracy and precision which fit for the purpose. The degree of precision and trueness must be specified.
Reagent Blank	Complete analysis made without the inclusion of sample materials for QC purpose.
Reference Material	Material one or more of whose analyte concentrations are sufficiently homogeneous and well established to be used for the assessment of a measurement method, or for assigning values to other materials. In the context of this document the term "reference material" does not refer to materials used for the calibration of apparatus.
Reference Method	Quantitative analytical method of proven reliability characterised by well established trueness, specificity, precision and detection power. These methods will generally have been collaboratively studied and are usually based on molecular spectrometry. The reference method status is only valid if the method is implemented under an appropriate QA regime.
Reference Procedure	Procedure of established efficiency. Where this is not available, a reference procedure may be one that, in theory, should be highly efficient and is fundamentally different from that under test.

Repeatability	Precision under repeatability conditions, i.e. conditions where independent test results are obtained with the same method on replicate analytical portions in the same laboratory by the same operator using the same equipment within short intervals of time. (ISO 3534-1)
Representative Analyte	Analyte chosen to represent a group of analytes which are likely to be similar in their behaviour through a multi-residue analytical method, as judged by their physico-chemical properties e.g. structure, water solubility, K_{ow}, polarity, volatility, hydrolytic stability, pKa etc.
Represented Analyte	Analyte having physico-chemical properties which are within the range of properties of representative analytes.
Reproducibility	Closeness of agreement between results obtained with the same method on replicate analytical portions with different operators and using different equipment (within laboratory reproducibility). Similarly, when the tests are performed in different laboratories the inter-laboratory reproducibility is obtained.
Representative Commodity	Single food or feed used to represent a commodity group for method validation purposes. A commodity may be considered representative on the basis of proximate sample composition, such as water, fat/oil, acid, sugar and chlorophyll contents, or biological similarities of tissues etc..
Ruggedness	Ability of a chemical measurement process to resist changes in test results when subjected to minor changes in environmental and method procedural variables, laboratories, personnel, etc.
Sample Preparation	The procedure used, if required, to convert the laboratory sample into the analytical sample, by removal of parts (soil, stones, bones, etc.) not to be included in the analysis.
Sample Processing	The procedure(s) (e.g. cutting, grinding, mixing) used to make the analytical sample acceptably homogeneous with respect to the analyte distribution, prior to removal of the analytical portion. The processing element of preparation must be designed to avoid inducing changes in the concentration of the analyte.
Screening Method	A methods used to detect the presence of an analyte or class of analytes at or above the minimum concentration of interest. It should be designed to avoid false negative results at a specified probability level (generally $\beta = 5\%$). Qualitative positive results may be required to be confirmed by confirmatory or reference methods. See Decision Limit and Detection Capability.
Selectivity	Measure of the degree to which the analyte is likely to be distinguished from other sample components, either by separation (e.g., chromatography) or by the relative response of the detection system.
Specificity	Extent to which a method provides responses from the detection system which can be considered exclusively characteristic of the analyte.

Standard Addition	A procedure in which known amounts analyte are added to aliquots of a sample extract containing the analyte (its initially measured concentration being X), to produce new notional concentrations (for example, 1.5X and 2X). The analyte responses produced by the spiked aliquots and the original extract are measured and the analyte concentration in the original extract (zero addition of analyte) is determined from the slope and intercept of the response curve. Where the response curve obtained is not linear, the value for X must be interpreted cautiously.
Tailing Factor	Measure of chromatographic peak asymmetry; at 10% peak height maximum, the ratio of the front and tail segments of peak width, when separated by a vertical line drawn through the peak maximum.
Test Portion	See "Analytical Portion"
Test Sample	See "Analytical Sample"
Trueness	Closeness of agreement between the average value obtained from a large series of test results and an accepted reference value.
Uncertainty of measurement	Single parameter (usually a standard deviation or confidence interval) expressing the possible range of values around the measured result, within which the true value is expected to be with a stated degree of probability. It should take into account all recognised effects operating on the result, including: overall long-term precision (within laboratory reproducibility) of the complete method; the method bias; sub-sampling and calibration uncertainties; and any other known sources of variation in results.

ABBREVIATIONS

C_b	See Annex IV	IEC	International Electrotechnical Commission
C_{max}	See Annex 4	JECFA	Joint FAO/WHO Expert Committee on Food Additives
C_{min}	See Annex 4	MRL	Maximum Residue Limit
CV_{Atyp}	See Annex 4	MRM	Multi-Residue Method
CV_{Ltyp}	See Annex 4	RRt	Relative retention value for a peak
CV_{SP}	See Annex 4	Rs	Resolution of two chromatographic peaks
GLP	Good Laboratory Practice	SD	Standard Deviation
GSM	Group Specific Method	$S_{y/x}$	Standard deviation of the residuals calculated from the linear calibration function
		WHO	World Health Organization

Progress in the development of principles for assuring reliability of analytical results

The first, and probably the most practical approach, to obtaining comparable results from chemical analyses was to standardise the analytical methods. The predecessor of FAO/WHO Codex Alimentarius Commission (CAC), the Joint FAO/WHO Committee of Government Experts on the Code of Principles Concerning Milk and Milk Products , established in 1958, was among the first to recognise the need for internationally harmonised methods.

It was soon recognised that methods which are to be used by many laboratories must first be tested in many laboratories. A number of harmonised protocols were adopted by the International Union of Pure and Applied Chemistry (IUPAC) to address this issue. These were developed through a series of meetings and discussions involving experts from the Analytical, Applied and Clinical Divisions of IUPAC and representatives of the main international organisations sponsoring collaborative studies, the International Organization for Standardization (ISO) and AOAC International (formerly the Association of Official Analytical Chemists, or AOAC).

- In addition to the use of reliable methods, the proficiency of the chemist and the testing laboratory are of vital importance in obtaining valid results. Again, the joint efforts of ISO, AOAC International and IUPAC resulted in harmonised protocols for internal quality control within laboratories[1] and for testing the proficiency of laboratories[2]. Various guidelines and standards were introduced by different national and international organisations, setting out the requirements for a laboratory to demonstrate that it is recognised as competent to carry out the specified tests. The certification of such competence is usually performed through third party conformity assessment, using concepts and principles defined internationally in standards issued by ISO. The basic principles of the ISO (and equivalent) standards can be summarised as follows:

- For a laboratory to produce consistently reliable data it must implement an appropriate programme of quality assurance.

- Analytical methods must be thoroughly validated before use. These methods must be carefully and fully documented, staff adequately trained in their use and control charts should be established to ensure the procedures are under proper statistical control.

By the mid-1990's the accreditation of laboratories and the regular organization of proficiency tests became widespread. In 1997, the CAC accepted the principle that only the certificates issued by accredited laboratories should be accepted in international trade disputes.

As a result of these well-defined requirements, and the regular application of internal quality control measures, the reliability of residue analytical data has improved substantially. The proficiency tests indicated that methods validated in one laboratory can provide results consistent with those obtained in laboratories useing methods which were studied collaboratively.

The principles for the choice of methods of analysis adopted by CAC have undergone several revisions. The latest principles[3] are as follows:

(a)   Official methods of analysis elaborated by international organisations occupying themselves with a food or group of foods should be preferred;

(b)   Preference should be given to methods of analysis the reliability of which have been established in respect of the following criteria, selected as appropriate:

     (i)   specificity
     (ii)   accuracy
     (iii)   precision; repeatability, intra-laboratory reproducibility and iter-laboratory reproducibility
     (iv)   limit of detection
     (v)   sensitivity
     (vi)   practicability and applicability under normal laboratory conditions
     (vii)   other criteria which may be selected as required

(c) ᐧ   The methods selected should be chosen on the basis of practicability, and preference should be given to methods which have applicability for routine use.

(d)   All proposed methods of analysis must have direct pertinence to the Codex Standard to which are directed.

(e)   Methods of analysis which are applicable uniformly to various groups of commodities should be given preference over methods which apply only to individual commodities.

The CAC ideally requires methods that have progressed fully through a collaborative study in accordance with the internationally harmonised protocol. However, such studies are not always practical for the validation of analytical methods for pesticide and veterinary drug residues, because of the numerous possible combinations of residues and matrices.

Based on the concerns raised by several delegations, the Codex Committee on Methods of Analysis and Sampling (CCMAS) agreed to propose to the Commission that new work on single laboratory, or "in-house", method validation be undertaken.

Following a recommendation of the Codex Committees concerned, a FAO/IAEA Expert Consultation on Validation of Analytical Methods for Food Control[4], was held in Vienna in 1997. The consultation concluded that "the preferred means to validate an analytical chemical method used to determine compliance with Codex limits is a full collaborative study using internationally accepted protocols and in which all participating laboratories operate under internationally accepted principles of quality assurance". In addition, it recognised that "due to decreasing resources available for such studies at the national level and other factors including, but not limited to, insufficient numbers of laboratories to participate and increased costs, full collaborative studies are less frequently undertaken".

In addition, this consultation also recommended that "in those cases where collaborative studies or other inter-laboratory studies are impractical or impossible to carry out, evaluations of analytical methods could be done in one laboratory, provided that the validation work is conducted according to the five principles":

- "Laboratories carrying out the validation studies operate under a suitable quality system based upon internationally recognised principles";

- "Laboratories have in operation a third party review of the whole validation process (eg. GLP registration, accreditation according to ISO/IEC Guide 25, or Peer Review)";

- "Analytical methods are assessed in respect to the Codex general criteria for selection of methods of analysis"[3], "with emphasis on the assessment of the limit of quantitation rather than the limit of detection";

- "The validation work be carefully documented in an expert validation report in which it is unambiguously stated for which purposes (matrices and analyte levels) the method has been found to perform in a satisfactory manner"; and

- "Evidence of transferability be provided for all methods intended for Codex use for food control purposes".

As a follow-up action, the FAO/IAEA Secretariat, through Training and Reference Centre for Food and Pesticide Control, initiated a project for the elaboration of Guidelines for the validation of analytical methods in a single laboratory, prepared a working document in cooperation with a number of experienced and practicing analysts and cosponsored the AOAC-FAO-IAEA-IUPAC International Workshop on Principles and Practices of Method Validation from 4-6 November, 1999, in Budapest, Hungary. The Workshop brought together analytical chemists and representatives of agencies, governments, standards organisations and accreditation bodies involved in method validation in general, and specifically for residue analysis of pesticides, veterinary drugs and mycotoxins in foods and animal feeds.

The Workshop discussed a number of issues related to the method validation, and recognised the concerns regarding validation of methods expressed by the Vienna consultation. It endorsed the need for guidance on validation of methods within a single laboratory, whether for use only within that laboratory or as a precursor to an inter-laboratory validation. The recommendations and concerns from the Vienna consultation and the Budapest workshop were considered by the Miskolc consultation in preparing the *'Guidelines for single laboratory validation of analytical methods for trace-level concentrations of organic chemicals'*.

The Miskolc Consultation recommended that:

Laboratories which generate data in support of food and feed inspection should ensure that analytical methods routinely used for determination of pesticide or veterinary drug residues are validated, at least to the minimum standards suggested in the Guidelines.

References

1. Thompson, M. and Wood, R. 1995. Harmonized Guidelines for Internal Quality Control in Analytical Chemistry Laboratories. Pure & Appl. Chem. 67: 649-666.

2. Thompson, M. and Wood, R. 1993. International Harmonized Protocol for Proficiency Testing of (Chemical) Analytical Laboratories. Pure & Appl. Chem. 65: 2132-2144.
3. FAO/WHO. 1997. Codex Alimentarius Commission Procedural Manual, 10th Ed., 55-64. FAO, Rome.
4. FAO. 1998. Validation of Analytical Methods for Food Control.. Report of a Joint FAO/IAEA Expert Consultation, Vienna, Austria, 2-4 December, 1997, FAO, Rome.

A Critique on Available In-House Method Validation Documentation

Paul Willetts and Roger Wood

FOOD STANDARDS AGENCY, C/O INSTITUTE OF FOOD RESEARCH, NORWICH RESEARCH PARK, COLNEY, NORWICH NR4 7UA, UK

1 INTRODUCTION

As a result of the recent adoption in June 1997 by the Codex Alimentarius Commission of the Guidelines for the Assessment of the Competence of Testing Laboratories Involved in the Import and Export Control of Foods and also of the European Union's Additional Measures Food Control Directive in 1993, countries are required to ensure that laboratories meet strict quality requirements. These quality requirements include:

- Being accredited to ISO Guide 25 or its derivatives,
- participating in proficiency testing schemes which comply with the Harmonised Protocol,
- using internal quality control procedures which comply with the Harmonised Guidelines and
- using "validated methods"

"Validated methods means methods which comply with the principles of the Codex Alimentarius Commission or the European Union; in essence, such methods will have been validated by collaborative trial. Thus, methods which are to be used to enforce food legislation are, wherever possible, fully validated with their precision characteristics determined by formal collaborative trial. Notwithstanding this requirement, in some areas of food analysis, e.g. the determination of residues of veterinary drugs, the validation of some methods by collaborative trial is impracticable due to the complex nature of the matrices and the associated costs. In such circumstances, reliance needs to be placed on the performance characteristics obtained within-laboratory and it is important, therefore, that this process of "internal
or in-house validation" is carried out in a manner consistent with fitness for purpose.

A number of organisations are concerned that there should be guidelines prepared to aid laboratories with the procedures that they should adopt if they wish to undertake formal and transparent in-house method validation of their methods. In particular, it is a new work item for the Codex Committee on Methods of Analysis and Sampling (first discussed November 1997).

It is to be developed by the IUPAC Working Party on Quality Assurance (i.e. the Group which has already prepared the Internationally recognised protocols/guidelines on collaborative trials, proficiency testing, internal quality control and the draft recovery factors guidelines).

In order to aid the development of the above work, the following summary of existing in-house method validation protocols has been prepared. This is being taken forward as a comparative exercise and will aid in the preparation of "best practice" Guidelines prepared under the auspices of the IUPAC Interdivisional Working Party on Quality Assurance Schemes.

2 DOCUMENTS BEING REVIEWED

In order to aid the preparation of agreed In-House Method Validation Guidelines a number of existing protocols/documents have been reviewed. They are:

- Method Validation - A Laboratory Guide (Eurachem/LGC/VAM)[1]
- Validation of Chemical Analytical Methods - (NMKL Procedure) - 1996[2]
- Intralaboratory Analytical Method Validation - A Short Course (AOAC INTERNATIONAL)[3]
- Validation of Methods Report - Inspectorate for Health Protection, Food Inspection Service, The Netherlands - 1995[4]

There are other texts available which are not reviewed here, notably:

- A Protocol for Analytical Quality Assurance in Public Analysts' Laboratories
- Method Validation Guidelines (by Tom Csalkowskl, Whatman Inc)
- A Step by Step Approach to Validating your Assay (D Green)
- AOAC Review of Peer-Verified Methods (1993)
- An Introduction to Method Validation (D L Massart)

The comparative review of the four main texts is given below:

---

[1]   "Method Validation - A Laboratory Guide", EURACHEM Secretariat, Laboratory of the Government Chemist, Teddington, UK, 1996.

[2]   "Validation of Chemical Analytical Methods", NMKL Secretariat, Finland, 1996, NMKL Procedure No. 4.

[3]   "An Interlaboratory Analytical Method Validation Short Course developed by the AOAC INTERNATIONAL", AOAC INTERNATIONAL, Gaithersburg, Maryland, USA, 1996.

[4]   "Validation of Methods", Inspectorate for Health Protection, Rijswijk, The Netherlands, Report 95-001.

3 METHOD VALIDATION - A LABORATORY GUIDE (EURACHEM/LGC/VAM)

The Eurachem draft document is a laboratory guide to method validation, produced by LGC with the support of the UK DTI VAM initiative. While it is generic in nature and not specific to the food sector, much of its content is relevant to the food sector. The aim of the document is described as being 'to explain some of the issues around method validation and increase the readers' understanding of what is involved and why it is important'. This information is presented in the context of the six principles of analytical practice identified in the Valid Analytical Measurement programme and in particular the principle that 'analytical measurements should be made using methods and equipment which have been tested to ensure they are fit for purpose'. The document is intended to give guidance on methods testing.

The document addresses in turn the following points:

- what method validation is
- why method validation is necessary
- when methods should be validated
- how methods should be validated
- using and documenting validated methods
- using validation data to design QC

Method validation is described as the set of tests used to establish that a method is fit for a particular purpose.

In presenting method validation within the context of good working practice, the need for the analyst to be competent and to use equipment appropriately is underlined. The inter-relationship of the process of method validation to the development of the method itself is also highlighted, this relationship being described as 'an iterative process' that continues until specified performance requirements have been met.

Methods must be validated, and used, appropriately in order to ensure that reliable analytical results are produced. The validation of a method is necessary when it is necessary to verify that its performance characteristics are adequate for use for a particular problem. It is recommended that the reliability of a result, and by implication the adequacy of the performance characteristics of the method, is expressed in terms of the uncertainty of measurement, 'quoted in a way that is widely accepted, internally consistent and easy to interpret'. Such a statement of the measurement uncertainty is said 'to not only provide information on the confidence that can be placed on the result but also provides a basis for the intercomparability with other results of the measurement of the same quantity'.

The question of what constitutes 'appropriate' validation is addressed by illustrating the following different situations where method validation is necessary:

In this way an attempt has been made to draw a distinction between the varying degrees of validation which may be considered appropriate in different situations. However the use, in this generic document, of the term 'full' to mean the highest degree of within-laboratory validation, does not correspond with existing practice in the food sector, where it is generally understood that 'full validation' is that achieved by collaborative trial.

Situation	Degree of validation or revalidation required
New method developed for particular problem	Full
Existing method tested for applicability to a particular problem	Full
Established method revised to incorporate improvements	Partial or Full
Established method extended or adapted to a new problem	Partial or Full
When quality control indicates an established method is changing with time	Partial or Full
Established method in a different laboratory	Partial
Established method with different instrumentation	Partial
Established method with different operator	Partial

The question of deciding what degree of validation is required, is returned to in the section of the guide, 'how methods should be validated.' It is suggested that starting with a carefully considered analytical specification provides a good base on which to plan the validation process. Only general guidance is otherwise provided on this point however, it merely being suggested that the analyst takes into account 'past experience, customer requirements and the need for comparability with other similar methods already in use.'

Nevertheless the responsibility for ensuring that a method is adequately validated is stated as resting with the laboratory that wishes to use a particular method for a particular specification.

The document however does raise some pertinent questions regarding the situation where it is inconvenient or impossible for a laboratory to enter into collaborative study to test a method, for example:

Can laboratories validate methods on their own, and if so how? Will methods validated in this way be recognised by other laboratories? What sort of recognition can be expected for 'in-house' methods used in a regulatory environment?

While not directly answering these questions the guide does suggest that in the absence of collaborative testing, it may be feasible to get some idea of the comparability of the method with others used elsewhere by analysing certified reference materials or by benchmarking the method under examination against one for which the validation is well characterised.

The guide gives details of the validation characteristics typically studied during within-laboratory validation and gives, without accompanying in-depth statistical treatments, some limited advice on how these characteristic parameters may be determined practically, addressing both qualitative and quantitative analysis as appropriate.

Notably ruggedness is included as one of the method validation characteristics. By its inclusion, the process of method validation, in addition to giving an indication of a method's performance capabilities and limitations in routine use, will indicate what might cause the method to go out of control. Validation data can therefore be used to design appropriate QC.

In summary the validation characteristics presented are:

- selectivity/specificity
- limit of detection
- limit of quantification
- recovery
- working and linear ranges
- accuracy/trueness
- repeatability precision
- reproducibility precision
- ruggedness/robustness

Once again the information contained in this generic document, at times, appears to conflict with existing practice in the food sector, where for example, reproducibility precision is expressed as that derived by collaborative trial.

The proper documentation of validated methods, however, which is recognised as essential in this document, does conform to established practice in the food area.

The main information given for each of the validation characteristics or parameters is as follows:

a) Selectivity

If interferences are present, which either are not separated from the analyte of interest or are not known to be present, they may:

- inhibit analyte confirmation
- enhance or suppress the concentration of the analyte

In these cases, carrying out further development may overcome these effects, thereby raising the confidence of the analysis.

b) Limit of Detection

It is recognised that a number of conventions exist for determining 'the lowest concentration of the analyte that can be conclusively detected by the method', based on various statistical justifications.

The approaches suggested involve either of the following:
- carrying out 10 independent analyses of sample blanks. The LOD is expressed as the analyte concentration corresponding to the sample blank + 3 sample standard deviations.
- carrying out 10 independent analyses of a sample blank fortified at the lowest concentration for which an acceptable degree of uncertainty can be achieved. The LOD is expressed as the analyte concentration corresponding to 4.65 sample standard deviations.

For qualitative measurements the concentration threshold below which positive identification becomes unreliable can be ascertained by carrying out approximately 10 replicate assessments at each of a number of concentration levels across the range.

c) Limit of Quantification

The limit of quantification is the lowest concentration of analyte that can be determined with an acceptable level of uncertainty, or alternatively it is set by various conventions to be 5, 6 or 10 standard deviations of the blank mean. It is also sometimes known as the limit of determination.

d) Sensitivity

This is the measure of the change in instrument response which corresponds to a change in analyte concentration. Where the response has been established as linear with respect to concentration, sensitivity corresponds to the gradient of the response curve.

e) Working and Linear Ranges

For any quantitative method there is a range of analyte concentrations over which the method may be applied. At the lower end of the concentration range the limiting factor is the value of the limit of detection and/or limit of quantification. At the upper end of the concentration range limitations will be imposed by various effects depending on the detection mechanism.

Within this working range there may exist a linear range, within which the detection response will have a linear response to analyte concentration. Note that the working and linear range may be different for different sample types according to the effect of interferences arising from the sample matrix.

It is recommended that in the first instance the response relationship should be examined over the working range by carrying out a single assessment of the response levels for at least 6 concentration levels. To determine the response relationship within the linear range, it is recommended that 3 replicates are carried out at each of at least 6 concentration levels.

f) Accuracy

Whether or not a method is accurate depends on the systematic errors which are inherent either within the method itself, in the way the method is used, or in the environment in which the method is used. These systematic errors cause bias in the method. Bias can be investigated using alternative analytical techniques, the analysis of suitable (matrix-matched) certified reference materials or where a suitable material is not available by carrying out analyses on spiked test portions.

Ideally the absolute value of bias can be determined by traceability of the measurand to the true value of an internationally accepted standard.

It is recommended that a reagent blank and standard (or reference material) are each analysed 10 times and the (mean analyte-mean blank) value compared with the true/acceptable value for the standard (or reference material). A comparison with similar measurements carried out using independent/primary method will indicate the methods' relative bias.

g) Precision

The two most useful precision measures in quantitative analyses are repeatability and reproducibility. Both are expressed in terms of standard deviation and are generally dependant on analyte concentration, so should be determined at a number of concentrations across the working range.

For qualitative analyses precision cannot be stated as a standard deviation but may be expressed as false positive and negative rates. These rates should be determined at a number of concentrations, below, at and above the threshold level.

It is recommended that 10 repeat determinations are carried out at each concentration level, both for repeatability and within-laboratory reproducibility.

h) Ruggedness

It is a measure of how effectively the performance of the analytical method stands up to less than perfect implementation. In any method there will be certain parts which, if not carried out sufficiently carefully, will have a severe effect on method performance. These aspects should be identified and if possible, their influence on method performance evaluated using ruggedness tests, sometimes also called robustness tests.

An example of 8 combinations of 7 variable factors used to test the ruggedness of the method is given.

i) Measurement Uncertainty

The ruggedness/robustness studies provide important information for the evaluation of the measurement uncertainty. The methodology for evaluating uncertainty given in the ISO Guide relies on identifying all of the parameters that may affect the result, (i.e. the potential sources of uncertainty) and quantifying the uncertainty contribution from each source. This is very similar to the procedures used in robustness studies where again all the parameters that are likely to influence the result are identified and then the control over these parameters that is required to ensure that their influence on the result is acceptable is determined. Thus if carried out with this in mind the robustness studies can provide information on the contribution to the overall uncertainty from each of the parameters studied.

4 VALIDATION OF CHEMICAL ANALYTICAL METHODS - (NMKL PROCEDURE)

The procedure produced by the Nordic Committee for Food Analysis is a sector-specific document which provides a comprehensive overview of the requirements for method validation in the food sector.

The introduction outlines the requirement in European Union regulations that, whenever possible, validated methods are employed in official food control. However, the authors clarify this position, stating that the requirement is also applicable in laboratories which are accredited, or intending to seek accreditation, irrespective of whether they are associated with official food control or not.

Validation is described as the measures taken in order to test and describe whether a method in respect to its accuracy, use, implementation and sources of errors operates at all times in accordance with expectations and laid down requirements.

Validated methods should be employed to ensure that the methodology fulfils certain quality requirements and, so that in turn, confidence is created in the analytical results reported by a laboratory.

It is emphasised that priority of use should always be given to a method which has been tested in a method-performance study (collaborative trail), wherever such a method is available. A laboratory proposing to use such a trailed method should demonstrate that the methodology is suitable for the intended purpose and that the performance characteristics stated in the method can be achieved in that laboratory, before its use.

Wherever such methods are not available, there is a need for other methods, which have been validated internally only, to be used.

As in the LGC document, it is stressed that the internal validation of methods should be conducted in the context of employing competent staff and having access to adequate/appropriate equipment.

The NMKL procedure addresses this latter situation of internal validation of methods at length. The procedure describes the parameters to be investigated, their definitions, the way in which the work should be carried out and criteria for acceptance of methods. The aim of the described approach to within-laboratory validation is to ensure that this important work, as carried out in different laboratories, should be performed as similarly as possible and according to defined guidelines.

Notwithstanding this aim for similarity of structured approach, the document makes it clear that in presenting guidelines for the validation of analytical methods, the requirements include consideration of the type of method employed and the way in which it is applied as well as the purpose of the work and the resources available to the laboratory.

Thus it would be more correct to summarise the intention of the authors as being to ensure that within-laboratory validation is carried out as similarly as possible under the same circumstances (i.e. presumably meaning the same outline approach to meet a stated level of measurement uncertainty).

Flexibility of approach is hereby required to take account of method type, fitness for purpose requirements and the possibilities open to the laboratory concerned. Provided that the validation process is thoroughly documented, the recommended acceptance criteria can and must be adjusted to the needs of the individual laboratory.

****** *Critique comment: It does not seem acceptable to adjust the overall recommended acceptance criteria, in terms of an overall stated level of uncertainty, if the required fitness for purpose is to be met. The authors of the document are presumably suggesting that it may, nevertheless be possible to adopt different approaches to validation in order to meet this overall level of measurement uncertainty.*

This flexibility of approach to validation is presented by identifying six different categories of analytical method in use in chemical laboratories. The categories, based on the available documentation on the degree of validation of a method, are as set out below:

Degree of external validation	**Recommended internal validation**
The method is externally validated in a method-performance study	Verification of trueness and precision
The method is externally validated but is used on a new matrix or using new instrument(s)	Verification of trueness and precision, possibly also detection limit
Well established, but not tested method	Verification, possibly a more extensive validation
The method is published in the scientific literature and states important performance characteristics	Verification, possibly a more extensive validation
The method is published in the scientific literature without presentation of performance characteristics	The method needs to be fully validated
The method was internally developed	The method needs to be fully validated

From the above, in contrast to the LGC generic document which merely detailed 'full' or 'partial' validation, more specific information is given regarding the degree of validation recommended, although it may be considered that the distinctions are rather arbitrarily drawn.

The term verification is introduced, its meaning being that a laboratory investigates and documents certain validation characteristics (only), for example trueness and precision, the values for which correspond to established performance characteristics of the method.

Validation or verification must always be carried out before taking a new method into routine use. Verification should be repeated in cases such as, for example, when:

- major instruments are replaced
- new batches of major reagents, e.g. polyclonal antibodies, are taken into use
- changes made in the laboratory premises may influence the results
- methods are used for the first time by new staff
- The laboratory takes an already validated and verified method into use after it has been out of use for a long time.

****** *Critique comment: It is noted that the NMKL document specifically states that validation/verification should be clearly distinguished from the development work, since changes can not be introduced into the method during the validation/verification process. This however, does at first sight contrast with the sentiment expressed in the LGC document that these processes are inter-related.*

As in the case of the LGC document, there is some potential for confusion between 'fully validated' and 'full internal validation' of a method.

Nevertheless NMKL gives a good presentation of the steps and validation characteristics from which an appropriate internal validation exercise can be designed.

The recommended steps and internal validation characteristics which serve as the 'set' criteria for the acceptance of methods are as follows:

1. Design a validation or verification protocol
2. Determine specificity and standard curve
3. Determine precision, expressed as repeatability and reproducibility
4. Determine trueness
5. Determine working range/measuring range
6. Determine detection limit
7. Determine limit of quantification/Limit of determination
8. Determine robustness
9. Determine sensitivity
10. Evaluate the result from steps 2 - 8 against the protocol
11. Document the work in a report

As shown in the list, before starting the experimental work in the laboratory, a protocol or plan for the validation or verification should be established. The validation or verification plan should in a concise manner describe which validation elements are to be evaluated and the order in which this is to be done, the protocol containing a compilation of all the requirements which the method must fulfil.

The protocol should be based on the following elements:

- The needs of the client
- What is analytically achievable
- Conditions of the laboratory, which have an impact on validation or verification, such as for example work environment and available equipment

The extent and the emphasis of the work depends on whether the method is a quantitative or a qualitative method as well as on the principle of the method (chromatography, determinations using antibodies etc.) and the circumstances i.e. the field of application. For example, analytical methods used to determine concentrations which are close to the detection limit of the method require more extensive validation than methods used to determine concentrations well above the detection limit.

In a similar manner, a simple verification may be sufficient in connection with in-house methods used in process control in industry, whereas methods to be used in the official control of food will for example, require full validation.

Performed validation and verification work must always be documented in such a way that the results can be checked later, and the work if necessary repeated.

After a method has been taken into routine use, its performance should be monitored according to the quality control measures as described in the quality manual of the laboratory, for example using suitable control charts.

In the NMKL document detailed coverage is given in the paragraphs devoted to each of the individual validation characteristics, for example specificity, precision etc. In the case of each characteristic, a definition, amounting to a comprehensive explanation, of the parameter is given, followed by a section containing practically-based details for the experimental determination of the parameter. Although detailed statistical treatments are not presented, the suggested approaches are more detailed than those given in the LGC document, and appear to be based on sound statistical and meteorological principles.

The information contained in the paragraphs on the individual validation characteristics is a notable strength of the NMKL procedure. In view of this and the status of the procedure as a recent document addressing internal validation of methods in the food sector, this material is summarised fully in this critique by referring to each validation parameter, or element, in turn, as set out below. *Critique comments are added where appropriate.*

a) Protocol for Validation or Verification

Concrete requirements should form the starting point of the protocol and should give answers to the following questions:

* *In what circumstances is the method to be used?*
For example official food control or production control. This has a bearing on the choice of analytical principle, and places requirements on reproducibility.
* *Is a qualitative or an exact quantitative result required?*
This consideration affects requirements on specificity, limits of detection and quantification etc.
* *In which chemical forms does the analyte occur?*
Bound or free? As various chemical compounds?
This has a bearing on the choice of sample matrices selected for the validation and on the design of recovery tests.
* *What is the matrix?*
* *Is there a risk of interference from the sample matrix or other components?*
Influences selectivity and specificity etc.
* *How much sample is usually available?*
Are the samples, for example, homogenous? Influences requirements on reproducibility and repeatability, sensitivity etc.
* *In which working range is the method to be used?*
Close to the limit of determination or at higher concentrations? Influences the requirements on linearity, detection and quantification limits, reproducibility and repeatability etc.
* *Can results be verified by using reference materials or by participating in proficiency testing schemes?*
This has a bearing on the requirements of trueness, reproducibility, robustness etc.
* *Which environmental requirements must be fulfilled?*
Personal safety? Effects of the environment, such as extreme temperatures, particles in the air etc. Use of chemicals toxic to the environment, such as ozone-depleting solvents etc.? Are the requirements on robustness influenced, and thereby indirect requirements on staff, equipment, reagents, premises etc.?
* *Which economical requirements must be fulfilled?*
Sets limits for the entire validation and verification work.

If the laboratory is developing a totally new method, the protocol should also include the development work and aspects of the choice of the analytical principle, purchase of reasonable equipment etc. In many cases these factors are already determined, and if the laboratory selects a method which has already been tested in an interlaboratory methods-performance study, the protocol will only describe the steps of the verification work. It is noted that if it is decided to omit some of these elements in an internal validation, then the reason for doing so should be documented. It is important that the protocol and the results of the experimental work are compared at all times. Only after all predetermined critera are met should the method be used.

The field of application is an important characteristic of most analytical methods. The evaluation of the field of application will depend on the type of method. For example, a method for the determination of vitamin D would require much more work in order to establish its suitability for all types of foods than would an AAS method for the determination of calcium. Often work may be saved by allowing the needs of the laboratory, not the limitations of the method, to dictate the field of application. If a method is used solely for the analysis of meat products, it is unnecessary to investigate its applicability to e.g. dairy products. However, it is important that the final report of the validation or verification describes the nature of the samples tested, and the basis for the determined field of application.

Some analytical methods are self-defining and their field of application implicit. Thus, for example the amount of Kjeldahl nitrogen equals the amount of nitrogen determined in accordance with a Kjeldahl determination, and the dry matter content and ash will usually depend on the method (drying at $105^{\circ}C$, vacuum drying at $70^{\circ}C$, freeze drying, ashing at $550^{\circ}C$ etc.).

b) Specificity

Definition. The ability of an analytical method to distinguish the analyte to be determined from other substances present in the sample.

Procedure. A blank sample and a sample to which a known amount of the analyte has been added may be analyzed in order to check that there are no interferences with the analyte from any expected compounds in the sample, degradation products, metabolites or known additives. In some cases, for example in the analysis of pesticides, a more concentrated extract of the blank may be analysed in order to demonstrate that no signals occur.

Specificity may also be examined by carrying out determinations in the presence of substances suspected of interfering with the analyte. The analyst should be aware that the analyte might be present in the sample in more than one chemical form.

By experience the analyst will often be aware of the kind as well as the nature of interferences to be expected in connection with specific methods. For example, spectral interferences occur in ICP determinations, whereas in chromatographic measurements several substances may have the same retention time. Such problems may have unfortunate consequences for the analytical results, and these techniques therefore require more extensive examination of specificity than techniques associated with none or only a few known interferences.

Self-defining methods need not be investigated for specificity. For example, certain substances will be lost when drying at $105^{\circ}C$, but that fraction is of no interest in cases where dry matter is defined as the matter remaining after such treatment.

c) Standard Curve

Definition. The standard curve reflects the ratio between amount/content of the analyte in a sample solution and the resulting measurement response.

Procedure. Measurement response should be determined using at least six different measurement points. Determinations should be made on reference samples or on blank samples to which the analyte has been added in evenly distributed concentrations covering the entire working range. The concentrations should be selected on both sides of the active working range.

The experiment should be repeated at least once. The results may be graphically presented, and the equation should be given for the linear regression as well as the correlation coefficient as a measure of the distribution (R > 0.999).

When linearity cannot be achieved, the standard curve in the relevant concentration range should be based on sufficient points to accurately determine the non-linear response function. The so-called curvature coefficient is a better measure of the linearity than the correlation coefficient. The curvature is estimated by adjusting the results to the function:

$$R = k \times c^{n}$$

where R = response corrected for the blank
  c = concentration
  k = estimate of the sensitivity
  n = curvature coefficient

The function is converted to the logarithmic form:

$$\log[R] = \log[k] + n \times \log[c]$$

A linear regression of log[R] against log[c] will then give n as the angular coefficient. The value of n should be above 0.9 and below 1.1.

d) Precision

Definition. Precision is the degree of agreement between independent analytical results obtained under specific circumstances. Precision depends only on the distribution of random errors, and is not associated with the true value. Precision is usually expressed as the standard deviation of the analytical result. Low precision gives a high standard deviation. Precision is an important characteristic in the evaluation of all quantitative methods.

The precision of a method depends very much on the conditions under which it has been estimated. Repeatability and reproducibility conditions represent clearly different conditions, while internal reproducibility lies in between these two extremes.

Repeatability or Reproducibility. An estimate of the repeatability of a method is obtained when analytical results are obtained on identical test portions, in the same laboratory, using the same equipment and within a short period of time. The reproducibility of the method can be estimated on the basis of results obtained when the method has been used to analyze identical test portions in different laboratories using different equipment. By internal reproducibility is understood the agreement of results when the analysis has been carried out in the same laboratory, using the same method, but performed at different times by different analysts using, for example, different batches of reagents.

It is important that it is exactly documented how, and on what materials repeatability and internal reproducibility have been estimated (reference materials, control materials, authentic samples, synthetic solutions). In cases where the method under study is intended to be used over a large concentration range, the precision should be estimated at several concentration levels, for example, at low, medium and high levels. It is recommended that the estimation of precision is repeated in whole or partly during validation work.

It should be kept in mind that the precision depends very much on the concentration of the analyte and on the analytical technique.

Procedure. Make at least ten determinations from the same sample material. The conditions may be those of repeatability, reproducibility or something in between. Calculate the standard deviation either from results of replicate single determinations or from duplicate determinations.

****** *Critique comment: This suggested approach seems more complicated, possibly unnecessarily, than approaches proposed by other organisations and authors. In many other documents the approach is based on replicate single determinations.*

The standard deviation of single determinations is an estimate of the distribution around the average, whereas the standard deviation derived from duplicate determination is an estimate of the variation between two single determinations.

<u>*Single Determinations.*</u> Carry out for example at least ten determinations, and calculate the standard deviation, s from the obtained individual results, x_i according to the formula:

$$s = \sqrt{\frac{\Sigma(x_i - \bar{x})^2}{n - 1}}$$

where x = average
$i = 1, 2, \ldots, n$
n = the number of determinations

<u>*Duplicate Determinations.*</u> A duplicate determination consists of two determinations a_i and b_i. Carry out two determinations and calculate the standard deviation, s_r based on d differences $(a_i - b_i)$ according to the formula:

$$s = \sqrt{\frac{\Sigma(a_i - b_i)^2}{2d}}$$

where   i = 1, 2, ..., d

      d = the number of duplicates

Precision may be expressed either directly as the standard deviation, as the relative standard deviation, RSD or as the confidence interval.

The relative standard deviation is given by: RSD (%) =s/x.100

The confidence interval is given by: x ± D where D = 3 × s

The coefficient 3 in the formula of the confidence interval is a good approximation for 10 to 15 measurements and a confidence interval of 99%. If the precision estimate is based on a different number of measurements, it is recommended that, instead of the coefficient of 3, a value from a two-sided t-test table is used, especially if the number of observations is lower than 10.

For information, Table 1 contains typical values of acceptable relative standard deviations for repeatability, based on the concentraions of the analyte (Pure & Appl. Chem. **62** (1990) 149-162).

Table 1 *Recommended Acceptable Relative Standard Deviation for Repeatability for Different Analyte Concentrations*

Analyte concentration	RSD (%)
100 g/kg	2
10 g/kg	3
1 g/kg	4
100 mg/kg	5
10 mg/kg	7
1 mg/kg	11
100 µg/kg	15
10 µg/kg	21
1 µg/kg	30
0.1 µg/kg	43

The above concerns quantitative determinations. In the case of qualitative methods, the precision must be approached in a different manner since a qualitative analysis in practice is a yes/no measurement on a given threshold concentration of the analyte. The precison of a qualitative analysis may be expressed as the ratio of true or false positive or negative results, respectively. These ratios should be determined at different concentrations of the analyte, at and above the threshold concentrations of the method. The result should be verified by using a separate reliable method, if available. If such an alternative reliable method is not available for comparison, a recovery test using suitable "spiked/not spiked samples" of suitables matrices may be used.

% false positive = (false positive × 100)/sum of the known negative

% false negative = (false negative × 100)/sum of the known positive

e) Trueness

Definition. Trueness is the agreement between the sample's true content of a specific analyte and the result of the analysis. In this context a result is to be understood as either the average of duplicate determinations or a single result, depending on what is normal for the method. A distinction should be made between different analytical methods, on the basis of whether or not there are available reference methods and/or certified reference materials as well as proficiency testing schemes.

Alternative Definition. The degree of agreement of the true content of the analyte in the sample and the average obtained as a result of several determinations.

The true content of the sample will always be unknown. In order to evaluate the trueness of a method laboratories should depend on accepted norms such as a certified content of a reference material, results obtained using a validated method or a result obtained in a proficiency test. In all three cases several laboratories, analysts and instruments have been involved in order to minimize the errors.

Determinations for which Certified or Other Reference Material is Available

Procedure. Investigate the trueness of the method by analyzing a certified reference material. If a certified reference material is not available, use other reference materials, or a material prepared in the laboratory (in-house material) containing a known amount of the analyte. The content of the analyte in in-house control materials must be thoroughly investigated, preferably using two or more analytical methods based on different physical chemical principles and, if possible, based on determinations carried out in different laboratories. An in-house material should, whenever possible, be calibrated against a certified reference material. Carelessness in the assessment of the analyte concentration of an in-house control material is a serious loophole in the quality assurance of the laboratory.

Commentary. A certified reference material or an in-house control material must have an analyte concentration which is similar to the concentration level occurring in authentic samples. Certified reference materials and other control materials provide an estimate of the performance of the method in the examined range only. The analysis of reference materials is of little value if the reference material represents a different analyte level. For example, it is not possible to test the trueness of a method in the range from 0.1 µg/kg to 0.1 mg/kg using a reference material having an analyte concentration of 1 mg/kg.

Attention should also be paid to the fact that certified reference materials and other control materials are not always typical materials. Compared to authentic samples they are easier to handle, for example to extract and to ash. They are also more homogeneous than ordinary food samples. This may mean that results obtained from reference materials are often "better" than those obtained from unknown samples, resulting in false security regarding the results. It is therefore not recommended to use results obtained on reference materials uncritically as evidence that the analysis of an unknown food sample has the same trueness. Often a reference sample included in an analytical run does not go through the same homogenization steps as the unknown samples. This means that it is difficult to detect whether the samples have been contaminated during homogeiztion.

Finally, it must be recognized that not all materials designated as "certified reference materials" have the same high quality. Some caution is recommended.

With reference to what has been said above, the analysis of certified reference materials does not alone verify the trueness of an analytical method. The analysis of reference materials should be supplemented with other quality criteria, such as for example recovery tests (spiking).

Determinations for which Reference Methods are Available

Definition. A reference method is a method which has been studied in a method-performance study.

Procedure. Examine the trueness of the method by analyzing the same samples with the method to be verified or validated and the reference method. If the reference method is not in routine use in the laboratory, it is not justidied to introduce the method only in order to evaluate a new method. In such cases it is recommended that samples are sent to a laboratory having the necessary competence regarding the reference method, preferably an accredited laboratory.

Determinations for which Proficiency Tests are Available

Procedure. Examine the trueness of the analytical method by using the method when participating in a proficiency test including samples corresponding to those for which the candidate method is intended. Be aware that documented trueness holds only for relevant analyte levels and matrices.

Determinations for which No Reference Materials or Proficiency Tests are Available

Procedure. Determine the recovery by spiking samples with suitable amounts of a chemical containing the analyte. Recovery tests offer a limited control of the systematic error by checking the recovery of the analysis. The technique is especially useful in connection with unstable analytes, or when only a limited number of determinations are to be made. It is recommended that recovery tests are carried out in the relevant concentration ranges. If the method is used on several levels, recovery tests should be carried out on at least two levels. A known amount of the analyte is added to a test protion of a sample having a known concentration of the analyte. The added amount should correspond to the level normally present in the sample material. The test portion to which the analyte has been added is analysed along with the original sample with no added analyte. It is recommended to carry out ten replicates weighings five of the material without the added analyte, and five with the added analyte. The averages of each series are calculated. The recovery (%) of the added analyte is calculated as 100 x the difference in averages divided by the amount of the added amount:

[(Analysed amount - Original amount in the sample)/(Added amount)] x 100

The obvious advantage of using recovery tests is that the matrix is representative. The technique may be used for all analytes and for most matrices, provided that the analyte is available in the laboratory as a stable synthetic compound. The greatest limitation is that there may be a difference in chemical form between the analyte in the authentic sample and in the added synthetic compound. Furthermore, it is almost impossible to check all relevant concentration ranges.

Where recoveries in the range of 80-110% are obtained, it is often sufficient to test for recovery as described above three times. The determinations should be carried out during a limited period of time. It generally holds that more determinations should be made the lower the analyte concentration is, because random errors increase with decreasing concentrations. Calculate the average and the standrd deviation of the results obtained. Possibly apply the t-test in order to demonstrate whether or not the obtained recoveries are significantly different from 100.

It is important to evaluate the trueness of all quantitative methods. This also holds for self-defining methods, such as fat etc. It will always be important that laboratories obtain results comparable to those obtained by other laboratories using the same method.

The requirements on trueness and precision will in both cases depend on the concentration level and on the aim of the determination. In the case of trace analysis, a deviation in trueness and precision of over 10% is often acceptable, whereas such deviations will normally be totally unacceptable in determinations of, for example, Kjeldahl nitrogen or dry matter.

f) Concentration or Measurement Range

Definition. The range of the analyte concentration which has been experimentally demonstrated to fullfill the quality requirements of the method (see also precision and trueness).

All methods have a limited sensitivity which restricts the concentration range in which the method is applicable. The lower limit for reliable quantification is defined as the limit of quantification and is discussed in secton 3.7. The upper limit of quantification is especially important when using methods including standardization. This is best illustrated by some examples. Spectrometric methods usually have a linear working range up to a certain concentration. At higher concentrations, the curve will bend towards the concentration axis. Unless the method is to be used for very limited purposes in a known contration range, it is necessary to study the extent of the linear range. In addition, it is worth while to determine to what extent the requiremets regarding precision and trueness are satisfied in this range. In practise, a limitation concerning the lowest part of the standard curve will be obtained in the form of a limit of quantification.

Ion selective electrodes show a response which against the logarithm of the concentration exhibits a linear relationship over a wide concentration range (often from mg/kg to molar concentrations), but the curve will bend towards the y axis when the concentration approaches zero. The limit of quantification may then be defined as the lowest region of the linear standard curve. Samples with very low concentrations may be lifted up to the linear range by adding known amounts of the analyte. During validation and verification it should also be ensured that requirements on precision and trueness are met within the linear range.

In some cases it may be practical to use a non-linear standard curve in order to expand the working range, or on the basis of the use of a detector having an implicit non-linear response (for example a flame photometric sulfur detector in gas chromatography).

The problem will resemble a linear standardization since it should be investigated in which concentration range the applied standardization relation is valid and also ensured that other quality requirements of the method are fulfilled.

Gravimetric and titrimetric methods (among these some important analytical methods for the determination of dry matter, ash, fat and Kjeldahl nitrogen) do not require evaluations of linearity, but the limit of quantification should be determined if the methods are used for the determination of very low concentrations.

g) Limit of Detection

Definition. Quantitative determinations: The amount or the content of an analyte corresponding to the lowest measurement signal which with a certain statistical confidence may be interpreted as indicating that the analyte is present in the solution, but not necessarily allowing exact quantification.

Qualitative determinations: The threshold concentration below which positive identification is unreliable.

Procedure. Quantitative determinations: Determine the limiting amount or concentration of an analyte from the analysis of a number (>20) of blank samples. The limit of detection is calculated as three times the standard deviation of the obtained average of the blank results (three standard deviations correspond to a confidence of 99%). Calculate the amount or content in the sample using a standard curve.

Some analytical methods will give no certain signal in the blank sample. In such cases it can be attempted to enlarge the instrumental noise and state the limit of detection as three times the standard deviation of the noise.

When enlarging instrument noise, extracts of the blank sample (concentrated blank samples) should be used. This extract should in addition be significantly more concentrated than extracts from normal samples in order to establish that there is no interference from any component.

The limit of detection is very important in trace elemental analysis. On the other hand it will often be unnecessary to estimate this limit when evaluating methods for the determination of principal components of foods.

Qualitative determinations: When evaluating qualitative methods, it is recommended that the range in which the method gives correct results (responses) is investigated. This may be carried out by analyzing a series of samples, i.e. a blank sample and samples containing the analyte at various concentration levels. It is recommended that about ten replicate determinations are carried out at each level. A response curve for the method should be constructed by plotting the ratio of positive results against the concentration. The threshold concentration, i.e. the concentration at which the method becomes unreliable, can be read from the curve. In the example in Table 2 below, the reliability of a qualitative method becomes less than 100% at concentrations below 100 µg/g.

Table 2 *Example of Results Obtained for a Qualitative Methods*

Concentration µg/g	Number of determinations	Pos./neg. results
200	10	10/0

100	10	10/0
75	10	5/5
50	10	1/9
25	10	0/10

h) Limit of Quantification/Limit of Determination

Definition. The limit of quantification of an analytical procedure is the lowest amount of analyte in a sample which can be quantitatively determined with a certain confidence. The limit can also be called the limit of determination.

Procedure. Using the method, analyze a number of blank samples (>20). The limit of quantification is calculated as 10 times the standard deviation of the average of the blank sample. The amount or the content of the samples is determined from the standard curve.

In cases where the noise of the instrument is used as the basis of the determination of the limit of determination, see the last two sections under 3.5.

When evaluating quantitavive methods, it is important to define and estimate the lowest concentrations which can be measured with a sufficiently defined precision. The procedure recommended above is based on the assumption that the the precision at the limit of determination is equal to a blank analysis (background noise will dominate). A measurement corresponding to 10 standard deviations will then have a relative precision of 30% since the absolute precision (99% confidence, 10 - 15 measurements) is estimated to be three standard deviations.

i) Robustness

Definition. Robustness is defined as meaning the sensitivity of an analytical method to minor deviations in the experimental conditions of the method. A method which is not appreciably influenced is said to be robust.

Procedure. Some form of testing of robustness should be included in the evaluation of analytical methods elaborated in the laboratory. In the literature it is recommended to perform testing of robustness prior to the conduct of the method-performance testing. The nature of the analytical method will define which parameters of the methods need to be studied. The most frequently occurring parameters tested, and which may be critical for an analytical method are: the composition of the samle (matrix), the batch of chemicals, pH, extraction time, temperature, flow rate and the volatility of the analyte. Blank samples may be used in robustness tests since these will reveal effects from e.g. the matrix and new batches of chemicals. Information gained in robustness testing may be used to specify the conditions under which a method is applicable.

j) Sensitivity

Definition. The sensitivity of a method is a measure of the magnitude of the response caused by a certain amount of analyte. It is represented by the angular coefficient of the standard curve.

Procedure. Determine the standard curve of the method using at least six concentrations of the standard. Calculate the angular coefficient by the least square method.

5 INTRALABORATORY ANALYTICAL METHOD VALIDATION - A SHORT COURSE (AOAC INTERNATIONAL)

The AOAC documentation comprises a short course manual devoted to the subject of 'intralaboratory analytical method validation'. The course material was prepared for a training programme in 1996. In the manual, intralaboratory method validation is introduced by presenting the topic within the broader context of method validation generally.

Validation is defined as the process of determining the suitability of methodology for providing useful analytical data. The process involves the production of a set of quantitative and experimentally-determined values, known as the performance characteristics of the method, for a number of fundamental parameters of importance in assessing the suitability of the method.

Overall, the objective of method validation is stated as being 'to have properly designed, adequately studied and well documented analytical methods capable of producing useful data for the intended purpose'.

The AOAC document highlights <u>the benefits</u> of conducting method validation studies and details <u>when</u> methods need to be validated, proposing standardised terminology for this purpose.

The benefits of method validation, in addition to the assessment, characterisation and optimisation of method performance, include:

- demonstration of ability to meet project requirements
- provision of information to assist the design of QA/QC strategy
- generation of baseline data for comparison to long-term quality control
- identification of instrument, equipment and supply needs and availability
- identification of personnel skill requirements and availability
- consideration of safety and hazardous waste issues
- facilitation of operational efficiency when the method is implemented

The process of validation normally takes place following a method development study but may also precede further development of the method, if the suitability of the method for a given purpose is not demonstrated during the validation stage.

It is implicitly recognised within this document that method validation may not be justified when the analytical measurement activity that is to be carried out is not a sustained or continual effort involving predictable matrices.

The course outline distinguishes between:

- interlaboratory validation studies
- intralaboratory validation studies,

Interlaboratory method validation studies are the ultimate test for determining how a measurement process can be expected to behave over time with different operators in a number of laboratory settings.

Intralaboratory method validation is carried out within a particular laboratory only and is conducted prior to any interlaboratory validation study.

The document appears to suggest that the intra- or within-laboratory validation of a method is required to be carried out, not only in the first instance by the laboratory which has developed the methodology for a particular purpose, but in addition, subsequently in other laboratories deploying the methodology. Intralaboratory validation of the method will also be required, dependent on the intended use of the data and the customers willingness to accept qualitative and/or quantitative uncertainties, when changes are made to the 'fitness for purpose' objectives of the previous validation; for example when a new application is undertaken for a currently validated method.

Intralaboratory validation may involve different options being followed, each with differing time and resource requirements. For example the approach which is chosen may involve:

- the use of reference materials
- comparison to another method
- the use of blanks and fortified samples
- the use of 'positives' and fortified samples

Of importance in this regard, is the scientific validity both of the approach taken and the performance characteristics generated, in particular whether proper statistical tools were used for experimental design and data analysis, also whether there are sufficient data to adequately describe the method performance characteristics.

The AOAC document emphasises the need for the validation study to be properly planned and documented. The production of a 'Standard Operating Procedure' for validating a method is recommended. In addition, the limitations of the study and the data uncertainties should be clearly explained and should accompany the data in the analytical report.

It is noted that the use of a validated analytical method does not preclude the need for proper measurement design and QA/QC samples during analysis. Moreover, the quality of analytical data obtained will be dependent on the reliability of the staff, facility, chemicals, supplies, equipment and instrumentation as well as the inherent limitations of the method itself.

In the initial planning stage of method validation, the suggested approach advocates setting data quality objectives (DQO) and analytical measurement objectives (AMO) to enable the requirements of the project to be met.

DQO's are generally quantitative specifications, agreed with the customer that establish the minimum quality of data that will be useful for a given investigation. Both sampling and analytical measurement activities are involved in establishing DQO's. In turn, AMO's are set for the particular analytical method application, quantitative objectives being based on the homogeneity of the samples supplied and the acceptable margin of error that will fulfil the DQO's. Different types of method application, for instance, monitoring, surveying and compliance/enforcement analyses, typically have different quantitative needs.

To determine whether an existing analytical method is suitable for the customers requirement, or whether it is necessary to search for a new method to validate, it is necessary to have knowledge of, at least, the following:

- the analyte(s) of interest, in particular whether this is well defined
- the type of matrix
- the concentration range of interest (screening may be required)
- the qualitative and quantitative needs, including level of detection and quantitation
- the stability of the samples and analytes

If the scope of an existing method does not specifically include the sample types and/or can not fulfil the data quality and measurement objectives, then it may be necessary to:

- search for another method that is more appropriate
- modify and validate a currently operational method
- develop and validate an entirely new method
- use the existing method, realising that the method performance can not be demonstrated to meet the needs of the customer

A candidate method should be critically reviewed for the performance characteristics needed for a given application. This may typically include examining:

- the concentration range of the analyte(s) in the samples
- instrument response profile and calibration
- the ability to discriminate between samples containing different amounts of analyte (sensitivity)
- selectivity
- bias in the data (recovery, calibration, means of bias identification)
- variability in the data (within-run and between-run precision, sample contributions)
- The level of detection and quantitation (definition and measurement approach)
- ruggedness testing

These aspects of method validation, summarised in the above list are then dealt with individually in greater detail, providing more information on the different technical components of a method validation study.

It is noted, however, that the process of validation may be accomplished in several ways. For instance, it may not be necessary to study all of the performance characteristics listed, professional judgement being required to select the appropriate performance characteristics, the exact approach for a given measurement objective, and the definition of "acceptable" vs. "unacceptable" performance of a method.

An example is given from the United States Pharmacopoeia (USP) Section <1225> method validation guidelines to illustrate the flexibility of approach, with respect to method type:

Type of Method

Performance Characteristic	Major Component	Trace: Above LOQ	Trace: Between LOQ and LOD
Precision	Y	Y	N
Accuracy	Y	Y	?
LOD	N	N	Y
LOQ	N	Y	N
Selectivity	Y	Y	Y
Linearity	Y	Y	N
Range	Y	Y	?
Ruggedness	Y	Y	Y

Y   =   important
N   =   not important
?   =   may be important depending on method application

Relevant aspects of the detailed classification and review of the performance characteristics that may be important for validating analytical methods, based on measurement objectives, are set out below.

a) Calibration Function

The response profile of an instrument over a range of analyte concentrations is determined using calibration standards presented in a suitable solvent or sample matrix. The response function takes the form $S = g(C)$ where S is the measured net signal and C is a given analyte concentration. The graphed version of the response function, known as the calibration curve, may exhibit a linear relationship.

The experimental determination of the constants (slope and y-intercept) of a intervals can be derived statistically for each constant.

- The calibration standards should be run in random order.
- The random error associated with the calibration can be minimised by replication of the calibration measurements. The spacing of the concentration levels and the replication pattern depends on the precision required over the range studied.
- The calibration is valid in the case of a standard prepared in solvent only if the analyte response from the calibration standard is the same as an equal amount of the analyte in the sample extract.
- The magnitude of the error associated with the analytical results is a function of several variables (eg precision of measurement, number of calibration standards,

concentration of analytes in the samples). For this reason, the random error of the calibration function should not be regarded as a performance characteristic.

- Use all data points, not averaged data, in the regression analysis. The correlation coefficient should not be used to determine the appropriateness of the linear model.

The approach chosen for calibrating instruments depends on the analytical technique, the instrument, and the measurement objectives. Typical approaches include:

Multi-point calibration - linear regression (method of least squares - unweighted). The model assumes that the variance of the response variables is constant across the calibration range (homoscedasticity)

Calibration standards are used to determine the response of the instrument over the concentration range of interest, however, a good assessment of this model requires at least 5 or 6 concentration levels and multiple measurements at each concentration level

The residuals plot is a valuable tool for assessing the appropriateness of the model for the data set.

Multi-point calibration - linear regression (method of least squares - weighted). The assumption that the variance of the response variables is constant across the calibration range is violated in many instances (heteroscedasticity)

A weighted least squares regression will provide a more accurate assessment of the slope, intercept and confidence intervals for the regression line, however, a good estimate of the variance at each concentration is required otherwise the weighted regression may be more biased than the unweighted regression

Single-point calibration. The response from only one standard solution is used as a reference for calculating the concentration of an unknown.

Assumes that the y-intercept is zero within error ($y = mx$) and the change in response is directly proportional to the change in analyte concentration.

Since the slope is a constant and the intercept is zero, the response factor (y/x) at any concentration should be a constant within error.

Valid if the confidence intervals for the regression line are acceptably narrow relative to the precision required for the assay.

Some method validation guidelines may specify a maximum acceptable CV for the response factors, but the variability of the response factors is not the whole story. In addition, the simple linear regression model assumes that the variance in y is constant across the calibration range.

Sometimes this model is assumed to be adequate if the response of the analyte is close to the response of the calibration standard (within 10% or 20%, for example).

The possibility of trends in the response factors should be investigated. A single point calibration may not be justified, and a measurement bias will occur if this approach is used inappropriately.

An ANOVA can be performed on response factor data to compare the means at each concentration level.

b) Sensitivity

Refers to the ability to discriminate between different amounts of analyte (that is how small a difference can be measured)

The sensitivity of the method depends primarily on the characteristics of the calibration function and the variability of analyte recovery over the concentration range of interest.

This term is sometimes used (inappropriately) when referring to the detection capabilities of the measurement process.

c) Selectivity and Specificity

Selectivity. The method responds primarily to the analyte of interest and is little affected by other substances.

Specificity. The method responds only to a specific form of the analyte and is not affected by other substances or form(s) of the analyte.

(For example, there may be situations when it is important to measure only a specific form of the analyte, such as a particular oxidation state of a metal, when other forms are present in the sample.)

Specificity is the limit of selectivity.

Interferences. An interfering substance is one that causes systematic error in the analytical results (of any magnitude) for at least one concentration of the analyte within the range of the method.

There are *qualitative* and *quantitative* aspects of describing method selectivity.

Qualitative certainty of selectivity for the analyte(s) of interest in the sample matrix can be achieved by the use of blanks, most conclusively matrix blanks, spikes and reference standards for comparative identification. In addition to demonstrating selectivity in this way, supporting qualitative data can be provided by spectral data (UV, mass, IR, atomic emission, etc.), elemental analysis and the use of alternative methods or techniques (different chemical separation techniques, different detection techniques, different sample preparation techniques) for qualitative confirmation.

If matrix blanks are available, analyse a minimum of 3 matrix blanks, 3 method blanks and the lowest concentration standard. Assess the magnitude and variability of the background signals from these blanks and the significance of any interferences or contamination relative to the lowest analyte level specified in the scope of the method. If unacceptably high background signals are observed, optimise reagent/equipment choice, analyte extraction, clean-up, and/or chromatographic separation efficiencies to eliminate the interference(s). Typically responses from interfering substances should be ‹1% of the response of the lowest concentration measured.

If matrix blanks are not available, analyse a minimum of 3 positive samples, 3 method blanks and a standard that corresponds to the analyte concentration in the positive sample. Ensure the absence of unacceptably high background signals from the blank by optimising reagent/equipment choice and, if possible, obtain qualitative data (e.g. UV spectra, mass spectra, IR spectra etc.) for the positive sample and the calibration standard to confirm that the analyte of interest is the only species being measured in the positive sample. Co-chromatography, in conjunction with peak shape/peak purity analysis, may in addition be carried out for chromatographic techniques. If the method detection technique does not provide qualitative data (e.g. FID or refractive index detectors), then the selectivity of the method must be demonstrated by another means such as:

(a) if a non-destructive separation and detection technique is used, by collecting the eluting fraction containing the analyte of interest and obtaining data, using other analytical techniques, to confirm the identity and purity of the analyte.

(b) if it is not possible to isolate and collect the analyte using the specified method, then an indirect demonstration of method selectivity may have to be based on the closeness of agreement of data with that obtained by one or more additional methods/techniques. Note, however, that this approach is not recommended for demonstrating selectivity unless one of the additional methods used is a standard method with known/demonstrated selectivity for the analyte in the matrix being studied.

Quantitative assessment requires extensive and fairly complex experimentation. For example, if the concentrations of the analyte and other substances can be varied at will, there are quantitative tests that can be performed to experimentally estimate the interference effects.

d) Bias, accuracy and precision

Bias. The systematic departure of the measured value from the accepted reference or true value (high or low).

Accuracy. The closeness of agreement between a test result and the accepted reference or true value. (Note: when applied to a set of test results, the term involves a combination of random components and a common systematic error.)

Precision. The closeness of agreement between independent test results obtained under stipulated conditions.

There are many sources of bias, and they may be additive.

Bias affects accuracy.

Accuracy can be assessed directly if you have a certified reference material that matches your sample.

Precision is usually expressed by the standard deviation of repeat measurements.

The precision of interest to analysts concerned with the on-going analysis of samples is comprised of a within-run and a between-run component.

If the between-run standard deviation is unacceptably larger than the within-run standard deviation, then the reason for the added variability should be investigated, identified and controlled if possible.

Sources of bias and variability from the laboratory include:

◊   Sample storage (contamination, physical or chemical degradation)

◊   Sample handling (contamination or alteration during preparation, e.g. grinding, blending, etc.).

◊   Sub-sampling

◊   Precision of weighing and volumetric devices

◊   Purity of solvents and reagents

◊   Extraction (low efficiency)

◊   Concentration (volatility, adsorption)

◊   Clean-up (analyte loss)

◊   Interferences (chemical and physical)

◊   Reference standard quality

◊   Instrument calibration

◊   Instrumental analysis (sample introduction repeatability, matrix effects on analyte response, change in operating conditions)

◊   Data analysis (misidentification of analyte, integration, rounding errors, instrument read-out error)

◊   Different analysts

◊   Environmental conditions in the laboratory (temperature, humidity, etc.).

The process of validating a method, with regard to bias and variability, may be accomplished in several ways including the following options, listed in order of increasing complexity. The approach chosen depends on the intended use of the method and the resources available. In all cases, bias should be assessed across the concentration range of interest and variability studies should be conducted within-run and between-run. It is recommended that for residue method validations, the variability of test results is assessed at 2 representative levels, as a minimum, which cover the working range of the method. At each of the concentration levels a minimum of 5 replicates should be analysed in each of 4 batches. It is considered reasonable to analyse one batch of samples every 1-2 days.

1.   Use of a reference material, and comparison of the reference material results obtained by the candidate method to the accepted reference value
-       May involve a one-sample t test, two-sample t test or the calculation of a confidence interval.
2.   Use of an alternative method, preferably a standard method known to be unbiased and precise, and comparison of the analytical results obtained by the candidate method to those obtained with the alternative method
-       May involve a two-sample t test, paired t test, F test for variances, least squares regression, ANOVA.
-       Unless a standard method is used, this approach does not unequivocally prove no bias, only lack of evidence of bias. The more methods used for comparison, the stronger the evidence.

3.   Use of representative blank matrices to conduct recovery studies with spiked matrix blanks.

-    May involve least squares regression, two-sample t test, F test for variances, confidence intervals, ANOVA.

4.   Recovery studies with positive samples and spiked positive samples: background subtraction or method of standard additions.

-    May involve least squares regression, two-sample t test, F test for variances, confidence intervals, ANOVA.

If a reference material is not available and there are no other methods appropriate for comparison, then bias and variability are investigated by spiking, either into a representative matrix blank or into a sample containing a low level of analyte (positive sample).

The problem, however, with using spike recovery data to assess analyte recovery is that typically the analyte is not introduced into the sample matrix in the same manner as the analyte is incorporated into the actual samples.

When carrying out the spiking/recovery experiments:

-    the spiking procedure must be precise and accurate
-    the spike should be added prior to any sample extraction procedures
-    the amount of analyte spiked should not be less than the amount estimated to be present in the sample, the background level being subtracted to give % recovery
-    depending on the variability of the background analyte signal and the spike level it may be necessary to prepare many replicates to get a good estimate of the average analyte recovery and the uncertainty in this value.
-    when spiking positive samples, qualitative data should be obtained to demonstrate that only the analyte of interest is being measured.

e) Level or limit of quantitation:

Defined by the maximum acceptable level of uncertainty in the measured values. Sometimes defined as RSD = 10% (higher levels of uncertainty are not considered to be associated with quantitative data).

It is necessary to determine what is or is not quantitative for a particular application.

Is also expressed as being equal to: blank signal + 10 standard deviations of a well characterised blank.

It is assumed that signals from blanks and samples are normally distributed.

f) Level or Limit of Detection:

Addresses the concentration at which the analyst is confident that the analyte is actually present, but it is necessary to define confident.

There are many definitions and many approaches for estimating this level.

Arbitrary cut-offs are chosen, resulting in left censored and ambiguous reporting of detection level data.

If the limit of detection is expressed as blank signal + 3 standard deviations of a blank, the risk of a false positive error is 0.13% and the risk of a false negative error is 50%.

If the limit of detection is expressed as blank signal + 6 standard deviations of a blank, the risk of a false negative error is 0.13%.

Estimates may be made based on the variability of blank data or from samples with low analyte levels - involves statistics of repeated measures, tests of significance, and decision errors.

To estimate the blank standard deviation from the peak-to-peak noise in an analog readout, it is necessary to divide the measured peak-to-peak noise by 5. The standard deviation of the blank is estimated by the root mean square noise, not the peak-to-peak noise.

It is noted that the levels of quantitation and detection are rarely critical performance characteristics since measurements are typically made at high analyte levels with methods capable of detecting and quantitating analytes at significantly lower levels.

g) Scope:

Defines the appropriate applications of the method (matrix, concentration range, type of study/monitoring effort) based on the performance characteristics studied during the validation process.

Describes the known limitations of the method.

h) Ruggedness Testing:

Demonstrates the sensitivity of the measurement process to small changes in conditions.

Ruggedness testing would typically be performed during a method development study, but if a method was developed in another laboratory, it is beneficial to conduct the appropriate tests for your own laboratory environment.

6 VALIDATION OF METHODS REPORT - INSPECTORATE FOR HEALTH PROTECTION, FOOD INSPECTION SERVICE, THE NETHERLANDS

A system for the validation of physicochemical and mechanical methods is presented. In the former case, the need to deal with a wide range of analyte/matrix combinations is introduced, it being necessary to guarantee the quality of the test results. The first requirement is the use of a validated method. When applying international guidelines for method validation, questions are encountered such as:

- Which parameters should be determined and how?

- On what scale should validation tests be carried out?

- Can the validation process be divided into different stages?

- What statistical methods should be used?

In discussing terms/parameters of relevance to method validation, the report draws on relevant EC legislation; Council Directive (85/591/EEC) of 23 December 1985 the Commission Decision of 14 November 1989 (89/610/EEC) and the Commission Decision of 26 September 1990 (90/515/EEC).

Next, all the methods used are catalogued and divided into groups. For chemical methods, this is done on the basis of the level of the analyte (main item) and the technique (subitem); for mechanical methods a distinction is drawn between destructive and non-destructive tests. The main parameters which at least must be determine to validate the method are indicated for each group.

The following terms/parameters used for validation are largely derived from Haagsma *et al.* (Ware(n)- Chemicus **21** 82-95 1991), supplemented with information from the above-mentioned EEC decisions, taking into account current practice of the inspectorates and existing agreements on validation.

1. Terms

These are given as follows:

a) **Analysis**
The procedure for detecting, identifying and/or determining an <u>analyte</u> in a sample.

b) **Analyte**
The component or group of components whose presence or absence in the <u>test sample</u> must be demonstrated or whose content in the sample must be determined.

After: Commission Decision 89/610/EEC.

c) **Content** (mass fraction of an analyte in a test sample)
The fraction of the <u>test sample</u> consisting of the <u>analyte</u>. The number is a dimensionless quantity. However, the content is usually given as a differential of two mass units: %, mg/g, µg/g or mg/kg, ng/g, etc. Use of the SI system is recommended: terms such as ppm and ppb are therefore not used.

d) **Detection**
Determination of the presence of the <u>analyte</u> as a chemical entity.

e) **Determination/Quantification**
The establishment of the absolute quantity of <u>analyte</u> (mass, volume or moles) or of the relative quantity of analyte (<u>content, mass fraction, mass concentration</u>).

f)   Laboratory Sample
A quantity of material intended for testing, in the form and condition in which it is supplied to the laboratory.

After: ISO 78/2, 1982.

g)   Mass Concentration
The quantity of analyte per quantity of solution, expressed in mg per litre of solution.
The term mass concentration is not used if the quantity of analyte is related to the mass of the sample, and the term content (mass fraction) is used instead.

h)   Matrix
All the constituents of the test sample with the exception of the analyte.

i)   Test Portion
The quantity of the test sample used for the actual analysis.

After: ISO 78/2, 1982.

j)   Test Sample
The part of the laboratory sample which has been treated in accordance with applicable regulations in such a way, that it can be used for the intended analyses.

After: ISO 78/2, 1982.

k)   Test Solution
The solution obtained by subjecting the test portion to the operations specified in the method of analysis. The test solution is used to detect or measure the presence or content of the analyte.

l)   Validation
The procedure which ensures that a test method is, as reliable as possible, applicable to analytical research. The validation process is a part of the verification process.
In a verification process a distinction is made between the validation process and the evaluation process. The validation process answers the question "How accurate is the method?" and the evaluation process answers the question "How precise is the method in the hands of the end user?".

The following distinctions can be made in this regard:

    Status 0 : method not validated;
    Status 1 : method validated in one's own laboratory;

Status 2 : method evaluated by means of an interlaboratory study.

A method used for inspection purposes which yields results allowing legal measures to be taken must at least be validated in one's own laboratory (status 1).

2. Parameters

These are given as follows:

a)   Accuracy
The degree to which a test result obtained with a particular method approximates the <u>true value</u>. See also: <u>systematic error</u>

After: NEN 3114.

b)   Calibration Curve
The graphic representation of the measuring signal as a function of the quantity of analyte.
If a calibration curve is used, the following information should be given:
-    the mathematical formula which describes the calibration curve;
-    numerical values of the parameters of the calibration curve with 95% reliability intervals;
-    the working range of the calibration curve corresponds to the <u>range of application</u>. The line is described on the basis of 5 standards, evenly distributed over the working range. The results are processed with J. Kragten's Statpak program;
-    acceptable ranges within which the parameters of the calibration curve may vary from day to day; for this purpose the 2σ limits from the above mentioned program are used.

c)   Intra-laboratory Reproducibility
The degree of agreement between results obtained with the same methods from identical test material under different conditions (different operators, same or different apparatus, same laboratory, at different times). The measure of this parameter is the standard deviation S_L. The intra-laboratory reproducibility R_L is calculated as follows:

$$R_L = 2.8 * S_L$$

After: ISO DIS 5725-3 (11).

For the determination of this parameter, see the CHEK report:

a)   derivation from a large number of paired determinations by two or more analysts.

$$2{,}8 * \sqrt{S\tfrac{2}{L}} \ = \ 2{,}8 * \sqrt{\frac{\sum\limits_{i=1}^{k} S_i^2}{k}} \ = \ 2{,}8 * \sqrt{\frac{\sum\limits_{i=1}^{k} d_i^2}{2k}}$$

where:
k = the number of paired determinations and
d = the difference within a pair of determinations

b)   derivation from the standard deviation (S_r) with the aid of an empirical formula:

$$R_L = 1.6 * 2.8 * S_r$$

d)   Limit of Detection
The lowest measured content from which the presence of the analyte can be deduced with reasonable statistical certainty. This is equal to three times the standard deviation of the measured unit of 20 representative blank samples. The certainty is then 93% and the probability of a false negative or a false positive result 7%.

If the recovery of the method is smaller than 100%, the limit of detection shall be corrected for this recovery. (After: Commission Decision 90/515/EEC and Royal Society of Chemistry (1987)).

A blank sample is a sample whose measuring signal does not rise above the noise.

Since it is difficult to obtain blank samples and determinations of the content close to the limit of detection are very uncertain, the following practical solution is used. Make a calibration curve in the relevant matrix. Determine the slope B and the residual standard deviation of the curve (S_{res}). The limit of detection is defined as:

$$C_{det} = 3\ 8\ S_{res}/B$$

When determining the limit of detection in this way, the statistical methods after Kragten may be used.

e)   Limit of Quantification
The lowest content of the analyte in the test sample measured according to the analytical method which can be determined with reasonable statistical certainty. The limit of quantification is considered to be equal to twice the limit of detection (the probability of a false negative or a false positive result has been reduced to above 0.1%). The following practical solution is used:

$$C_{quan} = 6 * S_{res}/B$$

See further under "limit of detection".

f)   Maximum Permissible Relative Standard Deviation (max. RSD$_r$)
When carrying out an analysis under repeatability conditions, the relative standard deviation must not exceed the following values:

Average value of fraction of analyte in sample	max. RSDr
10^{-9} - 10^{-8}	30%
10^{-8} - 10^{-7}	20%
10^{-7} - 10^{-6}	15%

10^{-6} - 10^{-5}	10%
10^{-5} - 10^{-4}	7.5%
10^{-4} - 10^{-3}	5%
10^{-3} - 10^{-2}	3.5%
10^{-2} - 10^{-1}	2.5%
10^{-1} - 1	2%

Derived from: Commission Decision 90/515/EEC and Pocklington (1990).

g)   Noise
The phenomenon where test apparatus gives signals the intensity and frequency of which fluctuate unpredictably, regardless of the presence of an analyte in the system. The rate at which these changes occur is much higher than the speed at which the system responds to changes in the quantity of analyte (speed of detector response) and generally amounts to a few to several dozen changes per second. The measure of noise is the difference between the maximum and minimum values of the measuring signal in the absence of the analyte, observed for 2 minutes or in accordance with the specifications accompanying the apparatus.

h)   Precision
The degree of agreement between test results obtained by applying the procedure several times under fixed conditions. See also: systematic error.

After: NEN 3114 (1990).

i)   Random Error
The difference between an individual test result and the average of 10 test results, obtained with the same method for the same homogeneous sample. The working range corresponds to the range of application.

After: NEN 3114 (1990).

j)   Range of Application
The content range, for which the analytical method can be used. This range is defined as the range from 0.75 to 1.25 times the legal norm (Cardone, 1983) for contents higher than or equal to 10^{-6}, and as the range corresponding to the norm ± the norm * (2 * max.RSD_r/100) for contents below 10^{-6}.

For a zero tolerance the range is defined as the limit of quantification + the limit of quantification * (2 * max. RSD_r/100).

k)   Recovery
The average fraction of an analyte recovered from an analysis, after addition of a known quantity of the analyte to the blank test sample or the test portion ($n \geq 10$) under well-defined conditions.

Recovery is expressed as a percentage; the procedure with which the parameter is determined must always be stated. The working range corresponds to the range of application. Recovery must not be based on an internal standard.

Concerning the recovery, the following limits apply:

Content analyte in sample	Recovery percentage
≥ 0.5 mg/kg	80 - 110%
1 to 500 μg/kg	60 - 115%
< 1 μg/kg	> 50%

Corrections must be made for recovery if it is constant and the relative standard deviation of the above-mentioned experiment does not exceed the values given under 'f' (Pocklington, 1990) and if it significantly deviates from 100%. For correction, the upper limit of the 95% confidence interval for recovery is used for this purpose in the case of a maximum norm for the analyte. The lower limit of the 95% confidence interval is used in the case of a minimum norm for the analyte.

l)   Relative Standard Deviation (RSD)
Alternative terms are: variation coefficient (VC) and coefficient of variation (CV).

The standard deviation is expressed as a percentage of the mean measured value.

$RSD = (s/x_{mean}) * 100\%$

where:
s = the standard deviation and
x_{mean} = the mean measured value.   After: NEN 3114 (1990).

The relative standard deviation RSD_r is determined under repeatability conditions. The relative standard deviation RSD_R is determined under reproducibility conditions.

m)   Repeatability (r)
The degree of agreement between successive results obtained with the same method with identical test material under the same conditions (same operator, same equipment, same laboratory, at the same time or after a brief interval).

A measure of repeatability is the standard deviation (S_r);
for test series that are not too small;
$r = 2.8 * S_r$ (reliability 95%).

The repeatability is determined by means of a test series of the size $n \geq 10$. The repeatability may depend on the content of the analyte. The relative standard deviation should therefore be included in repeatability studies.

After: NEN 3114 (1990) and ISO 5725-(1986).

n)   Reproducibility (R)
The degree of agreement between the results obtained with the same method with identical test material under different conditions (different operators, different laboratories, different apparatus, at different times). The standard deviation S_R is a measure of reproducibility. If the number of determinations is not too small, the following applies (with 95% reliability):

$$R - 2.8 * S_R$$

Reproducibility may depend on the content of the analyte. The relative standard deviation should therefore be included in reproducibility studies.

After: NEN 3114 (1990) and ISO 5725 (1986).

Reproducibility is determined in the second phase of the verification study by means of a method evaluating interlaboratory study, carried out by the Project Group on Collaborative Studies (PCS) of the Inspectorate for Health Protection. For a successful test, at least 8 laboratories should send in results and the number of outliers must not exceed 22%.

o)   Rounding Off
The omission of non-significant numbers. Numbers do not contribute to the information of their omission produces only a small change in the number of relation to the standard deviation which applies to that number and indicates the uncertainty of the numbers. Follow NEN 1047 (1990).

p)   Selectivity
The extent to which a given analytical procedure allows the analyte to be distinguished from all the other substances in a test sample. A general quantitative statement on this parameter demands an extremely complicated mathematical model, so that, in line with European and other guidelines published to date, a qualitative description is sufficient.

See note under specificity.

q)   Sensitivity
The change in the measuring signal resulting from a change in content of one unit (the slope of the calibration curve).

After: IUPAC, (1976).

r)   Significant Numbers
Numbers, whether decimal or otherwise, which contain a degree of information in agreement with the repeatability of an analytical method. Significant numbers are obtained by rounding off the test result.

s)   Specificity
The ability of an analytical method to measure only that which is intended to be measured. The method gives no response to any characteristic other than that of the analyte. Specificity is the limit of <u>selectivity</u>.

Note
As clarification of the definitions of specificity and selectivity: if a reagent forms a coloured complex with only one analyte, the reagent forms a coloured complex with only one analyte, the reagent is called specific for that analyte. If the reagent, however, forms many coloured complexes with compounds in the matrix, but forms a separate colour for each compound, the reagent is called selective.
The determination of an analyte can be interfered with in three ways:
1)   The matrix and/or other analytes influence the sensitivity of the analytic measurement: the matrix effect.

2)   Some elements in the matrix contribute to the response of the analytical measurement, without influencing its sensitivity for the component: the interference effect.
3)   A combination of a matrix effect and an interference effect.

Depending on the extent of above mentioned effects, arrangements should be made to correct for their responses.

t)   Standard Deviation, Measured (s)
A measure of the spread resulting from random errors in a series of independent test results ($n \geq 10$) with the same measurement expectations.
The measured standard deviation of a random sample from a population is an estimate of the standard deviation σ of the whole population.
The working range corresponds to <u>the range of application</u>.
The formula is:

$$s = \sqrt{\frac{\sum\limits_{i=1}^{n}(x_i - x_{gem})^2}{n-1}}$$

If measurement expectations are not the same (e.g. a mean is not being determined), the formula is:

$$s = \sqrt{\frac{\sum\limits_{i=1}^{k}w_i^2}{2k}}$$

where: w_i =   the spread within the series and
              k =   the number of series of n measurements (see 'j')

u)   Residual Standard Deviation (S_{res})
A measure of the deviation of the measured values from the calculated <u>calibration</u> curve.

v)   Systematic Error

The difference between the mean value obtained from a large number of analysis results and the true value. The accuracy of measurements may be represented (after: NEN 3114) as follows:

$$X \quad = \quad X_w \quad + \quad \delta \quad + \quad \epsilon$$

measured value	true value	systematic error	random error
		⌞————————⌟	⌞————————⌟
		trueness	precision

⌞——————————————————————⌟
accuracy

x)   Trueness

The degree of agreement between the average value obtained from a series of analyses ($n \geq 10$) and the <u>true value</u>. The difference rather than agreement is generally regarded as a measure of trueness, i.e. the . The difference rather than agreement is generally regarded as a measure of trueness, i.e. the <u>systematic error</u> (bias).

After: ISO DIS 5725-1 (1988).

Trueness should be determined with the help of one of the following techniques (listed in order of preference):

-       the use of certified reference material;

-       the use of a reference method/standard method with little or no <u>systematic error</u> (a status 2 method);

-       the use of spiked samples, based on blank samples.

The working range corresponds to the <u>range of application</u>.

It is necessary to correct for the systematic error if it is constant, if it significantly deviates from 0 to and if the relative standard deviation of above-mentioned experiments does not exceed the values mentioned under 6 (Pocklington, 1990).

For correction the lower limit of the 95% - reliability interval of the systematic error is used, for a maximum norm of the analyst. The upper limit of the 95% - reliability interval is used for a minimum norm of the analyst.

y)   True Value

The value which characteristics as <u>content</u> and is completely defined in the conditions in which the content is determined. The certified value of a reference material can be seen as an approximation of the true value.

After: NEN 3114 (1990).

Ruggedness test is not included as a parameter but is discussed separately from the validation parameters, as follows:

The reliability of a method often leaves much to be desired in the course of time if it is sensitive to minor changes in the procedure, such as variations in the content of the reagent, temperature, and reaction time. If this sensitivity to minor changes has not been examined in the normal development procedure for the method, a ruggedness test (1) can clarify the situation. This test allows information to be obtained on the method's sensitivity to minor changes in a quick and systematic manner.

For new methods it is **advisable** to carry out a ruggedness test if this can be fitted into the development process. It is **necessary** to carry out such a test before the second phase of verification (interlaboratory study) begins.

3.   Validation

3.1.   General

To establish which parameters need to be determined for a particular method, if it is desired to obtain status 1, the following procedure has been adopted.

Since it is impracticable to specify these parameters separately for all the several hundred methods used, it has been decided to cluster the methods. For the clustering of the chemical methods, the first item chosen is the mass concentration (in the test solution), and the subitem is the technique used. For mechanical methods a distinction is made between destructive and non-destructive tests.

3.1.1.   Physicochemical Methods

The parameters listed in table I need to be determined for validation purposes.

In general, no quantitative parameters need to be determined for a qualitative method, with the exception of the limit of detection and recovery.

For methods for which the technical possibilities of the test apparatus exceed the level of the analyte to be determined (i.e. it is possible to determine much lower levels with the apparatus), it makes less sense to determine the limit of detection and limit of quantification. This is only worthwhile for a sample the content of which equals the legal norm if the ratio between the measuring signal of the analyte in the test solution and the noise is less than 30.

Table I

METHOD VALIDATION PARAMETERS*
limit of detection
limit of quantification
intra-laboratory reproducibility
repeatability
trueness/recovery**
calibration curve***
relative standard deviation
selectivity
specificity
standard deviation, measured
range of application

   *   =   See 3.1.1

   **   =   The recovery is only determine if the trueness cannot be determined in the first or the second of the afore mentioned ways. The trueness is then identical to the definition of recovery.

   ***   =   Depending on technique and/or whether the limit of detection and limit of quantification are also determined.

3.1.2. Mechanical Methods

Non-destructive tests

The parameters listed in table II should be determined when validating non-destructive tests.

Destructive tests

In the case of destructive tests, the size of the laboratory sample partially determines the procedure to be followed. If the whole laboratory sample (one unit) is destroyed in the test, the test should be carried out simultaneously by two operators if a subjective operation or a subjective observation or a subjective observation of the test results is involved (e.g. determination of ignition delay of fireworks).

If there is sufficient laboratory sample available to carry out the test several times, the procedure for non-destructive tests is followed (e.g. determination of the flash point of a solvent).

Table II

PARAMETERS FOR NON-DESTRUCTIVE MECHANICAL TESTS
intra-laboratory reproducibility
repeatability
trueness
relative standard deviation
standard deviation, measures

3.2.   Methods Developed and Validated in Other Laboratories

If a method has been developed and validated in another laboratory and given status 1, the laboratory wishing to adopt the method should act as follows.

An analyst in the recipient laboratory should perform the method, preferably with a sample similar to the one used in the laboratory that developed the method. Any problems concerning the interpretation of the description, differences in equipment and procedures are noted and discussed with the laboratory concerned. The method is then validated in the recipient laboratory in accordance with the procedure "Validation of methods", based on the criteria referred to in this report.

This procedure also applies if the method to be adopted has validation status 2 (e.g. it is taken from the literature) and the laboratory adopting it did not itself participate in the relevant interlaboratory study.

If significant modifications are made to a validated method, the method should be validated again.

A method with status 2 can be applied without further measures, provided the laboratory itself participated in the interlaboratory study for the evaluation of the method and the result was not classed as an outlier.

3.3.   Methods for Screening

Methods for screening are used to carry out a fast pre-selection of non-complying samples.

In general, for these methods different requirements are set concerning the parameters to be determined for validation purposes.

An important aspect, however, is the distinctive capability: false negative results shall be avoided, as far as possible.

3.4.   Research and Development Work

Research and development work shall be validated to a limited extent, i.e. at least the repeatability and the trueness shall be determined. For the trueness the standard addition method can be used.

4. Reporting Validation Data

The conditions, results and raw data obtained during the validation process are set forth in a <u>validation report</u>, together with all the parameters measured.

The conditions and parameters are recorded on a <u>validation form</u>, which is attached to the SOP as a separate section and then submitted to the quality officer for approval. The latter may make use of the validation report when assessing the validation process.

Annex 2 gives an example of a validation form (after Beljaars).

Subject Index